Contemporary Issues in Construction in Developing Countries

Economic growth and socio-economic development are particularly important for developing countries, and the construction industry plays a central role in driving both of these.

Traditionally the issues faced have been assumed to be merely offshoots of those encountered in industrialised nations and are usually discussed only in this context. In addition, research in construction management and economics has generally failed to take proper account of the unique or highly emphasised characteristics of the industries in developing countries, or their economic and social environment. This volume challenges underlying assumptions and focuses on the distinct characteristics of construction in developing countries. In so doing it considers the issues from the perspective of the developing countries themselves to present a strong contemporary picture for researchers.

It forms a companion volume to *New Perspectives on Construction in Developing Countries* which provides an update on the generic subjects relating to the construction industry in developing countries, and covers new concepts and issues.

George Ofori is a Professor at and former head of the Department of Building at the National University of Singapore. Until recently he was coordinator of the "Construction in Developing Countries" Working Group of CIB (the International Council for Research and Innovation in Building and Construction).

About CIB and the CIB series

CIB, the International Council for Research and Innovation in Building and Construction, was established in 1953 to stimulate and facilitate international cooperation and information exchange between governmental research institutes in the building and construction sector, with an emphasis on those institutes engaged in technical fields of research.

CIB has since developed into a worldwide network of over 5000 experts from about 500 member organisations active in the research community, in industry or in education, who cooperate and exchange information in over 50 CIB Commissions and Task Groups covering all fields in building and construction related research and innovation.

http://www.cibworld.nl/

This series consists of a careful selection of state-of-the-art reports and conference proceedings from CIB activities.

Open & Industrialized Building *A Sarja*
ISBN: 9780419238409. Published: 1998

Building Education and Research *J Yang* et al.
ISBN: 978041923800X. Published: 1998

Dispute Resolution and Conflict Management *P Fenn* et al.
ISBN: 9780419237003. Published: 1998

Profitable Partnering in Construction *S Ogunlana*
ISBN: 9780419247602. Published: 1999

Case Studies in Post-Construction Liability *A Lavers*
ISBN: 9780419245707. Published: 1999

Cost Modelling *M Skitmore* et al.
(allied series: Foundation of the Built Environment)
ISBN: 9780419192301. Published: 1999

Procurement Systems *S Rowlinson* et al.
ISBN: 9780419241000. Published: 1999

Residential Open Building *S Kendall* et al.
ISBN: 9780419238301. Published: 1999

Innovation in Construction *A Manseau* et al.
ISBN: 9780415254787. Published: 2001

Construction Safety Management Systems *S Rowlinson*
ISBN: 9780415300630. Published: 2004

Response Control and Seismic Isolation of Buildings *M Higashino* et al.
ISBN: 9780415366232. Published: 2006

Mediation in the Construction Industry *P. Brooker* et al.
ISBN: 9780415471753. Published: 2010

Green Buildings and the Law *J. Adshead*
ISBN: 9780415559263. Published: 2011

New Perspectives on Construction in Developing Countries *G. Ofori*
ISBN: 9780415585724. Published 2012

Contemporary Issues in Construction in Developing Countries *G. Ofori*
ISBN: 9780415585716. Published: 2012

Culture in International Construction *W. Tijhuis* et al.
ISBN: 978041547275X. Published: 2012

Contemporary Issues in Construction in Developing Countries

Edited by George Ofori

LONDON AND NEW YORK

First published 2012 by SPON Press

2 Park Square, Milton Park, Abingdon, Oxfordshire OX14 4RN
711 Third Avenue, New York, NY 10017

Routledge is an imprint of the Taylor & Francis Group, an informa business

First issued in paperback 2018

British Library Cataloguing in Publication Data
A catalogue record for this book is available from the British Library

Library of Congress Cataloging-in-Publication Data
Contemporary issues in construction in developing countries /
edited by George Ofori.
p. cm.
1. Construction industry--Developing countries.
2. Building--Developing countries. I. Ofori, George.
HD9715.D442C67 2011
338.4'7624091724--dc23
2011012982

ISBN: 978-0-415-58571-2 (hbk)
ISBN: 978-1-138-38133-9 (pbk)

Typeset in Sabon
by Fakenham Prepress Solutions, Fakenham, Norfolk NR21 8NN

Contents

To my mother, Maame Yaa Darkwa.
She made tremendous sacrifices for me.

Illustrations

Figures

Tables

Contributors

George Ofori, PhD, DSc, is Professor at the National University of Singapore. He is a Fellow of the Chartered Institute of Building, Royal Institution of Chartered Surveyors, and Society of Project Managers. His research is on construction industry development, international project management, and leadership. He has been a consultant and adviser to several international agencies, and governments.

Asanga Gunawansa, PhD, works at the Department of Building of the National University of Singapore. His teaching and research areas include: construction law; arbitration; legal aspects of infrastructure development; and international environmental law. Before joining NUS, he worked as a legal officer for the United Nations Organisation.

John Hawkins is a social scientist who specialises in UK and international procurement policy and governance in construction and its links with social development. He is responsible for knowledge creation and transfer in civil engineering best practice at the Institution of Civil Engineers and is a Policy Advisor to the Department for International Development's Construction Sector Transparency Initiative. He is also the British Standards Institute Representative on the Working Group for an International Standard on Construction Procurement.

Koshy Varghese, PhD is with the Department of Civil Engineering, Indian Institute of Technology, Madras, India. He earned his doctoral degree at the University of Texas at Austin, USA in 1992 with a specialisation in Construction Engineering and Project Management. His research is in the area of Computer Integrated Project Delivery.

Richard Neale, DSc HonDLitt, initially worked for consulting civil engineers and contractors. He was appointed lecturer at Loughborough University in 1974 and Professor of Construction Management at the University of Glamorgan in 1996, where he was Head of School and Director of Research. He is now Professor Emeritus at that university. He has worked extensively for the International Labour Organisation (ILO) in developing countries.

Joanna Waters, PhD, worked as a Research Fellow and Lecturer at the University of Glamorgan for eight years, after which she joined Cardiff University. Now an independent consultant, she has worked for the ILO on a construction health and safety training package and is a guest lecturer at the University of the West of England.

Abdul-Rashid Abdul-Aziz, PhD, is Professor at the School of Housing, Building and Planning, Universiti Sains Malaysia, Penang. His research interests cover project management, construction industry development, and internationalisation. He is a member of the Institution of Surveyors Malaysia and the Royal Institution of Chartered Surveyors.

Izyan Yahaya is a fellow at the University of Science Malaysia. She has worked in Malaysia as a quantity surveyor. She is a Registered Graduate Quantity Surveyor of the Board of Quantity Surveying Malaysia. Her special area of interest and research is construction project management.

Patrick X.W. Zou, PhD, is Associate Professor and Director of the Construction Management and Property Program at University of New South Wales (UNSW), Australia. His research focuses on risk management and safety management, and he has received two international awards. He has also undertaken research on the scholarship of learning and teaching and received several fellowships and awards including the UNSW Vice-Chancellor's Award for Excellence in Teaching. He will take up the position of Head of Discipline of Building and Construction Management at University of Canberra in October 2011.

Dongping Fang, PhD, is a Professor and Head of Department of Construction Management, School of Civil Engineering, Tsinghua University, China. He is the founding Director of the (Tsinghua–Gammon) Construction Safety Research Center and Executive Director of the Institute of International Engineering Project Management at Tsinghua University. His research, teaching and consulting are in the area of construction safety, risk management, and international construction.

Shamas-ur-Rehman Toor, PhD, currently works with the Islamic Development Bank Group in Saudi Arabia. He was formerly a lecturer at UNSW, Australia. His research interests include leadership, project performance management and portfolio management. His papers have been published in several international journals.

Stephen Ogunlana, PhD, is Professor of Construction Project Management at Heriot-Watt University, UK. He was formerly at the Asian Institute of Technology, Thailand. He researches and provides consultancy services in project management, public–private partnerships, organisational learning, system dynamics simulation, human resource management and construction process improvement.

Gonzalo Lizarralde, PhD, is Professor at Université de Montréal. He has taught at the University of Cape Town (South Africa), McGill University (Canada), and Universidad Javeriana (Colombia). He has authored more than 30 articles about housing and project management. He is Director of the IF Research Group that studies processes relating to the planning and development of construction projects. He is a founding member of i-Rec, an international network on post-disaster reconstruction, and co-author of "Rebuilding after Disasters: From emergency to sustainability".

Sarah Dix, PhD, has a bachelor's degree from Yale University and an MPP from Harvard University. She is a political economist with over 15 years of field experience in Africa, Latin America and the Pacific. She has been a consultant and technical advisor on governance, and has extensive expertise in surveys. She undertakes research on corruption, democratic governance, and civil society. Among her works is a five-country study on post-war corruption, published by the United Nations Development Programme (UNDP) in 2010, and a chapter in *Corruption and World Order* (Brookings 2009). She was previously a lecturer at Harvard University.

P.D. Rwelamila, PhD, is Professor of Project Management at the Graduate School of Business Leadership, University of South Africa, Past President of the South African Council for Project and Construction Management Professions, and joint coordinator, CIB W107, on Construction in Developing Countries. He is a Fellow of the Chartered Institute of Building, American Association of Cost Engineers and American Institute of Constructors; and a registered Professional Construction Manager in South Africa. He has worked in several countries, including Tanzania, Kenya, Uganda, Botswana, Zambia, Australia, United Kingdom and Sweden. He has authored more than 200 peer-reviewed journal papers, chapters in books and papers in conference proceedings.

Florence Yean Yng Ling, PhD, undertakes research mainly in project performance, relational transactions and international construction. An Associate Professor at National University of Singapore, she teaches cost estimating and has won numerous teaching excellence awards. She was a consultant quantity surveyor in Singapore. She was a Visiting Scholar at the Department of Civil and Environmental Engineering Department of the University of California, Berkeley, in 2002.

Vivian To Phuong Hoang graduated from the National University of Singapore with a BSc (Building) (Hons) degree. She practises as a quantity surveyor at Rider Levett Bucknall in Singapore, handling cost estimating, procurement and contract administration.

Subashini Suresh, PhD, is a Senior Lecturer in the Department of Construction and Infrastructure, University of Wolverhampton, UK. She teaches site production, construction planning, and advanced project planning. Her research, conducted in the UK, USA and India, focuses on process improvement, knowledge management, leadership in change management initiatives and organisational competitiveness.

Abubakar M. Bashir is a research student in the Department of Construction and Infrastructure, University of Wolverhampton, UK. His research focus is on lean strategies for promoting health and safety initiatives in UK construction organisations.

Paul O. Olomolaiye, PhD, is Professor, Pro Vice Chancellor and Executive Dean, Faculty of Environment and Technology, University of the West of England, UK. His specialist areas include: education, research and knowledge exchange in construction management; regional regeneration through product and process improvement in engineering companies; productivity and performance strategies in construction; education, research and knowledge exchange in technology and environment; and widening participation and engagement in higher education.

Edmundo Werna, PhD, is an expert in urban development and the construction industry with more than 30 years experience in academia (in Brazil, Italy and the UK), the private sector, government and United Nations agencies. He currently works at the headquarters of the ILO. His preceding assignment was at the headquarters of the United Nations Volunteers programme, where he designed and implemented the programme's urban development agenda. He has been consultant to local governments, non-governmental organisations and international agencies such as European Commission, World Bank, UN-Habitat, UNDP and World Health Organization. He has won many prestigious international awards for his research output.

Emilia van Egmond, PhD, is a senior lecturer and researcher on Innovation and International Technology and Knowledge Transfer for Sustainable Construction at the Eindhoven University of Technology, the Netherlands which she joined in 1986. She has professional and academic experience in many countries in Africa, Asia, Europe and Latin America, and is a member of advisory and evaluation committees for the Dutch government and several international organisations. She was the director of a project financed by the Dutch government on Capability Building in Ghanaian Construction.

1 Reflections on the great divide

Strategic review of the book

George Ofori
National University of Singapore

Introduction

The construction industry plays an important role in economic growth and socio-economic development. The performance of the industry must be improved if it is to be able to fulfil the expectations of the governments and people in developing countries for enhanced economic prospects and improved quality of life. The companion volume to this book (*Construction in Developing Countries: New Perspectives*) discusses important aspects of the construction industries in the developing countries in the present. It focuses on the development of the industries, and improvement of their performance. What is this book about and what does it seek to achieve? What are "contemporary issues" where the construction industry is concerned? Why is this book important?

There are many differences between the countries generally designated as "developed" and those referred to as "developing". The common yardstick is gross national income per capita. Thus, the main difference is economic. There are also differences along a number of diverse indicators, such as those comprising the Human Development Index. It is pertinent to ask: Are there significant differences between the construction industries in developing countries and their counterparts in the industrialised nations? If so, how wide is this divide? What impact does this have on the application of knowledge on key aspects of the management and economics of the industries in one of the groups of countries on industries in the other group?

A great volume of research on Construction Management and Economics is undertaken in the developed countries. This has led to the development of a body of knowledge as well as a range of procedures, tools and frameworks. These research works are based on the nature of the construction industries in the developed countries and their operating environments. However, considering the inter-relationship between construction and the economy, as well as national development (Ofori, 1993; Hillebrandt, 2000), it can be expected that the nature of the construction industries in the developing countries is different from that of their counterparts in the developed nations. Leaving this issue aside for the time being, it can be explained that the operating environments of the industries in the two groups of countries are also different. The environment in the developing countries poses many difficulties to the firms and practitioners, and these are usually not consistent with the assumptions made in the research in most aspects of Construction Management and Economics undertaken in industrialised countries. For example, Alinaitwe (2009) found the following to be the top ten barriers to the implementation of lean construction in Uganda: (i) inputs exactly when required; (ii) infrastructure in transportation and communication; (iii) capability of teams to maintain alignment with other teams; (iv) certainty in the supply chain; (v) steady prices of commodities; (vi) reward systems based on teams' goals; (vii) buildable designs; (viii) participative management style for the workforce; (ix) parallel execution of development tasks in multi-disciplinary teams; and (x) accurate pre-planning. Thus, the recommendations that are made in research on contruction in developed countries (if any) and the conclusions that are reached are not always relevant to the developing countries.

It can be suggested that the body of knowledge on Construction Management and Economics cannot be considered to be adequate if its existing concepts and tools only apply in the developed countries. Moreover, researchers seldom distil the implications of their findings for the developing countries. Therefore, the developing countries would not be benefiting from the advancement in construction technology and knowledge on the industry and its processes and products resulting from studies in developed countries.

In the next section, the differences between the broad group of developed countries and developing nations with respect to their construction industries and processes are discussed in some detail.

The great divide?

The differences between the construction industries in the groups of developed countries and developing nations can by analysed using the criteria that comprise many of the existing frameworks for studying the construction industry.

Characteristics of construction as framework for analysis

The differences between the construction industries in the groups of developed countries and those in the developing nations are evident when the peculiar features of the construction projects, the industry and the constructed product are considered (Hillebrandt, 2000). The range of issues explored by Ofori (2009) in considering the need for leadership in the construction industries in developing countries is the first set of criteria used as the framework for the discussion here to establish the differences between the two groups of construction industries. The first peculiar characteristic of construction is that a significant proportion of the construction projects which the poorer countries require for their development (such as airports, factories, harbours, hotels, hospitals, ports, power generation and distribution, water purification and distribution installations) are the same in many ways as those found in the developed countries, being based on technical principles in design, and using materials imported from the developed countries. The projects are technically complex and large. They involve a combination of specialist skills to undertake them, and the ability to manage a wide range of risks. Whereas these resources and requirements can be taken for granted in the developed countries, they are a real challenge in the developing ones. As a result, many foreign design and construction companies and professionals are involved in these projects in those countries, and in fact, dominate that segment of the industries. This has its own issues and problems on which much research work has been undertaken (see, for example, Gunhan and Arditi, 2005; Ling and Hoang, 2010).

Second, the construction projects take quite a long time to complete, and involve a very large number of discrete activities. This increases the probability that time-related risks relating to such construction projects would occur (Ofori, 2009), and also exacerbates the communication and coordination problems. The special circumstances of the developing countries make these issues and problems further complicated. For example, the countries lack the infrastructure to facilitate scheduling and logistics. Next, the required professional skills are in short supply. The materials supply companies and plant-hire organisations are not well developed, and do not provide the extent (trade discounts, delivery services, credit periods) and quality of service (level of knowledge of the goods, promptness of delivery) that firms and practitioners in the industrialised countries are used to. Thus, both planning and scheduling of projects, as well as supply chain management, are greater challenges here.

Third, the teams involved in construction projects are not only large, but are also multi-disciplinary, and the members are from several different organisations. On today's large projects, they are also multi-cultural. This is most evident in the developing countries where, as discussed above, most of the large and complex construction projects are undertaken by foreign contractors which tend to form joint ventures with local firms, or engage subcontractors in the host countries. Thus, it is even more challenging to manage and mediate among the cultural diversity in developed countries and the developing nations whose citizens are involved in the projects.

Fourth, the projects and the constructed product involve serious implications for the health and safety of the workers involved, and of the general public. Thus, due care, diligence and expertise are needed to safeguard these. In developed countries, the law enforcement framework on occupational health and safety (OHS) effectively guides technical and business practices. This is not the case in the developing countries. Thus, there is a difference in safety performance and the incidence of health issues between the construction industries of the two groups of countries.

Differences in performance

Another way of considering the differences between the construction industries of the two groups of countries is the overall performance of the industries. First, there are more reports of project performance deficiencies such as cost and time overruns, poor quality of work, technical defects, poor durability, and inadequate attention to safety, health and environmental issues (see, for example, Ofori, 2007). Whereas the construction industries in all countries, including those in the developed nations (see, for example, Egan, 1999) are criticised with regard to their performance, those in developing countries face formidable challenges, some of which are not an issue in the developed nations, and these have an adverse impact on their performance. For example, in Malaysia, Abdul-Rahman *et al.* (2007) found that the level of management quality of contractors undertaking public design-and-build projects was unsatisfactory. The quality-related factors contributing to this situation were: budget constraints; time constraints; client's complexity; poor communication; and design variations.

Second, the management of projects in developing countries is fraught with many problems because of the nature of the industries and their operating environments. The importance of the effective management of the stakeholders of the construction project in developing countries is most clearly evident on the international projects that are common at the large-complex project end of the markets in the developing countries (Ofori, 2003). On such projects, the teams are invariably multi-cultural. This underscores the need for leadership skills.

Third, there are even greater adverse implications of poor performance on construction projects for long-term national socio-economic development in developing countries as the constructed product is critical to this process of development.

Finally, in the developing countries, the clients, end purchasers, users and other stakeholders of construction have no knowledge of aspects of construction. This implies a need for professionalism among the participants in the construction project, and a dedication to meet the objectives and aspirations of the stakeholders in the most innovative, imaginative and value-adding manner for the benefit of the client and all concerned.

Drivers of construction activity

The main drivers of construction activity can form an appropriate framework for distinguishing the construction industries in developing countries from their counterparts in developed countries. For example,

Flanagan (2005) identified and mapped business drivers of international construction over the next five years against global drivers (opportunities) that will have greatest impact. The latter were: demographics and ageing; climate change, sustainability and environmental pressures; urbanisation, growth of cities, and transportation; people, safety and health; rapid technological and organisational change; vulnerability, security, terrorism and corruption; globalisation of economies and businesses: information, knowledge and communication; and governance and legislation. Table 1.1 presents an attempt to use some key drivers (some of which are in Flanagan's framework) to distinguish the two groups of construction industries.

Table 1.1 Differences between developed and developing countries with respect to major driving forces of construction

Issue	*Considerations*	*Developed countries*	*Developing countries*
Environmental issues and climate change	• Construction has a wide range of adverse impacts on the environment; the industry is called upon to act (CIB, 1999; Du Plessis, 2002).	• Advanced legislation, technology, production procedures. • Benchmarking systems. • Market drivers and incentives.	• Not yet a major consideration in practice despite research findings that it is a critical issue (Du Plessis, 2002).
Safety and health in the population and workforce	• Poor safety and health in construction affects the industry's image and attractiveness.	• Advanced legislation, technology, production procedures. • Performance indicators. • Market drivers. • OHS no longer critical issues.	• Major national health problems, such as infant and maternal deaths, HIV/AIDS and malaria. • High accident frequency and severity rates, and fatalities on construction sites.
Population issues	• Global population growth is considered to be a matter of concern owing to its implications on consumption of resources and disposal of waste.	• Situation of declining and ageing population. • Construction workforce in short supply, and ageing (McGraw Hill Construction, 2008).	• High population growth, exacerbating poverty and unemployment, and putting pressure on infrastructure and social services, especially from rural–urban migration. • Increasing gap between construction needs and provision.

Issue	*Considerations*	*Developed countries*	*Developing countries*
Poverty alleviation	• Alleviation of poverty has been a major concern at the global level and in developing countries. It is the first MDG; the target is to reduce it by half by 2015 (UN, 2010).	• Not a critical issue at broad national level.	• A matter of great concern in all developing countries. All governments in these countries declare that they are committed to reducing poverty (UN, 2010).
International construction	• Large and complex projects everywhere usually have a significant international component. In the developing countries, they are undertaken by foreign design and construction firms. The latter usually work in joint ventures with local firms.	• Firms dominate global international construction market. • Strategy is to expand activities to lucrative overseas markets with appropriate economic and legal systems (Gunhan and Arditi, 2005).	• Firms have low market share, even at home, on costly and complex projects. • Strategy is to benefit from presence of foreign firms (Ofori, 2003).
Globalization	• The features of globalization include trade reforms and liberalization, and flows of capital, goods and services around the globe (UN, 2010). Such flows lead to demand for construction which firms, especially from developed countries, follow to undertake. A significant proportion of the materials, components and installations are also obtained from various countries.	• Countries have upper hand in current system of international trade regime and negotiations on reform (MDG Task Force, 2010). • Business opportunities in booming markets abroad, for direct investment and technical activities.	• Countries face an unfavourable situation and low bargaining position. Necessary for them to give ample incentives to attract investment.

Issue	*Considerations*	*Developed countries*	*Developing countries*
Technology development and innovation	• The continuous development of technology is critical in construction, to prepare for new, more demanding tasks and to enhance productivity, quality and other aspects of performance on the projects and in the products.	• Developers and originators of technology and knowledge. • Transferors of technology. • Innovation is a business driver for enhancing competitiveness so it is keenly pursued (Nam and Tatum, 1997; McGraw-Hill Construction, 2008). • Government incentives for technology development and innovation.	• Low level of technology development. • Transferees of technology and knowledge, with low bargaining strength in the transactions (Kumaraswamy and Shrestha, 2002). • Slow or non-existent diffusion of new technologies in the industry. • Low extent of innovation.
Information and communications technology (ICT)	• ICT is important for facilitating, expediting and simplifying work in all sectors, including construction.	• Sophisticated hardware and levels of knowledge. • Advanced development and strategic application.	• The level of sophistication is low and the need for development is basic. • It is important for the countries simply to keep up with the developed nations. Need for education, training, access to hardware, proprietary application software and efficient infrastructure.
Quality and productivity	• Performance on projects is of major concern in the construction industry. Quality and productivity are among the key indicators of performance.	• Given due attention for some decades now. Part of industry and project key performance indicators (Egan, 1999).	• Not given much attention. Owing to poor productivity, constructed items are relatively costly, considering the cost of labour and the economic situation. • Poor quality of construction has implications for durability, efficacy and ability to withstand the effects of disasters.

Issue	*Considerations*	*Developed countries*	*Developing countries*
Disaster prevention and reconstruction	• The frequency of disasters appears to have increased over the past decade. Destructive powers of disasters are also increasing. • The frequency and duration of conflicts, mainly civil wars, has also increased. • Specialist planning, design and construction skills are required in order to effectively address disasters and conflicts.	• Not a critical issue, as strong legislation, codes and enforcement lead to built items that are able to resist the effect of disasters.	• Many of the countries are prone to various disasters, and the impact is more severe in terms of damage and loss of life. Inadequate and inappropriate construction contributes to it.

The great divide established

The discussion so far shows that there are key differences between the construction industries in the developing countries and those in the industrialized countries, arguably from any perspective that one might adopt in considering them. If this is the case, then works in Construction Management and Economics, which are mainly undertaken in the developed countries, and are intended to be applied there, must be modified if their findings and recommendations are to be of relevance to, and applicable in, the developing countries. A number of scenarios might be considered if this is not done. First, the policies and initiatives will simply not work, as the pre-requisites for effective implementation do not exist. Second, the policies and recommendations will take a long time to succeed as hindrances must be addressed. Finally, the implementation does real damage because the wrong policies and recommendations are applied.

As, in the developing countries, resources for implementing the policies and programmes are limited, the need is great and time is of the essence, it is important that the knowledge that forms the foundation of the policies and programmes should be sound and practically and directly relevant. This is the reason for this book.

The book

In this book, the authors consider some "contemporary topics" in Construction Management and Economics that are not usually addressed from the perspective of the developing countries. Contemporary topics are those that have been of interest in recent years. These are on subjects ranging from technology (such as information and communications technology), through practices (such as occupational safety and health (OSH) and environmental issues, especially climate change) to business strategies (such as internationalisation). The topics in this book range from issues at the industry and company levels to those at the project, team and individual levels. In each chapter, the author(s) review the state of the art in the subject area concerned, and consider the applicability, in the construction industries of the developing countries, of the main trends in the field of knowledge and findings from recent research. Among issues discussed are the prerequisites that should be in place, and the possibility of further developing the subject in the developing countries in order to contribute to the greater body of knowledge.

The authors wrote the chapters by following an agreed framework. These included the provision that each chapter would contain: (a) a review of the relevant literature; (b) a discussion of current issues with relevant examples from several countries (and not only the developing ones) and inter-country comparisons; (c) proposals for policy development and action; and (d) future directions for research and practice. The authors were urged to bear in mind that the developing countries themselves are not a homogeneous group; their circumstances, as well as future prospects, differ. They were also requested to include in their conclusions and recommendations the policy implications of their discussions and the findings and conclusions for the governments of the developing countries, with respect to their efforts to develop the construction industries, and enhance their performance and prospects.

In the rest of this "strategic overview" of this book, the contents of the chapters are summarised, and then key issues in the chapters are related to each other and to major developments in the general area of Construction Management and Economics in a synthesis.

Strategic overview of the book

In this section, an attempt is made to introduce the contents of the book by providing a summary of each chapter.

PART I: INDUSTRY PRACTICES

In Chapter 2 of this book, **Asanga Gunawansa** suggests that, as far as developing countries are concerned, Public–Private Partnership (PPP) has become one of the most attractive sources of capital and technology for much-needed infrastructure development, as domestic finance and technical and managerial capabilities are inadequate. However, the success of PPP initiatives in developing countries depends on the extent to which an investor-friendly environment exists in the host countries. He notes that, whereas several developing countries have introduced changes in the policies and regulations to, among many things, facilitate PPP projects, the progress has been slow. Some of the factors hindering change include political and economic instability, public opposition to the payment of market prices for previously free or subsidised facilities, and conflicts between central governments and provincial and municipal governments with respect to responsibilities for the development and maintenance of infrastructure. In the chapter, Gunawansa explains PPP and considers what it means to the governments of developing countries. He assesses the major risks associated with PPP projects in developing countries, and outlines risk mitigation measures. He ends the chapter by making proposals on how a developing country can develop a policy and regulatory framework conducive to the application of PPP.

In Chapter 3, **John Hawkins** discusses the social development opportunities that can be derived during the procurement of construction projects and how they can contribute towards economic growth and the achievement of the Millennium Development Goals (MDG) in developing countries. He suggests that designers and planners should consider utilising the procurement process in the management of social impacts that are a consequence of the construction process, and apply pro-active stakeholder engagement to inform, and hear from, all the people affected by the project. Hawkins notes that many developed countries have used procurement to promote social opportunities. He gives a number of examples. However, the ability of the developing countries (where the need might be even greater) to do the same has been constrained by the tendering procedures of the donor agencies which favour adherence to the "conventional" requirements of project delivery (time, cost and quality). Also, in developing countries the broader policies of governments have been outweighed by a focus on conventional micro-project requirements. With this background, Hawkins examines why it is important for developing countries to include social development opportunities in the delivery processes of their construction projects. He suggests that this issue is best considered during the procurement process so that it would simply become a project objective. Procurement specialists then need to ensure that these opportunities are clearly identified in the tendering and contract documents. Potential suppliers will then be able to develop a methodology to achieve the objectives and price them accordingly. Hawkins concludes his discussion by outlining actions that the governments of the developing countries can take to realise the proposals he makes.

In Chapter 4, **George Ofori** notes that, whereas public-sector construction projects are intended to benefit society by providing vital infrastructure and social facilities which improve the economic and social well being of the populace, and enhance their quality of life, studies show that mismanagement and corruption are common on such projects in all countries.

This leads to cost increases, delays and poor quality both of work and the durability of the completed asset. He discusses the nature and consequences of mismanagement, corruption and other aspects of malpractice in construction, and the many initiatives that have been introduced to address the problems and reduce their frequency and impact. He focuses on the Construction Sector Transparency initiative (CoST). The main principle of CoST lies in the question: "Are we getting what we paid for?" The idea of CoST is that, by disclosing key information on public-sector construction projects, mismanagement and corruption will be reduced, if not eliminated. Ofori also considers other major international anti-corruption initiatives. He makes proposals for addressing the situation, and considers the post-CoST future.

PART II: INDUSTRY PERFORMANCE ON PROJECTS

In Chapter 5, **Koshy Varghese** discusses the application of Information and Communication Technology (ICT) by construction firms in developing countries. He observes that advances in ICT have been drivers and enablers in upgrading practices and raising productivity in most industries worldwide. The exponential growth of computing capability has facilitated global communications and the modelling of complex systems. However, the construction industry lags behind other sectors in the harnessing of ICT. This "lag" has been attenuated in developing countries, because of many non-human factors such as policy restrictions, financial limitations and general resistance to change. Tracing the evolution of ICT applications in construction, Varghese presents the recent and future developments of such applications, discusses the levels of adoption in developing countries and proposes approaches by which barriers to the adoption of ICT can be removed, or their impact minimised. In the discussion, he uses some key ICT roadmaps formulated in Europe and the US to benchmark the current status of usage in developing countries. Finally, he considers the current and future actions required to enhance the adoption of ICT in the construction industries in developing countries.

In Chapter 6, **Richard Neale** and **Joanna Waters** note that the aims of "international development" and those of "effective occupational safety and health" present a humanitarian paradox within the construction industry. They observe that, whereas economic development and rapid industrialisation have brought many opportunities to improve the quality of life and well-being of the people in developing countries, in which the construction industry plays a major role, inadequate provision for OSH means that companies, workers and society as a whole face high costs in terms of the risks, injuries and fatalities which occur. Neale and Waters explore this paradox and argue that OSH is critical for construction workers and for the wider development and well-being of a country. They discuss barriers to instituting effective OSH in developing countries. They present a conceptualisation of OSH unique to developing countries, and explore the appropriateness of transferring models of OSH from developed countries. They note that a total cultural shift is needed to cultivate an 'holistic safety culture' in developing countries. They present suggestions for good practice and highlight ten key issues to focus on.

PART III: STRATEGIES

In Chapter 7, **Abdul-Rashid Abdul-Aziz** and **Izyan Yahaya** note that sub-contracting and joint venturing offer advantages. They enable construction firms to cope with the uncertain and uneven nature of demand for their activities, and to pool resources. They also offer opportunities for construction companies from developing countries to gain from technology transfer from their foreign counterparts. However, the demerits include the need for close monitoring of the sub-contractors, which might produce poor-quality output, or have low efficiency. The envisaged technology transfer might also not take place. Moreover, there can be collusion in bidding among the strategic partners. Abdul-Rashid and Izyan use the transaction cost approach to analyse the strategic alliances in construction such as joint ventures and subcontracting. They suggest that the success factors of sub-contracting and joint venturing primarily relate to the partners and the process. They highlight the need for trust and commitment from all the parties and suggest that policymakers must try to augment the positive aspects while attenuating the demerits. For example, the welfare of workers, particularly those of labour-only sub-contractors, should be addressed, and training programmes delivered. The transfer of technology from foreign contractors to their local counterparts should be appropriately managed by both foreign and local companies, and by the host government.

In Chapter 8, **Patrick X.W. Zou** and **Dongping Fang** discuss the risks, opportunities and strategies in international construction, and present an empirical study that they undertook on the construction market in China. They note that international construction markets and projects are complex due to the differences between countries in relation to government policies; legal, financial and tax systems; techniques; and culture. China's large and growing construction market offers opportunities for international property development, design or construction management enterprises. However, there are associated risks. In the field study, Zou and Fang used a postal questionnaire-based survey of practitioners in Australia on the risks they perceive in the Chinese market. The key risks found included: the repatriation of capital and earnings; intellectual property protection; debt collection; and lack of knowledge of the local business environment. The most favourable market entry mode was as a wholly foreign-owned enterprise; and the effective business development strategies were to: provide a niche and superior product or service; build a network of contacts in China; arrange for staff from the head office to manage projects in China; focus on client satisfaction; and obtain political backing in China. They suggest that the results and outcomes are applicable to other developing countries, and offer proposals to international companies and the governments of China and other developing countries.

In Chapter 9, **Shamas-ur-Rehman Toor, Stephen Ogunlana and George Ofori** observe that the features of the construction process and the construction project make leadership essential in the construction industry. Construction projects are expensive and technically demanding, and project teams are usually large and diverse. The process is long and involves a large number of discrete and inter-related tasks. The implications of poor performance on projects for a nation and its citizens can be severe. Thus, it may be argued that "effective leadership" is one of the main answers to the problems of the construction industry especially those in the developing countries. Toor, Ogunlana and Ofori consider leadership needs in the construction industry in these countries. They suggest that leadership development should be actively pursued in these countries, and present an agenda for leadership research in construction in the developing countries.

PART IV: RECONSTRUCTION PROGRAMMES

In Chapter 10, **Gonzalo Lizarralde** observes that, when natural hazards collide with social and physical vulnerabilities, disasters occur and housing deficits increase. Therefore, housing solutions in developing countries must account for regular housing deficits as well as for the sustainable reduction of vulnerabilities. Stakeholder participation is important in the achievement of these two objectives. However, regular and post-disaster housing solutions frequently pay little attention to the post-occupancy phase when considering the role of stakeholders, although most of the customisation and upgrading actually take place after completion. Lizzaralde analyses six recent housing projects in four developing countries, and suggests that housing solutions can benefit from two decisions: coordinating the role of different public and private participants (including the end users) *during* and *after* the construction project; and replicating the response of the informal sector to incremental housing development. He concludes that housing projects require careful planning of the post-occupancy phase, which must include anticipating the role of the informal sector in incremental construction.

In Chapter 11, **Sarah Dix** observes that more than 50 large-scale violent conflicts have taken place in many developing countries over the last two decades. These events have taken a heavy toll in the form of the destruction of the social fabric of the communities, and the physical infrastructure. The construction industry can contribute to the rebuilding of areas that have experienced wars and major internal unrest. Doing this stimulates the local economy, and provides jobs to the members of the local communities, including former combatants. This expedites their re-integration into society after war ends. However, mismanagement, waste and corruption are common features of post-war reconstruction programmes. Dix explains that the post-war environment is different from that which prevails after natural disasters or in other developing countries in "normal" circumstances. She then examines the challenges of post-war construction, including the accountability deficit. She presents the experiences of civil society watchdogs and community-driven reconstruction in a number of post-war countries to show their potential. She notes that the interests of national and international political actors pose challenges to the development of new state–society relations. Dix points out that this array of issues also faces the construction enterprises undertaking projects to improve the physical infrastructure in the post-war situation. Finally, she suggests that attention should be given to the building-up of the capacity of the local construction industry in the post-war reconstruction programme.

PART VI: PROJECT MANAGEMENT

In Chapter 12, **P.D. Rwelamila** notes that studies show that, in the developing countries, where the need for success is greatest, and failure has the most adverse effect in terms of meeting basic needs of large numbers of people, performance on construction projects is commonly unsatisfactory. He undertook a study to consider construction project performance in developing countries and the factors which influence it. Rwelamila examines the key frameworks for assessing project performance that are proposed in the literature and draws out a number of key performance indicators which are suitable for application to construction projects in developing countries. He uses these to assess the performance on projects in three countries, using published data. He offers recommendations on ways and means of addressing the challenges relating to construction project performance in developing countries.

In Chapter 13, **Florence Ling** and **Vivian To Phuong Hoang** investigate project risks faced by foreign firms on construction projects in developing countries, using Vietnam as a case study. The objectives of their study are: to investigate project risks faced and the risk response activities adopted; and to develop a risk management framework to help foreign firms to manage project risks in developing countries. They interviewed 18 experts from abroad who have managed construction projects in Vietnam. The major risks they identified include those relating to design, construction, finance and culture. Of the ways to respond to the risks in developing countries, the two most important are: proper selection of team members; and adequate planning, controlling, monitoring and supervision of consultants, contractors and subcontractors. Ling and Hoang believe that foreign firms undertaking construction projects in developing countries can make use of their findings to identify their project risks and determine the appropriate risk response measures to give their projects a higher chance of success. They discuss the policy implications of their findings in developing countries.

In Chapter 14, **Subashini Suresh, Abubakar M. Bashir** and **Paul O. Olomolaiye** suggest that the principles of lean construction have been successfully adopted to minimise construction waste by construction industries in some developed countries. Although several techniques have been used to minimise wastage of resources on projects in developing countries, the lean concept is still rarely adopted by construction practitioners in those countries. They introduce the concept of lean construction as a strategy for eliminating waste in the construction process in developing countries. They consider its origin, principles and tools, and barriers to its implementation. Suresh et al. discuss the adoption of lean construction in several developing and developed countries. They present case studies of four projects in Nigeria. The cases examined the effectiveness, level of leanness and implications of the techniques adopted by clients to minimise wastage of resources during the design and construction stages. The case studies indicated that, besides being ineffective, the supposedly lean techniques used by the construction practitioners have negative impacts on project quality. They develop a Lean Protocol and recommend that practitioners in the developing countries should have a holistic understanding of lean construction principles and consider the application of the protocol's tools.

PART VII: INDUSTRY DEVELOPMENT

In Chapter 15, **Edmundo Werna** points out that green jobs are the result of the employment opportunities which are created as actions are taken to improve environmental protection. The adverse environmental impacts of construction range from various forms of pollution to climate change through emissions of global warming gases. He notes that both new construction and refurbishment of the existing stock of buildings provide the opportunity to improve the environment in many ways, for instance in reducing energy consumption. These activities create jobs, and might also require the development of new professional and technical skills. Werna defines "green jobs" and highlights the importance of green jobs in construction to the developing countries. The discussion is undertaken in the context of a comprehensive view of the "world of labour". He next turns to green construction in developing countries. He discusses the core subject of the chapter, green jobs in construction. He presents several cases studies on the policies, programmes or particular projects on green jobs in the construction industries of a number of developing countries. Finally, he proposes some policy directions.

In Chapter 16, **Emilia van Egmond** first highlights some major challenges which developing countries face. They include: the high proportion of the urban population in these countries who still live in slums which lack durable housing, adequate living space, improved sanitation and or clean water; the need to safeguard the natural environment and resource base; and how to address water scarcity, deal with waste, and mitigate against greenhouse gas emissions. She notes that this poses a challenge to the construction industry and also offers an opportunity for innovation and the development of more efficient (green) technologies that will contribute to sustainable development. Van Egmond presents case studies of construction technology development and innovation in some developing countries. In these cases, she addresses two main questions: how, and in which form, were the technology developments and innovations accomplished and applied; and why were they not taken up on a larger scale although they appeared to provide better results than the existing technologies in terms of performance? She makes some observations on the way forward with respect to technology development and innovation in construction in developing countries.

A synthesis

There is a paradox between the relatively low level of development of the construction industries in the developing countries, as well as their practices, tools and techniques, on the one hand, and the immensity of the problems which have to be addressed on projects and within the industry on the other. It is ironic that, contrary to the expectations of most researchers, the analytical techniques and tools which are available need to be more robust if they are to be successfully applied in the developing countries.

Relationships can be traced through all the topics discussed in the chapters in this book. A few of these are outlined by taking some of the topics at a time.

Developing countries have been urged to consider using the PPP approach to procurement to close the gap between needs and provisions in terms of both volume and quality in their infrastructure and other public services (World Bank, 2005). The calls were intensified as the approach seemed to be working in some developed countries, and the few developing countries such as China and Malaysia that had adopted it. Whereas PPPs appear to have great potential in the developing countries, it is pertinent to consider some questions. What are the peculiarities of PPPs? In what circumstances are they most likely to succeed? Are the developing countries ready to adopt this relatively novel approach? What are the risks to these countries? How can they be addressed? These are questions to which the developing countries would like to have answers. This is because a PPP programme would involve a great deal of risk on the part of the government. On such projects, implementing initiatives to promote social opportunities, as well as to attain transparency, would help to mitigate against some of the political risks. To these ends, stakeholder management should be a key activity.

The employment generation potential of construction is commonly highlighted. That construction projects are location specific is seen as a way of spreading the practical realisation of this possibility throughout the country. In order for the construction industry to deliver on its employment generation potential, appropriate techniques and methods should be applied on projects. The selection of these inputs is facilitated by the procurement approach adopted, as it can specify the materials which should be used, and, to a large extent, how the work is to be done. It is also relevant to link this to the importance of OSH in construction as action is taken to attain the labour intensity of the projects. Moreover, it would also be appropriate if the opportunities were to create green and decent jobs. Indeed, the creation of green jobs has been considered as a major national objective in most developed countries (Booz Allen Hamilton, 2009). In taking measures to revamp their economies after the 2008 global economic and financial crisis, the US and many European countries proposed green programmes, from which large numbers of jobs would be created.

Successful transfer of technology and knowledge to construction companies in the developing countries will foster innovation and help the firms to improve their performance for the benefit of all the stakeholders in the projects such as governments and society as a whole. An example of these technologies is ICT. For instance, its successful application will enable risks to be identified and appropriately managed. This will then lead to enhanced performance on construction projects.

Leadership is necessary if projects are to be effectively managed in developing countries, and performance on the projects enhanced. This brings up the importance of project leadership in comparison with project management. The former is not focused on anywhere, although it has been shown that it is more important as a determinant of success on projects than the latter (see, for example, Russell et al., 2006). Leadership is also key to the relief of the hindrances to efforts to improve performance in the industry such as effective application of ICT, and the cultural change that will be necessary to enable OSH practice in the developing countries. Thus, leadership development would be of great benefit to the developing countries.

In post-disaster and post-conflict situations, good performance on projects is critical, in order to minimise cost and stretch budgets to accomplish greater volumes of output over the shortest possible time, factoring in social needs to provide jobs and utilise displaced former combatants and other unskilled people. In these situations, effective stakeholder management is also critical if the needs of the beneficiaries are to be met. Moreover, the construction activity in a post-disaster or post-war situation is among the most challenging, for example, as logistics is a major problem owing to the damage to the infrastructure. At the same time, in such a situation, construction activity is of an importance that goes far beyond the provision of physical items, or even jobs, to contribute to the rehabilitation of ex-combatants, and the rebuilding of the fabric of the society. Some pertinent questions are: What are the most suitable scheduling systems? What technologies could be used and which procurement approaches would be best? How should the community be engaged?

So what is to be done? With their difficult operating environments, the developing countries can offer the crucible for challenging studies which will lead to breakthroughs in scheduling, cost modelling, risk analysis and management techniques, and benchmarking. Other relevant issues include human resource management and communication, strategic alliances, supply chain management as well as contractual matters and those relating to inter-personal conflict. These are pertinent because many of the larger projects essential to the development of the countries are undertaken by foreign companies with workforces from several different countries. Effective stakeholder management is another subject that offers opportunities for advancing knowledge, owing to the large number and variety of stakeholders of different backgrounds in each project.

The discussion in this chapter shows that it is necessary for research in Construction Management and Economics to consider the circumstances of the developing countries, either in the design of the research, or in the interpretation of the findings. Of even greater importance is that research should be undertaken from the perspective of the developing countries on every new topic that emerges in the field of construction. This has been a driver of the chapters of this book.

References

Abdul-Rahman, H., Rahim, F.A.M., Danuri, M.S.M. and Low, W.W. (2007) A study on quality management during the pre-construction stage of design-and-build projects. Proceedings, CME 25 Conference, Reading, 16–18 July.

Alinaitwe, H.M. (2009) Prioritising lean construction barriers in Uganda's construction industry. *Journal of Construction in Developing Countries*, Vol. 14, No. 1, 15–30.

Booz Allen Hamilton (2009) *Green Jobs Study*. Washington, DC.: US Green Building Council.

Du Plessis, C. (2002) *Agenda 21 for Sustainable Construction in Developing Countries*. Rotterdam: International Council for Research and Development in Building and Construction and United Nations Environment Programme.

Egan, J. (1999) *Rethinking Construction*. Report of the Construction Industry Task Force. London: Department of the Environment, Transport and the Regions.

Flanagan, R. (2005) The synergy between business and global drivers in future planning for construction enterprises. In Kahkonen, K. and Porkka, J. (eds) *Global Perspectives on Management and Economics in the AEC Sector*. Technical Research Centre of Finland and Association of Finnish Civil Engineers, Helsinki, pp 1–15.

Gunhan, S. and Arditi, D. (2005) International expansion decision for construction companies. *Journal of Construction Engineering and Management*, Vol. 131, No. 8, 928–37.

Hillebrandt, P. (2000) *Economic Theory and the Construction Industry*, 3rd edition. Basingstoke: Macmillan.

Kumaraswamy, M.M. and Shrestha, G.B. (2002) Targeting technology exchange for faster organisational and industry development. *Building Research and Information*, Vol. 30, No. 3, 183–95.

Ling, Y.Y. and Hoang, V.T.P. (2010) 'Political, economic, and legal risks faced in international projects: case study of Vietnam'. *Journal of Professional Issues in Engineering Education and Practice*, Vol. 136, No. 3, 156–64.

McGraw-Hill Construction (2008) *Key Trends in the European and US Construction Marketplace – SmartMarket Report*. Bedford, MA.

MDG Task Force (2010) *The Global Partnership for Development at a Critical Juncture – MDG Task Force Report*. New York: United Nations.

Nam, C.H. and Tatum, C.B. (1997) Leaders and champions for construction innovation. *Construction Management and Economics*, 15 (3), 259–70.

Ofori, G. (1993) *Managing Construction Industry Development: Lessons from Singapore's experience*. Singapore: Singapore University Press.

Ofori, G. (2003) Frameworks for analysing international construction. *Construction Management and Economics*, 21(4), 379–91.

Ofori, G. (2007) Construction in developing countries. *Construction Management and Economics*, 25(1), 1–6.

Ofori, G. (2009) Leadership and the construction industry in developing countries. Paper presented at CIB W107 on Construction in Developing Countries, Penang, 5–7 October.

Russell, J.S., Doll, N., Orner, K. and Sullivan, G. (2006) Leading with heart in the construction industry: preparing level 5 leaders at UW-Madison. Paper presented at Center for Project Leadership Forum, New York.

United Nations (2010) *The Millennium Development Goals Report 2009*. New York: United Nations.

World Bank (2005) *Infrastructure and the World Bank: A progress report*. Washington, D.C.: World Bank.

Part I

Industry practices

2 Public–private partnerships for major infrastructure projects in developing countries

Asanga Gunawansa
National University of Singapore

1. Introduction

Public–private partnerships (PPP) is the name given to arrangements between public-sector entities and private-sector ones to develop and/or deliver services to the public, a model that is widely used for developing infrastructure facilities. It is a procurement method that involves private-sector supply of infrastructure assets and services.

PPPs are implemented for a wide range of social and economic infrastructure projects. PPPs have a successful track record in developing traditional infrastructure projects such as airports, power facilities, telecommunication facilities, roads, tunnels, bridges, water and sanitation projects, which have traditionally been provided by the public sector. PPPs are also used to implement a wide range of social infrastructure projects such as hospitals, schools and prisons that were also considered as falling within the realm of public services. Thus, PPP is often identified as an arrangement in which the private sector supplies infrastructure assets and services traditionally provided by governments.

PPP is seen as a different concept from outright privatisation. It provides a structure which exploits some of the lessons countries have learned from privatisation. Privatisation was seen as too blunt and abrupt a technique to get the private sector involved in the provision to the public of services that were previously strictly kept within state control. The difficulty of selling the idea of privatisation to the public who have got used to free or heavily subsidised services from the public sector was one of the key reasons for the failure of privatisation in many countries. Compared to total privatisation, the slower, balanced process of PPP is felt to enhance value to the government and provide a long-term continuing participation of the government in functions of public interest. Furthermore, PPP provides the flexibility governments lacked in privatisation as it allows governments to engage the private sector to develop and operate facilities whilst having some control in the projects as partners. Moreover, if structured properly, PPPs should create a fair distribution of wealth, avoiding the financial windfalls that are often-criticised elements of privatisation.

The needs of public-sector as well as private-sector entities to enter into PPPs can differ from project to project, and from one jurisdiction to another. For example, the need of a cash-strapped developing country to enter into a PPP to develop a project to provide clean water or electricity to its citizens will be different from the need a developed country may have in considering a similar project. As a result, different definitions have been given to PPPs in different countries. As far as private-sector entities are concerned, they may enter into PPPs for the purpose of gaining access to new markets, or in search of new profit-making ventures. For some private-sector entities, it may be a way of exporting their technology. For some others, entering into PPPs may be an extension of their corporate social responsibility in assisting public-sector entities to provide better facilities to the public (Gunawansa, 2000).

The concept of the public sector and private sector partnering for the development and provision of services is also known by different names in different jurisdictions. In the UK, the term used is private finance initiative (PFI); in Australia, the expression is privately financed projects (PFP); and in the US, PPP is commonly used (Yescombe, 2007). The other popular terms used for public–private partnerships include private participation in infrastructure (PPI); and private sector participation (PSP).

Sometimes, the names of PPP projects are based on the procurement model used for developing such projects, or the project financing model used for funding them. For example, names used to refer to PPP contracts include build–transfer–lease (BTL); build–own–operate–transfer (BOOT); build–operate–transfer (BOT); build–own–operate (BOO); and operation and management (O&M).

Statistics from the World Bank and the Public–Private Infrastructure Advisory Facility (PPIAF) PPI Project Database show that the total investment commitments to infrastructure projects with private participation in developing countries, by subsector in 1990–2008 were: telecommunications, 54 percent; electricity, 23 percent; water, 3 percent; roads, 8 percent; seaports, 4 percent; airports, 3 percent; and railways and natural gas, 2 percent each.

Definition of PPP

The partnership of modern legal systems is based upon the *societas* of Roman law. There is no specific definition given to *societas* by any of the Roman jurists. However, the Roman understanding of what a partnership should be is sufficiently expressed in the definitions given by several jurists of other civil law jurisdictions. For example, according to German jurist Samuel von Pufendorf, a "partnership is a voluntary contract between two or more competent persons to place their money, effects, labour, and skill, or some or all of them, in lawful commerce or business, with the understanding that there shall be a communion of the profits thereof between them". According to the French jurist, Pothier, a "partnership is a contract whereby two or more persons put, or contract to put, something in common to make a lawful profit in common, and reciprocally engage with each other to render an account thereof" (Story, 2007).

Today, the concept of "partnership" is understood as a legal relationship of voluntary association of two or more persons for the purpose of gain. Many countries have defined the concept in various statutes. Although these definitions may identify different types of partnership, limit the number of partners and specify their rights and obligations under a partnership arrangement, the general idea given to the concept is not far from what has been described above.

Thus, it may be concluded that a public–private partnership is a contractual arrangement in which a public-sector entity and a private-sector entity come together to engage in a commercial activity with the expectation of mutual gain. However, as observed by Khanom (2009), there is no precise and commonly accepted definition of PPP; and the difficulty in defining PPP arises as a result of the diverse interests and objectives which the public and private parties have in entering into PPPs.

One definition of PPP, as embraced by The Canadian Council for PPP, provides:

> A cooperative venture between the public and private sectors, built on the expertise of each partner, that best meets clearly defined public needs through the appropriate allocation of resources, risks and rewards.

In the UK, Her Majesty's Treasury (1998) defines PPP as:

> An arrangement between two or more entities that enables them to do public service work cooperatively towards shared or compatible objectives and in which there is some degree of shared authority and responsibility, joint investment of resources, shared risk taking and mutual benefit.

In Singapore, the Ministry of Finance (2004) defined PPP thus:

> PPP refers to long-term partnering relationships between the public and private sector to deliver services. It is a new approach that Government is adopting to increase private sector involvement in the delivery of public services.

In India, the Department of Economic Affairs of the Ministry of Finance (2005) defines PPP as follows:

> The Public–Private Partnership (PPP) Project means a project based on a contract or concession agreement between a Government or statutory entity on the one side and a private sector company on the other side, for delivering an infrastructure service on payment of user charges.

The Canadian definition focuses on the cooperative venture between the public and private parties and the appropriate allocation of resources and risks. This gives the indication that PPPs are considered as partnering arrangements between parties with equal bargaining power. Similarly, the UK definition focuses on compatibility between the parties and sharing of responsibilities, risks, resources and profits. The Singapore definition focuses on the PPP as a long-term relationship between the public and private sectors that enables the public sector to involve the private sector in providing services to the people. This definition does not give any indication as to the real need for the public sector to enter into PPPs. Moreover, in Singapore, PPP is also seen as a way to bring in specialist private-sector expertise to stimulate an exchange of ideas and bring more international players into the domestic market (KPMG, 2007). The Indian definition focuses on the fact that the government gives a concession to the private sector to develop a project and provide services in return for the payment of user charges. This gives the indication that the public sector's engagement in the partnership is limited to the granting of the concession, due to financial constraints and lack of modern technology. The private sector is required to finance and develop the project and offer services in return for payments for services provided.

Despite these subtle differences in the various definitions, most versions of PPP as a procurement mechanism are very similar, although the degree of control shared by the partners, and several other characteristics of the partnership, may receive different emphasis from definition to definition. The role of the public sector varies from being the concession granter that authorises the private sector to participate in the business of providing public services, to acting as promoter or facilitator of a project, or a joint venture partner or shareholder of a private sector-led company that develops and operates a project. Sometimes the public-sector entity is also the purchaser of the services delivered by the project.

The role of the private sector also varies, depending on the type of procurement model used for project development and operation by the parties. Thus, depending on the type of agreement, the private-sector entity can be the project promoter and an equity holder of the project company. In addition, it can also play the role of the construction contractor, management contractor or the project operator.

Main characteristics of PPP

In PPP, a public-sector entity usually takes the decision to develop a project and specifies the outputs or services required. The responsibility for financing the project often lies with the private-sector developers, although sometimes public-sector financing will be made available for the project. The private-sector project developers also often have to bring other resources, including the technology and management expertise required for the development. Given the large sums usually needed to develop infrastructure projects, the equity contribution of the shareholders comprising mainly private entities may vary from 20 to 40 percent. In the circumstances, the larger share of the development cost of the project often has to be borrowed from lenders. Thus, PPP projects are usually financed with a combination of debt and equity in a ratio that sees most of the development finance being provided in the form of debt (Rondinelli, 2002).

A private consortium would often be responsible for the equity financing, design, construction, operation and maintenance of a facility. In addition, it may also have to secure the necessary debt portion of the project development cost. The project consortium will often be a limited liability, special purpose vehicle (SPV) created solely for the purpose of developing the project.

The PPP contract between the public agency and the consortium would usually be for a period of between 10 and 30 years. Unlike traditional procurement methods, the public sector may not own the facility during this time period. Thus, a PPP allows the public sector to move away from directly owning and operating facilities that provide services to the public. It is often only after the contract period that the facility returns to public ownership.

In a PPP, the private consortium is normally required to recover its investment through income earned by operating the facility. However, the public sector may sometimes compensate the consortium with service payments, rights to levy tariffs or fees against the public users, or a combination of these. Moreover, for a project that produces a public utility service, an off-take contract may sometimes be signed between the consortium and the public agency, whereby the latter agrees to purchase the output of the facility at an agreed price and volume on a long-term basis (Yescombe, 2002). This type of off-take arrangement is mostly agreed in connection with projects in developing countries, where the governments are unable to guarantee that the end-users will use and pay for the services provided by the consortium (Gunawansa, 2000).

Another characteristic of PPP projects is that lenders usually finance projects without requiring security outside the project assets. This feature of PPP projects is popularly known as the "non-recourse" nature of financing projects, or "off-balance sheet" financing. The advantage of non-recourse financing is that it allows the shareholders of the project consortium to keep financing and project liabilities outside their non-project-related assets. This reduces the impact of the project on the cost of the shareholder's existing debt and the shareholder's debt capacity, releasing the debt capacity that the project would have taken up for additional investments (Delmon, 2009). In non-recourse project financing, the lenders' recourse in the event of debt default will be limited primarily or entirely to the project assets and any performance guarantees and bonds provided to secure the debts.

However, it should be pointed out that, in the case of developing countries, most projects are financed on "limited recourse" basis. This is because, due to the presence of financial, political, market and various other risks in such countries, lenders are reluctant to finance projects without security from the investors and/or guarantees from the public-sector project sponsor. Thus, limited recourse financing may be structured to provide lenders with recourse to the assets of the shareholders of the project consortium in certain specific situations (Delmon, 2009). In the case of developing countries, on the basis that project failures are often blamed on the failure of public-sector entities to fulfil their contractual undertakings, lenders often seek direct sponsor guarantees to support the release of debt financing for infrastructure projects (Gunawansa, 2000).

The contractual structure used in PPP projects would normally consist of a concession or a licence issued by a public-sector entity to a private-sector organisation to develop the project. The private-sector entity, through a combination of shareholder agreements with other private-sector partners (the equity portion of the project development cost) and loan agreements with lenders, would then raise the funds required for the project (Savas, 2000). As stated above, in order to keep their project assets off balance sheet, the developers are likely to raise the funds from the lenders on a non-recourse or limited recourse basis. The developers will be given a reasonable period of time (usually 10 to 30 years) by the public-sector entity to manage the project, once developed, in order to enable them to recover the development cost, pay off the debt, and earn profits (Gunawansa, 2000).

Typically, the PPP payment mechanism provides the government with the power to withhold or deduct payments if the quality of service provided by the private-sector consortium is lower than agreed. The government may also reserve the right to step in and regain control of the asset in the event of repeated default in service provision by the private-sector consortium.

Another key feature is that, although PPP is seen as an escape route for public-sector entities as far as financing and developing certain infrastructure services for the public are concerned, the duty of providing the public services will often remain with the public sector under the applicable laws of the country. For example, water is now recognised as a basic human right, and provision of water is considered as a public duty of the State by many governments. Thus, although cash-strapped governments may enter into PPPs for the development of water-related infrastructure and delivery of water to the people, public entities may not be allowed to renounce their duty to the public. In the circumstances, public entities often continue to regulate the prices and the quality of services provided to the people by project companies created under PPP arrangements.

Key project participants

What is unique in PPPs is that, in addition to the main project participants, as far as the partnership is concerned, namely, the public-sector entity and the private-sector consortium, several other parties play major roles in the project. These parties include international financing agencies and domestic and local banks which finance the debt portion of the project development cost as lenders, designers and contractors which work on the project, suppliers, underwriters (insurers), the operator of the facility, and off-take purchasers which pay for the services provided by the project. These different players are brought together by their public or commercial needs and they team up for the purpose of achieving their individual goals by co-operating with each other. Figure 2.1 shows how these different parties with their diverse interests combine to form a project company to develop an infrastructure project.

Each of the parties shown in Figure 2.1 has a critical part to play in a PPP project, often at different stages of the project's life. Because PPP projects are often financed on the basis of non-recourse or limited-recourse to the project sponsor, various project-related risks must be efficiently allocated to parties which will assume recourse liability and which possess adequate resources or access to credit to accept the risk allocated. The allocation of risks can vary from transaction to transaction. It is largely dependent on the bargaining power of the project participants and the ability of the project to cover risk contingencies with the underlying cash flow and reserve accounts. The major risks associated with PPP projects and the risk mitigation measures are discussed in more detail below.

Public entity

The role of the public entity in a particular project will depend on the nature of the project. Given a choice, a public entity in a developing

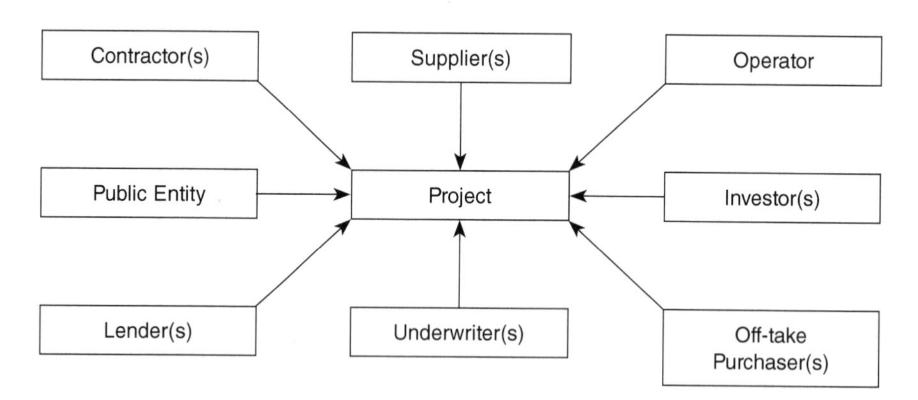

Figure 2.1 Key parties in a PPP project

country would prefer to act as the regulator of an infrastructure project and have the minimum level of involvement in the funding, management and operational aspects of the project. However, the host government is usually expected to play a greater role, as explained earlier. At a minimum, the host government is likely to be involved in the issuance of a concession to develop the project and, in addition, in the issuance of consents and permits on a periodic basis throughout the duration of the project. In some instances, the host government may also be required to enter into a support agreement with the private-sector consortium that confirms the continued commitment of the government towards the development of the project. In some cases, a public-sector entity in the host government may be the purchaser or off-taker of the products of the project. Furthermore, in some instances, a public-sector entity of the host government may even be a shareholder of the project company, sometimes by virtue of holding a golden share in consideration of the concession granted or the land provided for the development of the project.

Investors

A project needs a champion who can take the initiative to drive the project from the start-up stage to its completion. In a PPP, as noted above, since the private sector plays a major role in promoting, financing, developing and operating a project, it is important to have a private-sector investor who takes the lead in bringing in other investors as shareholders and initiates the negotiations with the public sector for developing the project. In PPP, such a leader can be a single entity or a joint venture.

Investors will have to resort to their own resources for the equity injections required for developing the project. Depending on the project, there may also be investors who come in either for the anticipated returns or for a combination of purposes, including the possibility of being suppliers to the project. Such investors may provide either pure equity or quasi-equity financing such as subordinated loans with an equity kicker (Denton Wilde Sapte, 2004). Thus, usually, investors take a stake in the project company, which is normally an SPV established to own and/or operate and manage the project.

Investors can come into a project in two different ways. Sometimes they act as sponsors of a project by bidding for a project for which a host government has already called for tenders or expressions of interest. There are other instances where the investors identify a commercially viable project and make representations to the host government to obtain a concession to develop it. It is not uncommon for the investors to find local partners who may or may not be government entities. In developing countries, the current trend is for a government entity to join the project promoter in forming the project company. There are two reasons for this trend. First, the host government may not want to transfer total control of an infrastructure facility to a private-sector consortium. Second, the investors may search for local participation in equity with the aim of reducing investment-related risks, and also to secure local commitment in order to enhance the chances of the success of the project.

Lenders

The largest share of project financing for PPPs normally consists of debt, which is usually provided by lenders who have no direct control of the management of the project. Lenders invest in debt, either on their own or as part of a group of banks providing syndicated loans. The lenders usually come into the project after the preliminary negotiations of the contract between the project sponsor and the government are concluded, and at the stage during which the host government will be interested in examining the financial plan of the sponsor. Although the lenders do not participate in the initial negotiations prior to the preparation of the main contracts for signature, they play an active role once the main negotiations between the parties begin.

For a party that may not have any management control over the project company, except for interests over its project assets, the role played by the lenders during negotiations relating to the project is very significant. In this author's view, the lenders hold the key to successful negotiations as they finance the major portion of the project development cost.

In their typical role as bankers, lenders are reluctant to assume risks, and thus the major interest of the lenders during the project negotiation process would be to ensure that project risks are minimised and the remaining risks are efficiently allocated among other parties. Thus, lenders can delay the process of negotiations, as they will object to risks that have not been hedged adequately by their standards. There are many guarantee structures to get around this, but the transaction costs associated with these could be very high, in both time and money (Denton Wilde Sapte, 2004).

As noted earlier, due to the limited or non-recourse nature of project financing in PPPs, the lenders will have to be cautious and ensure that necessary protections are in place for their investments as, usually, the sponsors will not be liable for debt beyond the assets of the project company. Accordingly, the lenders will try to ensure that the project contracts include a security package that will provide assurance to the lenders that their loans will be paid effectively, efficiently and as scheduled. A typical security package will include a mortgage on available land to the project and its fixed assets; sponsors' commitment to the project including a share retention agreement and a project fund agreement; and assignment of major project agreements, including construction, supply and, if any, off take contracts to the lenders. Moreover, lenders may require financial covenants ensuring prudent and professional project management, and assignment of insurance proceeds in the event of a project calamity (Scriven et al., 1999).

In practice, most lending institutions insist on direct agreements governed by the law applicable to the financing documents, from each of the parties contracting with the project company. The main purpose of these direct agreements is to ensure that the project contracts will not immediately fall away if the project company is in default of its obligations under them, and that, if the lenders enforce their security, the project can continue either under the control of the lenders or, if the lenders transfer it, under the control of a third party purchaser. Moreover, the law governing the financial arrangements will most likely be one chosen by the lenders, and they will seek, as far as possible, to insulate themselves from the effects of local law in relation to financing arrangements. However, this will need clever negotiation on the part of lawyers representing the lenders as it will usually be necessary for the law governing some of the securities, for example, land, to be that of the country in which the secured assets are located (Scriven et al., 1999).

Construction contractor

A PPP project can be negotiated between a host government that grants the concession, an investor or a group of investors willing to accept the concession and develop the project, and the lenders who are willing to finance a major portion of the project cost. However, if the project is to be successful, it is also necessary to have a qualified contractor which will undertake the construction of the project. Sometimes the contractor may also be the sponsor of the project or one of the investors in the project consortium.

In PPP projects, careful consideration is given to the selection of the construction contractor to ensure that the chosen contractor is capable of completing the construction within the agreed time period and the allocated construction cost, and of meeting the commercial requirements of the project. Failure in any of these three aspects would affect the project consortium's expected gains from the operation of the facility as it has a specified time period for the operation of the project. The construction contractor will be concerned with the underlying financing structure of the project, including whether the sponsor has arranged for sufficient funds to pay the contractor for the work to be performed.

Operator

Most PPP projects use sophisticated and modern technology in the provision of public utility services. The operator is charged with the responsibility of operating such a facility. Frequently, the construction contractor and the operator of the facility are a single entity. Sometimes, the operator is also part of the private-sector consortium that developed the project. Usually, the operator will be expected to sign a long-term contract with the project consortium for the operation and maintenance of the project.

The operators tend to accept little risk in the form of up-front capital or expenditure. An operator simply anticipates making a profit from operating the infrastructure more efficiently than a comparable government-run facility (Hoffman, 1989).

The successful operation of the facility is vital to the generation of the cash flow necessary for the economic viability of the project. Thus, a key issue will be whether the operator of the facility is prepared to guarantee certain operation levels, for example in terms of production or efficiency, or whether it is only prepared to commit to more general operating obligations such as a duty to operate the facility in accordance with good industrial practices. The sort of undertaking that should be obtained from the operator will usually depend on the type of facility to be provided by the project. If the facility is new and uses new and untested technology, the operator will be hesitant to give guarantees on specific levels of achievements. On the other hand, if the project involves taking control of a facility previously provided by the public sector, it will be in the interest of the host government to insist on improved levels of achievement, as, otherwise, the very purpose of private-sector participation in the project will be futile.

Suppliers

The suppliers to PPP projects range from companies which supply goods such as fuel, raw materials and equipment, to those which provide services such as insurance, banking,[1] electricity and telecommunication to the project. The suppliers have the objective of delivering to the project the necessary goods and services in exchange for a price. The other project participants will seek firm price, quality and delivery commitments with a minimum of uncertainty in the price, terms, and time of supply from the suppliers. On the other hand, the suppliers will want to maximise profit and may sometimes even form part of the project consortium in order to ensure continued demand for their goods and services.

Output purchasers (off-takers or end-users)

At the heart of the viability of a PPP project is the revenue stream generated by the commercial operation of the facility. In the circumstances, the off-takers that enjoy the end product of the facility are essential for the success of the project. Infrastructure services provided by a PPP project may be purchased by a single customer (an example would be power supplied by an independent power producer to an electricity utility) or by many users (such as the users of a toll road).

1 The reference made to banks here is to commercial banks which will provide banking services to the project. Thus, they are different from the banks that will provide the necessary debt for project financing transactions.

The off-takers desire as firm a price and quality as the market will permit, and a reliable source of output at an acceptable price. On the other hand, the project promoters will expect a steady market for the output in order to service the loans and generate profits. Unless the project involves the construction and operation of a facility in which sufficient income returns can be projected with reasonable accuracy, such as in the case of a port facility with an already established clientele, the investors and lenders will be more comfortable in entering into an off-take contract with some state-sector utility purchaser. For example, in most cases, a power project in a developing country includes a power purchase contract between the project company and the state-owned power supply agency. The agreed price and the agreed period will usually depend on the project's life term as specified in the project contract, and would usually allow sufficient time for the payment of debts and equity and also for the return of reasonable profits to the investors.

Insurers and underwriters

Insurance companies and underwriters will be brought into a PPP project to cover those risks that the parties accept but wish to mitigate. Lenders will wish to ensure that the insurer is of reputable standing, and would be able to pay quickly after a claim is made. From this standpoint, local insurers in developing countries may not be acceptable to lenders due to their limited resources and lack of exposure to complex project financing transactions. The lenders will wish to ensure that all insurance is maintained with insurance companies they have approved beforehand. In circumstances where having local insurers is a legal requirement (even when they do not have adequate financial standing or reputable claims settlement history), lenders may mitigate the risk of failure to perform by such an insurer by requiring (if the project company has not already done so) that the risks are reinsured with international insurance companies of adequate calibre and by requiring an assignment of such reinsurance proceeds.

While it is in the best interest of lenders to, and investors in, a project to insure the project against various risks, the host governments too will have an interest in ensuring that sufficient insurance cover is obtained. This is because the failure of a project would have a drastic impact on the ambition of the government to eventually inherit a fully operational project. Besides, project failure will also expose the host government to both political and public criticism and will discourage prospective future investors.

The insurance companies may be motivated in providing insurance to PPP projects due to the high premiums they can charge. However, as the payments involved will be huge in the event of any of the covered incidents happening, there are very few local and international insurers that can provide sufficient cover for various risks involved in infrastructure projects.

The basic contractual structure of a PPP Project is shown in Figure 2.2.

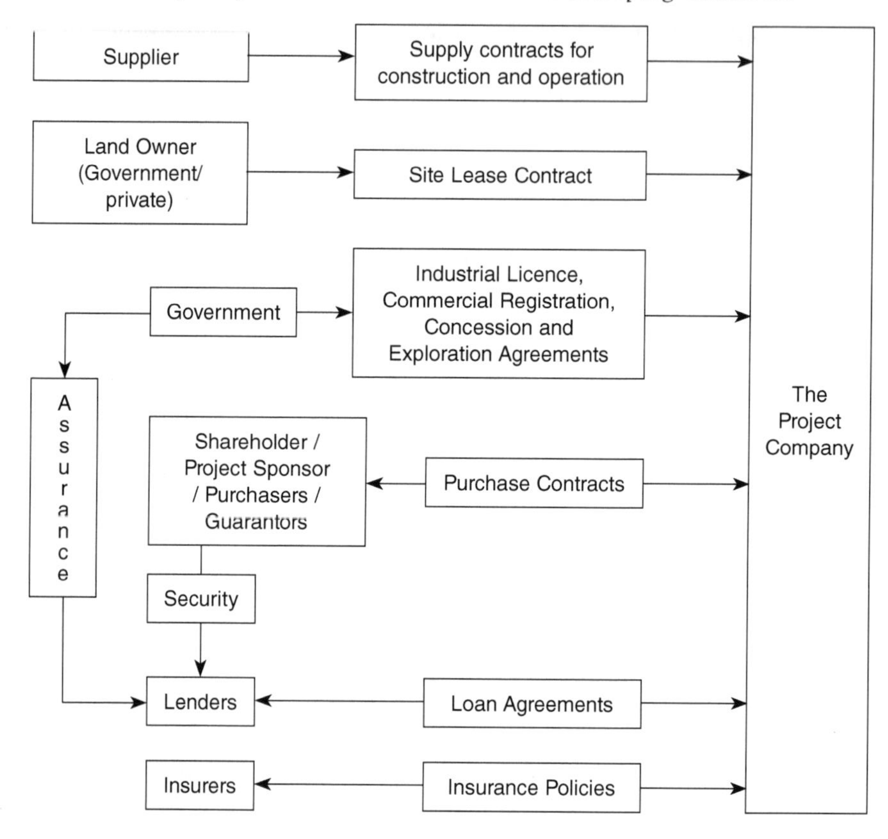

Figure 2.2 Basic contractual structure of a PPP project

Advantages and disadvantages of public–private partnerships

PPPs bring together governments on the one hand and investors and lenders on the other. Theoretically speaking, PPPs enable host governments to attract private capital investment without guaranteeing payment of project costs and without completely rearranging internal economic frameworks through direct privatisation. From the lenders' point of view, project financing allows them to lend on a project-specific basis to more efficient private-sector entities instead of lending money to public-sector organisations, which, over the years, and in many developing countries, have shown that they are not capable of developing and operating efficient infrastructure facilities. Moreover, PPPs enable the lenders to avoid lending money to public-sector entities in developing countries in situations that would present an otherwise unfavourable credit risk due to political instability or unrest, or other similar non-economic factors (IFC, 1999). For private investors, project developers, project operators and construction contractors, project financing opens new markets.

Table 2.1 identifies the significant advantages and disadvantages of PPPs to the key parties involved in the projects.

2. Development of PPP as a procurement mechanism for infrastructure projects

Although private investment in public infrastructure can be traced back to the 18th century in European countries (Kumaraswamy and Morris, 2002), according to Harris (2004), the increasing adoption of PPPs by countries in the late 1990s was due to the success of PPPs in the UK. It was the development and refinement of PFI by the UK in 1992, as one of a range of government policies designed to increase private-sector involvement in the provision of public services, which led to the renewed international interest in PPPs. Since then, many countries around the world have either embarked on, or considered, a PPP programme of their own (Harris, 2004).

Table 2.1 Advantages and disadvantages of PPPs to the key parties

Advantages	*Benefiting Party*
Moulding a project in a form which is compatible with government policies Maximising national sovereignty by ensuring that ownership of the project remains with the public sector and the project is handed back after the concession period is over Receiving subsidised or risk-free participation Sharing in the rewards of value-added services Training of labour and inheritance of modern technology Minimising any perceived adverse effects of foreign direct investment Improving predictability and stability of operational conditions and acceleration of infrastructure provision Reduction of project costs and time, while enhancing its overall efficiency and effectiveness Shifting some of the project risks to the private sector PPP may provide greater economic benefits than other forms of public procurement methods	Host Government / Public-Sector Entity
Access to new and previously restricted markets and sources of new profit Minimising political risk and better allocation of risks Availability of tax or other investment incentives	Investors and Lenders

Disadvantages	*Affected Party*
Exposure to risk of incompatibility with private-sector participants Need to grant long-term concessions over national assets Need to contribute capital or other assets Need to provide undertakings, guarantees and buy-back agreements to provide comfort to the investors and lenders Need to offer tax and other incentives Possibility of political and public criticism resulting from giving access to previously subsidised public utility sectors to the private sector Exposure to business risks "Soft" value of host country's capital contributions Less efficient decision-making and financing structures	Host Government / Public Entity
Exposure to risk of loss of confidential commercial information and know how Exposure to risk of incompatibility with government bureaucrats Exposure to unexpected regulatory and legal changes due to long life term of projects	Investors and Lenders

However, before the rebirth of PPP in the UK as a much developed and refined procurement mechanism that enabled the private sector to partner with public-sector entities in developing infrastructure, such mechanisms had been used in some parts of the world. For example, according to Kettl (1993), in the US, "every major policy initiative launched by the federal government since World War II, including Medicare and Medicaid, environmental cleanup and restoration, antipoverty programs and job training, interstate highways and sewage treatment plants, has been managed through public-private partnerships."

Development of PPP programmes

Initially, most PPPs were negotiated individually, as one-off deals. Thus, although some countries have had a taste of PPPs, it was only in 1992 that the first systematic programme aimed at encouraging PPPs was introduced when the Conservative government of Prime Minister John Major introduced the PFI arrangement in the UK (Allen, 2001). The 1992 programme focused on reducing public-sector borrowing, and developing infrastructure with private participation. Since then, many countries in various parts of the world, both developed and developing, have introduced their own versions of PPP programmes. These range from specific legislation to public policy statements and procurement guidelines.

As noted earlier, two alternative mechanisms that are considered for the engagement of the private sector are total privatisation of public facilities and PPPs (Ford and Zussman, 1997). The former enables governments to transfer the total responsibility of developing, managing and providing services to the people to the private sector. The latter enables governments to invite private-sector entities to finance and develop infrastructure projects without the state losing control over the regulatory aspects of service delivery, including the pricing of the services provided by the infrastructure facility (Savas, 2000; Gunawansa, 2000; Abdel-Aziz, 2007). The total privatisation of public infrastructure facilities that have provided services to the public at prices heavily subsidised by governments was considered politically controversial. Moreover, governments were hesitant to subject certain facilities to total privatisation due to reasons such as national security. Thus, PPP became the popular option.

Therefore, the concept of PPP was adopted by many countries and country-specific PPP programmes were introduced as they provided an opportunity for efficient allocation of project risks whereby risks could be borne by the parties best able to manage them. Moreover, procurement of PPPs became popular as they enable the provision of value for money to the public by drawing on the expertise of the private sector, and benefitting from its technology and management skills (Algarni et al., 2007; ADB, 2008). Moreover, as far as the developing countries are concerned, international financial agencies such as the International Monetary Fund (IMF) and the Asian Development Bank (ADB) also put pressure on governments to introduce PPP programmes as they consider the championing of the financing of development projects by private-sector entities to be a more viable option compared to the provision of direct soft loans to developing countries for infrastructure development (Gunawansa, 2000).

Need for private investment in infrastructure development

Infrastructure is essential for economic growth, which in turn is essential for poverty alleviation. According to the World Bank, expanding infrastructure to meet growing demands will absorb trillions of dollars of investment over the coming decades in the developing and transition economies (IEG, 2007). As far as developing countries are concerned, infrastructure has remained the largest component of public investment programmes, accounting for between 2 and 6 percent of gross domestic product (GDP). Nearly half of the project lending provided by international financial institutions to developing countries goes to infrastructure. The Organisation for Economic Co-operation and Development (OECD) estimates that developing countries might have to invest over US$700 billion a year in infrastructure in the coming decade in order to sustain rapid growth rates, and that this figure will rise to $1 trillion a year by 2030 (World Bank 2005).

Historically, public infrastructure projects have been primarily developed by public-sector entities using traditional procurement methods, such as Design–Bid–Build and Design and Build. Such projects were financed from allocations made from public funds or moneys borrowed or received under bilateral arrangements with developed countries, or funds received under lending facilities administered by international financial institutions such as the World Bank and the ADB. Once the projects were developed, the services were provided to the people, and the projects were managed, maintained and regulated by public-sector entities. Thus, the role of the private sector was limited to the design and construction of the facilities.

However, as governments started to face the challenge of stretching scarce public funds to deal with the increasing demand for new and modern infrastructure facilities, whilst also performing the other fiscal duties expected of modern governments which seek to safeguard the welfare of their citizens and improve their quality of life, the need to find alternative mechanisms to finance and develop public infrastructure facilities was recognised. As a result, PPP, under which the public-sector entities could partner with private-sector entities in developing, managing and providing services to the people, has became a popular option for many countries.

3. What PPP means to developing country governments

Until about the mid-1970s, the provision of utility services and administration of physical infrastructure in developing countries were tightly guarded by the state sector, with government departments or statutory corporations enjoying a monopoly. Private investor participation was restricted in the public utility and physical infrastructure sectors due to nationalist sentiments and concerns about foreign economic and political influence over strategic public utilities and state assets. That many developing countries had been under the colonial rule and power of Western European states until the early or middle part of the last century may have contributed towards this fear, as the colonial experience has left a legacy of concern that private-sector investments from foreign nations may serve as a modern form of economic colonialism in which foreign companies might exploit the resources of the developing countries (Gunawansa, 2000).

The need for PPP

In recent years, restrictions on private-sector investments in infrastructure development have been substantially reduced in many developing countries. This change has been largely the result of multi-lateral treaties such as the World Trade Organisation (WTO) and external pressures from the international financial agencies to liberalise economies and pave the way for economic development with market-based mechanisms. The globalisation and political development in the area of international relations have also contributed towards removing or radically altering some perceptions about the state's control and distributional functions with respect to infrastructure facilities.

Moreover, the generally poor performance of public utilities, and changing views on the role of the state in the economy, have contributed to the public provision of infrastructure falling from grace. Furthermore, the growing demand for increased as well as quality infrastructure services has not allowed developing countries to curtail the need for infrastructure development even when budgetary constraints limit the scope of public funding. This provided further impetus for the change in the approach of developing country governments towards private-sector participation. From another angle, financial deregulation in the capital markets has introduced new suppliers of equity capital to cross-border investment, thus making room for private-sector provision of utility services. Finally, technological developments in the area of communication have reduced the capital intensity and lead times involved in the provision of infrastructure services, thus expanding the potential for competition in activities that were once dominated by state monopolies.

In summary, the changes discussed above have resulted in one of the swiftest and most dramatic changes of context for utilities and infrastructure industries in developing countries. Intense global competition among large multi-national companies (MNCs) (in terms of both operations and ownership) with deep roots in the capital markets has replaced a landscape of national, often over-regulated monopolies in fragmented markets, financed primarily through national budget allocations, mostly deficits. With MNCs competing with each other to gain access to new and developing markets, with developing countries becoming increasingly more willing to adopt open market economic policies, and with local investors being willing to join hands with foreign investors, private-sector participation has grown as one of the most important forms of capital flow for developing countries.

Many argue that PPPs create "win-win" options for developing countries as well as for investors, lenders and other project participants. The main arguments that are put forward to support the benefits for developing countries include the inability of public finance to meet the growing needs for infrastructure development and the limited or non-recourse nature of project financing which reduces or removes the burden of servicing debt and equity from the developing countries. The main argument supporting the view that project financing techniques are equally beneficial to the investors, lenders and other project participants is that markets previously under state monopoly could now be accessed by the private sector. Theoretically speaking, PPPs have, to some extent, removed from the developing countries the burden of servicing debt and financing equity for developing projects by transferring that burden to a purpose-specific project company, i.e. an SPV, that is set up to run the facility during its agreed life term. The non-recourse or limited-recourse nature of project financing might ensure that the lenders and investors do not look beyond the project assets and project income for their loan payments and profit earnings. Thus, engaging in PPPs appears a viable option for developing countries.

However, in practice, the theoretical benefits of PPP only look convincing on paper as, in some developing countries, due to the unique nature of the risks associated with such projects, many development projects are financed with techniques that are not limited or non-recourse in nature. For example, many low-income countries with political and/or economic instability continue to be burdened with the obligation of servicing debt and equity with government pay-back guarantees, if they are able to attract investors. Moreover, although it is theoretically correct to say that PPP techniques have opened the doors for the private sector to areas that were previously closed, in practice the situation is somewhat different. Due to policy and regulatory inadequacies as well as the hostile reaction of end-users towards private-sector involvement in certain utility service sectors, project developers, investors and lenders do not find it easy to enter many developing countries. These risks are discussed in more detail below.

The real beneficiaries from PPPs have been the developing countries which have shown the potential for rapid economic growth, those which have large markets, and countries with comparatively risk-free investment environments. Assessments by the Independent Evaluation Group (IEG) of the World Bank (IEG, 2007) and results from both the IEG's and IFC's development outcomes tracking system (DOTS) reveal that countries with better investment climates were more likely to achieve the Millennium Development Goals (MDGs). The findings also show that a better investment climate is associated with better development results. Thus, countries such as China, India, Mexico and Brazil are amongst the biggest beneficiaries of the PPP model. According to the World Development Report 2005 (World Bank, 2005b), investment climate improvements in the 1980s and 1990s caused private investment as a share of GDP to nearly double in China and India. a recent report by the Commission on Growth and Development (2008) indicated that sub-Saharan Africa has seen its fastest growth in decades, owing partly to better microeconomic policies.

As far as PPP in infrastructure development assisting developing countries to eradicate poverty is concerned, Fan and Chan-Kang (2004) report good returns in India. They estimate a reduction in the poverty head-count of 10 people per 1 kilometre of road extension in low-potential rain-fed areas, and economic rates of return in the hundreds or even thousands of percent. Similarly, for China, they estimate high economic returns to road investments. Thus they conclude that, among infrastructure investments, roads had the greatest impact on reducing poverty. Considering another sector, the International Energy Agency (2004) estimates that energy, as a factor of production, accounted for 13 percent of China's GDP growth over 1980–2001, 15 percent of India's, 30 percent of Mexico's, and 77 percent of that of Brazil. The recent projects in these sectors in China and Mexico have been largely financed and developed under PPP schemes.

The World Bank (2009) also notes that, in 2008, about half of the World Bank Group's Investment Climate Advisory Services (ICAS) portfolio of projects supported 108 reforms, many of which contributed to improvements to business environment indicators across 23 countries. These reforms included the enactment of best practice laws, improvement of processes that resulted in cost savings to the private sector, and implementation of other proven best practices in areas such as taxation, trade logistics, alternate dispute resolution, business entry and other operating regulations.

PPP models widely used by developing countries

The most sought-after project financing techniques for PPPs aimed at infrastructure development are the BOO and BOT models. These two models involve a consortium submitting a proposal to finance, design, build, operate and sometimes transfer the project back to the host government after the expiry of the agreed operation period. Projects are financed with a combination of debt and equity from several key players including international financing organisations and private-sector project developers, thus making these models complex arrangements in which several parties with diverse interests come together to develop projects.

Key features of the BOO model

The BOO model is mainly used to develop projects such as limited-term power generation facilities which are not meant to be permanent fixed assets of the host country. These contracts (which usually last for 10 to 15 years) are more like stop-gap measures to facilitate the fulfilment of medium-term requirements in the provision of infrastructure facilities to the people. In the circumstances, contracts of the BOO type are not used on projects such as highways and port construction, as such projects would continue to exist and provide services even after the concession period is over. A good example of a typical BOO project is a temporary power generation facility which is expected to contribute to the national grid in order to address the host nation's immediate and up to medium-term requirements. Unlike in BOO, ownership of the project does not rest with the private entity in BOT. Only the development and operational rights of the project will be given to a private entity for an agreed period of time. During this time (usually a 20–30 year concession period) the private entity is expected to recover the project investment, operation and maintenance costs, and a reasonable profit. The project has to be transferred to the host government at the expiry of the agreed concession period.

Key features of the BOT model

It is important to note that the term 'BOT' itself is sometimes used generally to identify most of the project financing models, including BOO and variants of the BOT model that are currently in use. These include: Build Lease Transfer (BLT), Build Transfer (BT), Build Transfer Operate (BTO), Rehabilitate Lease Operate (RLO), and Build Own Operate Subsidise Transfer (BOOST). However, from a legal and contractual perspective, BOT refers to a contractual arrangement under which a private-sector consortium is given a concession to build a project, operate it for an agreed time period and then transfer it back to the host government.

BOT contracts are mainly used for infrastructure projects that involve high investment costs and modern technology, and thus are beyond the capacity of host countries to finance and develop. Typical projects which are developed using this model include highways, ports, dams, telecommunication facilities and water sanitation facilities which usually become permanent fixed assets of the host country when the project construction and operation periods are completed. The operation period is long enough to allow the investors to pay off the costs and debts and realise a profit, and is typically 20–30 years.

The government retains the ownership of the infrastructure facility, and most of the time it becomes both its regulator and the customer during the project operation period. In a typical power generation project where the facility is expected to contribute towards the national grid, the power purchaser is the government or a government-controlled entity. On the other hand, if the infrastructure facility is a highway, the users of the highway then become the direct customers by paying a toll to the project company. In most BOT agreements in developing countries, the government also inherits the facility as a fully operational item of infrastructure after the concession period.

BOT agreements tend to reduce market and credit risks for the private sector because, most of the time, the government is the only purchaser of the services provided by the facility. This reduces the risks associated with insufficient demand and ability to pay. Where the government is not the direct purchaser (such as on a highway facility), the project developers sometimes negotiate government payback guarantees in the event the volume of usage falls below the minimum target levels identified during the project negotiations. The negotiating power of the governments in such situations is limited, as the private-sector investors will avoid BOT arrangements where the government is unwilling to provide assurances that their investments will be paid back.

The government often remains responsible for establishing performance standards and ensuring that the project developer meets them. Thus, in essence, the government's role shifts from being the provider of the service to the regulator of its price and quantity. Such regulation is particularly critical in the utility sectors such as power and water supply, given that they are public goods and their delivery systems are natural monopolies. The fixed infrastructure assets are entrusted to the project developer for the duration of the contract, but they remain government property.

The initiative to develop a project can be taken by either the government or a project developer. When the government takes the initiative, the decision is usually based on the national demand for new or increased infrastructure services. The government may identify a specific project that needs to be developed and then follow the established government procurement procedure to call for proposals to develop the project. From the government's standpoint, the necessary preparation for an infrastructure development project would include the assessment of: existing infrastructure facility, capital and tariff regimes; current service coverage; general customer satisfaction; and current balance sheet (revenues versus costs). Based on such an analysis, the government should outline broad goals for improvements. These include setting coverage objectives and service standards, as well as ensuring transparency, efficiency and customer satisfaction. If it appears that private-sector participation might be an option, the next step should be the development of a multi-disciplinary review team. The review team should then conduct a more thorough evaluation of the current system and evaluate the technological, financial, social, political and legal feasibility of various solutions. It is important that the review team identifies and actively involves key stakeholders such as local residents, development committees, community organisations and NGOs in a meaningful way throughout the process. It is also important that this process be transparent and that the views of all stakeholders are actively solicited.

Most developing countries have competitive and open procurement practices. Thus, the bidding process for infrastructure development projects is usually competitive and transparent. It is important to note here that the United Nations Commission on International Trade Law (UNCITRAL) has model laws and guidelines concerning public procurement and infrastructure development contracts (examples are the UNCITRAL Model Law on Procurement of Goods and Construction, and the UNCITRAL Model Legislative Provisions on Privately Financed Infrastructure Projects). Many developing countries have modelled their own legislation or guidelines on the UNCITRAL models.

When the initiative to develop a project is taken by a private-sector entity, such a potential developer will have to convince the host government of the project benefits. Thus, the project developer will have to be equipped with information such as initial risk analysis and project feasibility study. A financial plan should also be ready to convince the host government that the developer has the capacity to raise the required equity and debts contributions. If the host government is convinced, the developer would be allowed to do the initial preparatory work such as preliminary negotiations with lenders, and preparing an acceptable project plan so that negotiations for a BOT agreement can begin. These initial works are usually carried out by the project developer under a Memorandum of Understanding that it has signed with the host government. The procurement laws in most countries provide exceptions to the open tendering procedures that allow direct negotiations with project developers in situations such as this.

4. Major risks associated with PPP projects in developing countries

The risks associated with PPPs are sometimes more complex than those associated with traditional forms of procurement. PPP mechanisms used in developing countries have not been sufficiently tested, as they are relatively new. There are not many projects around that have run their full lifecycle to provide examples of success or failure. In the circumstances, the manner in which risks are identified and shared between governments, investors, lenders and other project participants is still being explored on a trial and error basis. Harmonising the diverse interests of all project participants whilst trying to ensure project success is a challenge for all project participants (Beidlemen et al., 1990).

4.1. Traditional risks

There are a number of ways to categorise risks associated with PPP projects.[2] For the purpose of the discussion in this chapter, risks have been categorised first as traditional and non-traditional risks. The traditional risks have been further categorised as commercial and non-commercial risks. A further sub-categorisation has been introduced to commercial and non-commercial risks, presenting them as project-specific and non-project specific risks.

Figure 2.3 presents a summary of commercial and non-commercial risks associated with PPP projects. Tables 2.2 to 2.5 provide short descriptions of each category of traditional risk presented in Figure 2.3.

4.2. Some unique features of PPP risks in developing countries

Most of the risks associated with PPP projects are not new, as they have been associated with infrastructure development projects in both developed and developing countries for several decades. However, in the case of developing countries, project participants take a more cautious approach as the investment environments in such countries are considered to be volatile compared to those in the developed countries. Moreover, some of the risks associated with PPP projects have assumed a new outlook in developing countries in the wake of initiatives to attract more foreign investment and also due to the changing political, economic and social conditions.

2 For a detailed description of various risks associated with financing infrastructure projects, see: Tisley 2000. Also see: Nevitt and Fabozzi 2000.

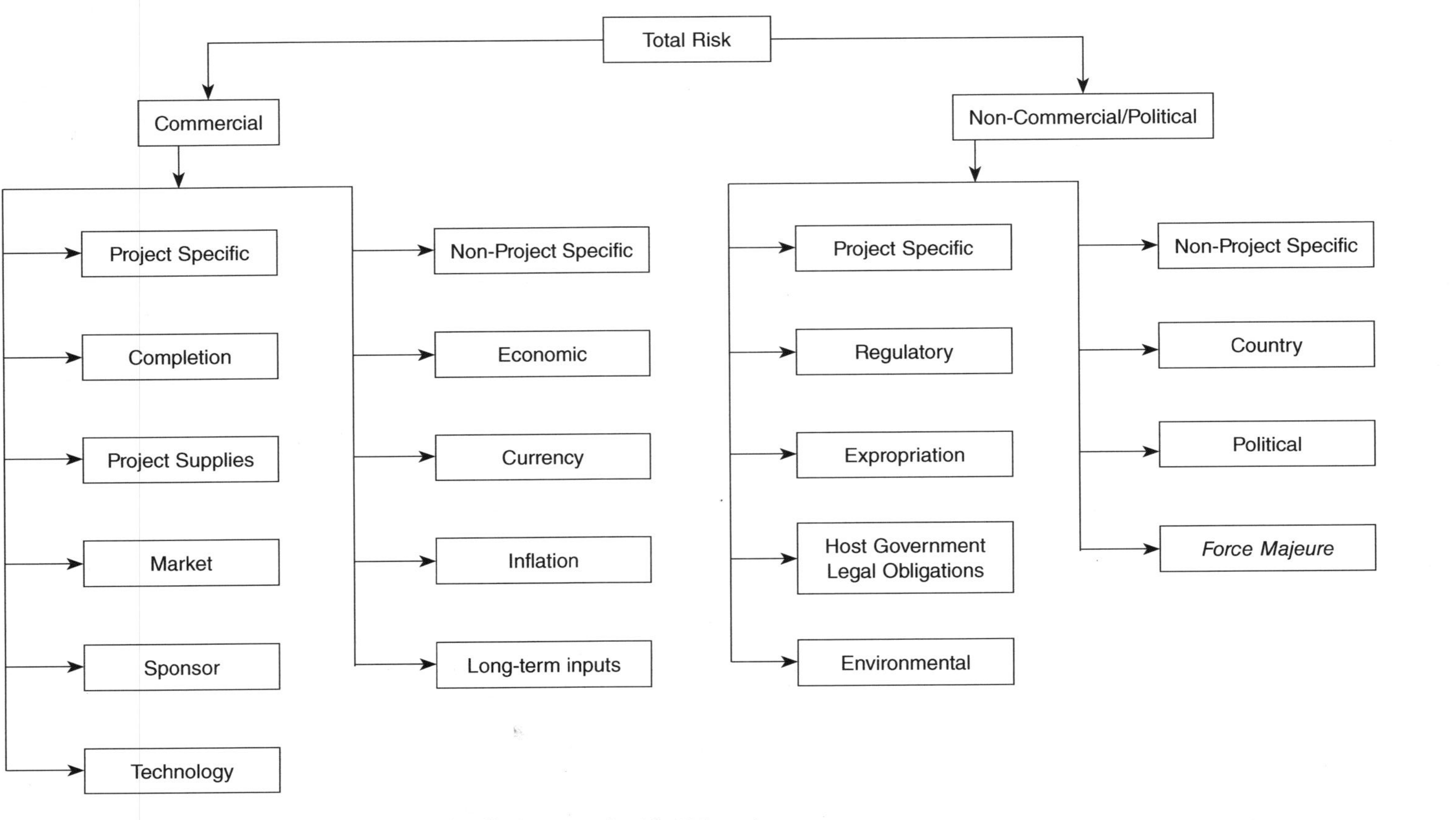

Figure 2.3 Commercial and non-commercial risks associated with PPP projects

Table 2.2 Project specific commercial risks

Risk	*Description*
Completion	The risk here is that the project may never be completed or that there may be delays in completion. Delays may occur due to various reasons such as failure of technology, cost overruns, required variations to the design, shortage of supplies and shortage of workers. Delays may also occur due to inefficiency of the contractors or the lack of commitment and supervision by the SPV, or both. There may also be time overruns on projects owing to delays relating to permits and approvals from the host government.
Project supplies	This risk usually occurs when the completed project does not get the necessary supplies to fulfil its full operational capacity. For example, if the project is a hydro power generation facility, the absence of rain would cause a supply shortage which would adversely affect the project and the operation of the facility.
Market	Inadequate demand or the reduction in demand for the services to be provided by the project is one of the major causes of revenue and profitability problems on most infrastructure projects. For example, developing a toll road or a pleasure port would not be worthwhile unless there is adequate demand for the services provided by such facilities.
Sponsor	Project sponsors of a PPP are the equity investors. The risk here is that the sponsors fail to provide the promised equity contributions and as a result the project falls short of the total amount of funds required for the development.
Technology	On a complex PPP project, failure of a new technology used in the project or facility is a possible risk.

Table 2.3 Project specific non-commercial risks

Risk	*Description*
Regulatory	On most PPP projects, the public sector retains the regulatory powers relevant to the services to be provided by the facility created by the project. Thus the regulatory risk is that the host government may introduce new regulations that would adversely affect the project and the operation of the facility, and the expected profits. An example would be the imposition by the host government of new taxes, planning controls or price controls.
Expropriation	On PPP projects, nationalisation of the project or public-sector takeover of the project or facility is a possible risk.
Host-government obligations	Statutory and regulatory changes by the host government (which might be) in breach of its contractual obligations are the risks involved here. For example, a new law might be introduced to remove a concession which had been promised to the investors. The imposition of exchange controls preventing foreigners from liquidating local financial assets would fall into this category.

Risk	*Description*
Environmental	There are two types of risks involved here. First, the failure by the SPV to take into consideration the relevant environmental laws and regulations and perform the necessary environmental impact assessments can expose the project to prosecution, civil litigation and criticism. Second, failure to take into consideration the environmental changes that might affect the long-term operations of the facility and introduce adequate adaptation measures might expose the project to operational delays and financial losses. An example would be developing a public housing facility in a seaside location without taking into consideration the adverse impacts of rising sea levels.

Table 2.4 Non-project specific commercial risks

Risk	*Description*
Economic	The risk here concerns the economic viability of the host state in which the project is developed. Economic stability may be adversely affected due to internal factors such as unsound economic policies and external factors such as a global financial crisis. Each of these situations might have an adverse impact on PPP projects.
Currency	Currency risk arises when the expected revenue, expenses, capital expenditure and loans are in more than one currency. Often PPP projects are financed and developed meeting the capital expenditure in international currencies such as US dollars, euros or British pounds sterling. However, revenue is collected in local currency. Thus, fluctuations in currency exchange rates could adversely affect the project.
Inflation	The risk here is the adverse impact on project completion and operation of the facility, of failing to take into consideration the effect of inflation in construction and operational costs. Failure to consider inflation as a factor might result in project sponsors and lenders having to commit extra funds to the project.
Long-term inputs	Availability and price of non-project specific long-term inputs can adversely affect PPP projects. Availability of labour at agreed or predictable salary levels and adequate sources of general supplies and services such as energy, water and telecommunication at steady prices are key long-term inputs that may affect a PPP project.

Table 2.5 Non-project specific non-commercial risks

Risk	*Description*
Country	Country risk associated with PPPs concerns the ability of a country to honour its undertakings to investors and lenders. Thus, the legal and regulatory regime of a country and issues such as public acceptance of the project, applicable laws and the availability of a sound system of justice and dispute resolution are relevant considerations.
Political	There is a wide range of risks which revolve around the possibility of governmental action which may adversely affect the development or the performance of the facility. Thus, political risk includes expropriation or nationalisation of projects, change of government, stability of political climate, and host-country commitment to PPPs.
Force majeure	A party's performance of contractual obligations might be rendered impossible by events not within the reasonable expectation or control of the parties. Such events could include natural disasters as well as acts of terrorism, riots and wars. Force majeure events can severely affect, or even halt, the project during different phases of the project lifecycle.

The most significant changes have occurred in connection with political risks. The demand for decentralisation of power and political autonomy in some developing countries has given a new dimension to the political risks associated with PPP projects. The other political risks that have changed their traditional outlook in recent times include those of political instability due to internal conflicts and external factors.

Risk of decentralisation

In an era where PPP is used as an innovative method to bolster and stimulate sagging economies in developing countries, one of the major obstacles to development is the growing demand for decentralisation of administration. Some policy-makers and observers consider decentralisation of state power as the path to better administration of the state and improving the quality of life of the people. In countries where significant control of the economy is concentrated in the centre without any substantial financial and/or administrative autonomy for the local authorities, there is a growing demand for decentralisation of political power (UNFPA, 2000). Some countries have already introduced decentralised systems of administration. However, in these too there is a demand for more delegation of powers.

In developing countries such as India, Mexico, Nigeria and the Philippines, decentralisation is based on the political and legal structures created by the constitution or other statute. In these countries, the states or provinces form a federation, which generally has its own elected government with a wide range of fiscal and programming powers and responsibilities. In contrast, countries such as Bolivia, Ghana, Sri Lanka and Viet Nam are unitary states, although they have political sub-divisions at the local, provincial and higher levels. In these countries, decentralisation often takes a more administrative and operational character, regulated through decrees or directives from the central government. In the latter group of countries, decentralisation is also defined by the extent to which fiscal powers have been decentralised. In most countries the central authority represents the highest level of governance with first priority over fiscal resources. It is only when the federal or central authorities agree to share their resources that true decentralisation can take place (Litvack et al., 1999).

As far as PPP is concerned, the main obstacle that decentralisation poses is the possibility of conflict between the central administration and the state, provincial or local authorities in sharing of power and responsibility in connection with projects. In addition to slowing down the progress of development activities, the lack of definition of functions may also cause confusion and uncertainty in the minds of project sponsors, lenders and other key project participants. Thus, although decentralisation may be a good thing for developing countries, when it comes to sharing political power, if policies are not clear and the respective functions and powers of the administrative authorities are not clearly defined, it is likely that decentralisation will not complement initiatives to develop PPPs. Another obstacle to PPP is that, whilst most central governments will have the capacity to ensure that, even after development projects are completed with private sector participation, the end product is offered to the public at affordable rates, state, provincial or local governments in decentralised developing countries may not have such capacity. Moreover, a central government will have the capacity to provide sufficient comfort to project sponsors and lenders by extending government guarantees or other securities to secure projects. However, state, provincial or local governments once again may not have such capacity. Furthermore, project sponsors would prefer to work with the central administration rather than with the local authorities.

The following two examples, Case Studies 1 and 2, provide insight into the confusion that can be caused when the specific roles of the central administration and the state, provincial or local authorities are not properly spelled out in a decentralised system of administration.

Case Study 1 The dispute between Metalclad Corporation and the United Mexican States

The dispute between Metalclad Corporation and The United Mexican States arose from the construction of a landfill in Guadalcazar in the central Mexican state of San Luis Potosí, by an enterprise owned and controlled by Metalclad. The construction was designed for the confinement of hazardous waste from the area. Approvals having been obtained at the federal and state levels, construction of the landfill was completed in March 1995. However, demonstrations that took place at the inauguration of the landfill kept it from opening.

In November 1995, Metalclad concluded an agreement with federal environmental agencies setting forth the conditions under which the landfill would operate. However, in December 1995, the local municipality issued a denial of a construction permit for the landfill that had been requested 13 months earlier. The municipality then challenged the agreement Metalclad had concluded with federal agencies and obtained a judicial injunction, which prevented the operation of the landfill until May 1999.

At the hearing before the International Court for Settlement of Investment Disputes (ICSID),[3] it was argued on behalf of the United Mexican States that the local government actions are generally not subject to the standards of protection required from the North American Free Trade Agreement (NAFTA)[4] member states towards the property of, and investments by, entities from other member states. It was argued that Article 105 of the NAFTA does not use the term "local governments" in describing the extent of the obligations set forth in the Agreement. According to this argument, the NAFTA parties deliberately excluded the term "local governments" from Article 105 to signal a departure from otherwise applicable customary international law, which provides that a State is liable for the acts of all its political subdivisions, including local governments.

The US argued *inter alia* that there was no general exclusion from the NAFTA standards for local government action. It was the US belief that the Parties intended that, except where specific exception was made, actions of local governments would be subject to the NAFTA standards.

The Tribunal held that the actions of the Mexican State and municipal authorities entailed a breach by Mexico of its obligation to afford Metalclad's investment treatment in accordance with international law, including fair and equitable treatment, under NAFTA Article 1105. The Tribunal further held that, by permitting the actions

3 Metalclad Corporation v. The United Mexican States, ICSID Case No. ARB (AB)/97/1.

4 North American Free Trade Agreement (NAFTA) is a trade agreement, negotiated among three federal governments, namely the United States of America, Canada and Mexico, which came into operation on 1 January 1994. It is the first agreement ever to include services as well as goods. The full content of the NAFTA agreement is available online at: http://www.nafta-sec-alena.org/english/nafta/nafta.htm/

of the municipality, Mexico had taken measures tantamount to expropriation of Metalclad's investment under NAFTA Article 1110, since those actions "effectively and unlawfully prevented the Claimant's operation of the land-fill."[5]

This example shows that, in a decentralised system, local governments may sometimes act in violation of international, regional or bi-lateral treaty undertakings made by the central government and that this would result in central governments having to defend themselves against actions brought by investors or home countries of investors under agreed contract or treaty terms.

Case Study 2 Dispute concerning the alienation of tribal lands by the State of Andhra Pradesh

This dispute on the alienation of tribal lands by the state of Andhra Pradesh in India concerned a move by the state government to alienate lands traditionally held by various tribes to private companies in violation of the Fifth Schedule to the Indian Constitution that provides protection to the *Adivasi* (indigenous) people living in the Scheduled Areas.[6]. Samatha, a non-governmental organisation, filed action against the state government of Andhra Pradesh alleging, *inter alia*, that essentially the Fifth Schedule is a historic guarantee to indigenous people on the right over the land on which they live and that the state government has violated this right.

The Supreme Court of India, in a landmark judgement delivered in July 1997,[7] observed that it was clear from the inception of the colonial administration that the tribal areas were treated distinctly from other areas. Moreover, the tribes were protected from exploitation. Their rights and title to enjoy the lands in their occupation and their autonomy, culture and ecology were preserved; infiltration of the non-tribal people into tribal areas was prohibited. The court held that Chapter VI, Part X of the Indian Constitution deals with "Scheduled

5 The award gives an account of the facts on which the Tribunal found a breach of the obligation to grant fair and equitable treatment. The award thus contains one of the first rulings ever to apply the standard of fair and equitable treatment under a treaty governing investment matters.

6 This Fifth Schedule is still under threat of being amended to effect transfer of tribal lands to non-tribal people and corporate bodies. There have been several mass campaigns in recent times protesting against any move to amend this schedule and alleging that the very survival and culture of the 80 million-strong tribal population of India is under threat.

7 Samatha vs. State of Andhra Pradesh 1997 8 SCC 181.

Tribes and Tribal Areas" and that all lands leased to private mining companies in the scheduled areas were null and void. Furthermore, it was held that the transfer of land in the Scheduled Area by way of lease to non-tribal people, corporation aggregate stands prohibited. The court also directed that the conference of all Chief Ministers, Ministers responsible for the Ministry concerned and Prime Minister, and Central Ministers concerned should take a policy decision for a consistent scheme throughout the country in respect of tribal lands, need they be subjects of development.

This example shows that certain actions that may be taken by the state or provincial governments to attract investors in a decentralised administrative system can sometimes violate the constitutional and statutory protections guaranteed by the central government to its people.

Risk of political instability

A favourable investment climate is based on many factors. Political stability is one of the most important of these factors. The risk of political instability in developing countries can discourage investors and lenders from participating in PPPs. It may also influence the abandonment of projects that are already underway. Investors and lenders would be discouraged from financing the projects by the threat of political upheaval, and by the prospect of a new regime changing its policy towards investors, for example, by imposing punitive taxes on them or by expropriating their capital assets.

Political instability in developing countries can be caused by internal factors such as civil wars and movements for separation of states. There can also be external factors such as hostilities with neighbouring countries. The countries that have suffered most from lack of investment due to political instability are in Africa and Asia. Somalia and Sierra Leone are two examples of African countries that have not been able to attract significant foreign investments during the last three decades due to internal political instability. Afghanistan is an Asian example. In the absence of inward foreign investment and the present lack of a financially and technologically capable private sector, these countries have hardly benefitted from PPPs.

During the Siyaad Barre regime from 1969 to 1991, Somalia had a fairly well functioning state apparatus. It deteriorated over time due to civil wars among clans that led to a collapse of the state and the fleeing of Siyaad Barre (Braathen et al., 2000). Thereafter, the army dissolved into competing groups loyal to different clan/tribal leaders, resulting in further political instability and volatility of the economy. Since then, there have hardly been any significant investment projects in Somalia (Collier and Hoeer, 2000). Recently, PPP has been considered as an option for developing water infrastructure in Somalia. This too has happened with the United Nations Children's Fund (UNICEF) taking the role of the facilitator and the funds coming from the European Union, the Swiss National Committee for UNICEF and the Government of Denmark (UNICEF, 2010).

In Sierra Leone, multi-party democracy was disrupted by a rebellion led by the Revolutionary United Front of Valentine Strasser, a movement of exiles. Strasser's *coup d'état* in 1992 was met by revolts of the Revolutionary United Front, and civil war broke out two years later. With the participation of the United Nations, the situation has been defused and settled. As a result of the near decade-long political instability, the country has not yet been successful in attracting any significant investments. However, Sierra Leone seems to be going in the right direction as seen from the Government of Sierra Leone's (2009) Resource Mobilisation Strategy in which the Ministry of Finance and Economic Development declared, *inter alia*:

> The Government's policy is to attract large private-sector proposals with Build–Operate–Transfer (BOT), which will certainly require a special BOT Law, and Public Private Partnership projects. There are opportunities for realising these options in the power, water, road, port, airport and telecommunications sub-sectors; projects range from green-field to concessions and divestitures.

Afghanistan is also an example of a country that has failed to benefit from PPPs due to political instability created by both internal and external factors. After the troops of the former Soviet Union withdrew from Afghanistan in 1989, the government lasted only a few years, and in 1992 the *Mujahideen* took control of the state (Rothstein, 2006). The *Mujahideen* rule did not last for long as, in 1996, the Taliban, an Islamic fundamentalist movement formed in 1994, captured Kabul and subsequently overran about 90 percent of the country (Rubin, 2002). In the years that followed, there was a military struggle between the Taliban and the opposing Northern Alliance. In October 2001, the war against terrorism in Afghanistan began when the US and British governments launched military operations against the Taliban and the Al-Qaeda movement, in response to the September 11 2001 attacks on the US. As a result, until recently, there has been hardly any foreign investment going into Afghanistan from private entities for development projects.

Need for risk mitigation for project success

As noted earlier, PPP projects usually have a long life, ranging from 10 to 30 years. Thus, during the project negotiation stage, it is important to identify the risks a PPP project might have to face. The private-sector participants in PPP projects are mostly profit driven and would want to minimise the risks they have to bear on long-term projects. As far as the lenders are concerned, since they finance the projects mostly on a non-recourse or limited recourse basis, minimisation of project risks is one of the primary requirements. For the public-sector entities, in addition to attracting private-sector finance, technology and project management and operation skills, another reason for engaging in PPPs is to share the risks.

Failure to identify risks and improper allocation of risks to parties could lead to frustration of projects, and even breach of contracts. Thus, risk sharing is the other distinguishing characteristic of PPP; its success depends on the appropriate allocation of risks. Therefore, the project risks need to be identified carefully and allocated to parties that are best suited to manage them at the lowest possible cost. However, this is complicated and difficult to deal with since many project participants typically assume several project roles.[8] Moreover, due to their diverse interests, the capacity of the parties to absorb project risks will be different. Thus, the risks involved in the project should be spread among the various parties, with the party that can most efficiently and cost-effectively control or handle each risk usually assuming it.

However, it should be noted that, although risk sharing among the key project participants is the ideal scenario for PPP projects, in the case of developing countries, most of the time, the public-sector entity involved in the project and the host state end up as the highest risk takers. The demand for government guarantees by the investors and lenders before investing in certain developing countries identified as high-risk investment destinations is the key reason for this. As there are over 140 developing countries in the world[9] and most of them are competing to attract investors, it is not difficult to understand that some developing countries offer to provide government guarantees and become the chief risk takers in projects to win over the investors.

Thus, country risk is important in influencing the investment in, and development of, PPP projects. This can be better perceived by Sovereign Credit Rating (SCR), which indicates the risk level of the investing environment of a country, and is used by investors looking to invest in a country. It takes political risk into account. SCR helps investors to understand the risk exposures and uncertainty, default records and access to international bond markets for a specific country. Normally, projects do not receive a rating higher than the host country's SCR and, as a result, countries with low SCR struggle to attract investors. A project may receive an improved rating from

8 For example, the construction contractor may also be one of the shareholders of the Project Company and/or operator. The concession granter (public entity) sometimes may have a direct or indirect link with the output purchaser who might have a contract with the project company to purchase the project output.

9 See the list of all countries: Human Development Index, in the Human Development Report released by the United Nations Development Programme on 5 October 2009, at http://hdr.undp.org/en/media/HDR_20072008_EN_Complete.pdf/

agencies, for example, if a financially sound off-taker participates in the project. In the circumstances, apart from seeking government guarantees, tight off-take agreement, reliable supply agreement and setting up escrow accounts are some key measures employed by PPP project participants to deal with country risk.

Discussing in detail all the possible and available risk mitigation measures for PPP projects is beyond the scope of this chapter. Table 2.6 presents a summary of the key mitigation measures for major risks associated with PPP projects in developing countries.

Table 2.6 Risk mitigation measures for PPP projects

Risk	*Mitigation measures*
Completion	The contractor should be required to provide a completion guarantee that specifies a completion time and a minimum efficiency rate. This should be supported with a performance guarantee. The risk of delays in construction can also be mitigated by incorporating specific clauses in the construction agreement that would make the contractor liable to pay liquidated damages for any delays within its control. Delays within the control of the public entity – for example, delays in obtaining necessary approvals and licences for the project – can be mitigated by requiring the host government to issue the approvals and licences in an efficient manner and requiring the host government to provide guarantees to the effect that such approvals and licences will be duly issued.
Project supplies	Long-term supply contracts with the suppliers and maintenance of a list of suppliers instead of relying on a single supply source are methods often employed for mitigating the risk of project supplies.
Market	One method adopted to mitigate market risk is to secure a market for the output through the conclusion of long-term sales contracts with customers, when that is possible. Another widely used risk mitigation mechanism is to enter into an agreement with the host government to underwrite revenue streams by means of "take-or-pay" or "take-and-pay" contracts.
Sponsor	Obtaining bid bonds and project completion guarantees and performance bonds from the sponsors are widely used mitigation measures. In addition, providing minimum shareholding requirements for the major sponsors and also making the contractual mechanism flexible for new shareholders to be drafted in when the original sponsors are facing financing challenges are contractual measures that could be taken to mitigate the risk. Providing step-in rights for the public entity and the lenders to assume control over the project is an option of last resort.

Risk	*Mitigation measures*
Technology	Lenders may not favour the use of new technology in the project since the borrowers will rely on cash flows from the project to service debt. The public-sector entity should also consider insisting on the use of proven technology for the project. If new technology is used, then adequate warranties should be obtained from the technology providers. In addition, lenders and the host government may require the project sponsor to provide a guarantee for the technology used. A prudent contractor would include a construction contingency in the total cost of the project, and build in some excess capacity to allow for technical failures that may prevent the project and facility from reaching the required capacity.
Host-government obligations: Regulatory; Expropriation; and Political	In order to reduce political risks, it is necessary to insist that host governments establish legal safeguards to protect investment projects from arbitrary decisions, especially in respect of future legislation which would affect the project or the status of concessions, permits or licences granted to the SPV. Often, guarantees are obtained from the host government which give the assurance that there will be no expropriation of the project, or change in the status quo of the applicable laws and regulations under which the project was negotiated. Offering a public entity of the host government an equity stake in the project is a good approach to mitigate political risks. Most project sponsors obtain political risk insurance when investing in developing countries which are considered to be politically unstable.
Economic	Guarantees obtained from the host government concerning agreed payments for project outputs by means of "take-or-pay" or "take-and-pay" contracts constitute a widely used measure for mitigating the risk of economic downturns affecting project revenue.
Currency	All projects involve local costs. Thus, the overall currency risk can be reduced to some extent by covering the local costs with local funding. Mixing local and foreign funding of the project in this way will ensure that the project does not rely excessively on foreign funds. Indexing the project output price to the exchange rate is another method that is available. However, this arrangement can expose the project to dramatic changes in the exchange rate. Maintenance of reserves in the country or off-shore in a special escrow account to deal with currency exchange fluctuations is another option. Entering into rate protection agreements (forward exchange contracts) is another option.

Risk	*Mitigation measures*
Inflation	The usual method of avoiding the risk of inflation is to have fixed-price turnkey contracts where the risks involved are borne by the contractor. The other mitigation measures available include the injection of additional equity and debt by the project sponsors and lenders, and the inclusion of a renegotiation clause in the contract to restructure the cost and pricing mechanism. Having an escrow account with funds provided by the sponsors for the completion of the project is another measure.
Long-term inputs	Long-term supply contracts with suppliers of services such as energy, telecommunications and water could be taken to mitigate the risk of inadequate supply of key services.
Environmental	Proper environmental impact assessment and careful evaluation of the existing legislation in the host country on environmental protection would help to mitigate the environmental risks. Obtaining comprehensive insurance coverage against environmental risk is a common practice.
Force majeure	The unpredictability encountered on projects and the fact that force majeure events cannot be blamed on project participants makes this one of the most difficult risks to mitigate. Various forms of commercial insurance are available against force majeure risk. They cover all stages of a project and cover both asset loss and business interruption. However, the cost of such insurance is high.

5. Developing policy and governance architectures to enable win-win options

Many argue that PPPs offer "win-win" options for developing countries as well as for investors, lenders and other project participants. The main points put forward to argue the benefits for the developing countries are: that they are incapable of meeting the growing demand for infrastructure development without engaging the private sector; and that, as PPP projects could be financed on "non-recourse" or "limited recourse" bases, the burden of paying for the cost of projects, or servicing project development debts, is removed from the public sector. The main argument supporting the view that it is equally beneficial to the investors, lenders and other project participants is that segments of the construction market that were previously under state monopolies can now be accessed by the private sector. This creates business and profit-making opportunities.

However, these are theoretical benefits that might not be realised in practice. As pointed out above, due to the unique nature of some risks associated with PPP projects in developing countries, many PPP projects are not strictly financed following the "non-recourse" or "limited recourse" methods. For example, many low-income countries with political and or economic instability continue to be burdened with the obligation of providing government pay-back guarantees to the investors and lenders.

Moreover, although it is theoretically correct to say that PPPs have opened the doors to the private-sector entities to enter previously inaccessible markets, the reality is somewhat different. Due to policy and regulatory inadequacies, and hostile reaction of end-users towards the private sector's role in services previously provided by the state, project developers, investors and lenders do not find their entry into many developing countries easy.

In the circumstances, to ensure that PPP creates a "win-win" situation for all project participants, it is important that all risks associated with projects are identified at the earliest possible stage, and measures are taken to mitigate them by allocating them to the project participants best equipped to absorb them or deal with them. It is also important to ensure that projects are actually developed in the interest of the public and not owing to any rent-seeking motives of the law-makers, officials or project developers. In other words, it is important that PPP projects are launched to become essential engines for sustainable development.

In order to ensure that the risks associated with PPP projects are duly assessed, investment promotion policies of developing countries as well as PPP agreements should be subjected to a conceptual shift from the mere promotion and protection of investment towards promotion and protection of sustainable investment. The tools needed to do this include: environmental and social impact assessments; environmental management systems; corporate codes of conduct; measures for the protection of the rights of end-users and parties who may be adversely affected; stable political and regulatory environments; and effective and efficient dispute resolution mechanisms. The first five points relate directly to the importance of promoting sustainable development, and are inter-related. The last point is worthy of further explanation. As PPPs may bring together parties from different jurisdictions with diverse interests, it is fundamental to ensure that effective and efficient mechanisms acceptable to all parties are in place which can be applied for resolving disputes during the development, construction and operation phases of projects.

References

Abdel-Aziz, A.M. (2007) Successful Delivery of Public–Private Partnerships for Infrastructure Development, *Journal of Construction Engineering and Management*, 133(12): 918–31.

ADB (Asian Development Bank) (2008) Public–Private Partnership Handbook, Manila: ADB; pp 12–15.

Algarni, A.M., Arditi D. and Polat G. (2007) Build-Operate-Transfer in Infrastructure Projects in the United States. *Journal of Construction Engineering and Management*, 133(10): 728–35.

Allen, G. (2001) *The Private Finance Initiative*. London: House of Commons Library, Economic Policy and Statistics Section, http://www.parliament.uk/commons/lib/research/rp2001/rp01-117.pdf (accessed 23 October 2010).

Beidlemen, C.R., Fletcher D. and Veshosky D. (1990) 'On Allocating Risks: The Essence of Project Finance, *Sloan Management Review*, Vol. 31, pp 47–55.

Braathen, E. Bøås, M. and Sæther, G. (eds) (2000) *Ethnicity Kills? The Politics of War, Peace and Ethnicity in Sub-Saharan Africa*. London: Macmillan.

Collier, P. and Hoeer, A. (2000) *Greed and Grievance in Civil War*, Policy Research Working Paper 2355. Washington DC: World Bank.

Commission on Growth and Development (2008) The Growth Report: Strategies for Sustained Growth and Inclusive Development, http://cgd.s3.amazonaws.com/GrowthReportComplete.pdf (accessed 21 October 2010).

Delmon, J. (2009) *Private Sector Investment in Infrastructure*, Washington DC: The World Bank and Kluwer Law International, pp 49–52.

Denton Wilde Sapte (2004) *A Guide to Project Finance*. London: Euromoney Publications.

Department of Economic Affairs, Ministry of Finance of India (2005) Scheme for Support to Public Private Partnership in Infrastructure, http://www.pppinindia.com/pdf/PPPGuidelines.pdf (accessed 23 October 2010).

Fan, S. and Chan-Kang, C. (2004) Returns to Investment in Less-Favored Areas in Developing Countries: A Synthesis of Evidence and Implications for Africa, *Food Policy*, 29(4): 431–44.

Ford, R. and Zussman, D. (1997) Alternative Service Delivery: Sharing Governance in Canada, Toronto: Institutes of Public Administration of Canada (IPAC).

Government of Sierra Leone (2009) *Resource Mobilisation Strategy*, http://www.sierraleoneconference2009.org/docs/resource_mobilisatn_strategy.pdf (accessed 22 October 2010).

Gunawansa, A. (2000) Legal Implications Concerning Project Financing Initiatives in Developing Countries. *Attorney General's Law Review*, July 2000.

Harris, S. (2004) *Public Private Partnerships: Delivering Better Infrastructure Services*, Working Paper. Washington DC: Inter-American Development Bank.

Her Majesty's Treasury (1998) Partnerships for Prosperity The Private Finance Initiative. London HM: Treasury.

Hoffman, S.L. (1989) *A Practical Guide to Transactional Project Finance: Basic Concepts, Risk Identification, and Contractual Considerations. Business Lawyer*, Vol. 45: pp 181–232.

International Energy Agency (2004) *World Energy Outlook 2004*, Paris: OECD/IEA.

IEG (World Bank Independent Evaluation Group) (2007) The Nexus between Infrastructure and Environment, http://siteresources.worldbank.org/INTOED/Resources/infrastructure_environment.pdf, (accessed 22 October 2010).

IFC (International Finance Corporation) (1999) *Project Finance in Developing Countries*, Lessons of Experience Series No. 7. Washington DC: IFC.

Kettl, D.F. (1993) Sharing Power: Public Governance and Private Markets. Washington DC: Brookings Institution.

Khanom, N.A. (2009) Conceptual Issues in Defining Public Private Partnership, paper presented at the Asian Business Research Conference 2009, Dhaka, Bangladesh.

KPMG (2007) *Building for Prosperity: Exploring the Prospects for Public Private Partnerships in Asia Pacific*, KPMG, http://www.kpmg.com.sg/publications/Industries_PPPinAsia2007.pdf (accessed 23 October 2010).

Kumaraswamy, M.M. and Morris, D.A. (2002) Build-Operate-Transfer-Type Procurement in Asian Megaprojects. *Journal of Construction Engineering and Management*, 128(2): 93–102.

Litvack, J., Ahmad, J. and Bird, R. (1999) *Rethinking Decentralization at the World Bank*, Washington DC: World Bank.

Ministry of Finance, Singapore (2004) Public Private Partnership Handbook, Version 1, http://app.mof.gov.sg/ppp.aspx (accessed 1 July 2010).

Nevitt, K. and Fabozzi, F.J. (2000) *Project Financing*, 7th edition. London: Euromoney Books.

Rodinelli, D.A. (2002) Public-Private Partnerships. In Kirkpatrick, C., Clarke, R. and Polidano, C. (eds) *Handbook on Development Policy and Management*. Cheltenham: Edward Elgar, pp 381–88.

Rothstein, H.S. (2006) *Afghanistan: and the troubled future of unconventional warfare*. Annapolis: US Naval Institute Press.

Rubin, B.R. (2002) *The Fragmentation of Afghanistan*. New Haven: Yale University Press.

Savas, E. (2000) *Privatisation and Public–Private Partnerships*. New York: Chatham House.

Scriven, J., Pritchard, N. and Delmon, J. (1999) *A Contractual Guide to Major Construction Projects*. London: Sweet & Maxwell.

Story, J. (2007) *Commentaries on the Law of Partnership as a Branch of Commercial and Maritime Jurisprudence with Occasional Illustrations from Civil and Foreign Law*, Seventh edition. Clark, New Jersey: Law Book Exchange Ltd, pp 1–6.

Tisley, R. (2000) *Advanced Project Financing*. London: Euromoney Books.

UNFPA (United Nations Population Fund) (2000) *UNFPA and Government Decentralization: A Study of Country Experiences, Evaluation findings*. Office of Evaluation and Findings, June 2000, Issue 30.

UNICEF (United Nations Children's Fund) (2010) *In Somalia, a Public–Private partnership provides safe water to thousands*, http://www.unicef.org/infobycountry/somalia_55308.html (accessed 23 October 2010).

World Bank (2005a) *Infrastructure and the World Bank: A Progress Report*, http://siteresources.worldbank.org/DEVCOMMINT/Documentation/20651863/DC2005-0015(E)-Infrastructure.pdf (accessed 21 October 2010).

World Bank (2005b) *World Development Report 2005: A better investment climate for everyone*, http://siteresources.worldbank.org/INTWDR2005/Resources/complete_report.pdf (accessed 16 May 2010).

World Bank (2009) *Doing Business 2009*, http://www.doingbusiness.org/Documents/FullReport/2009/DB_2009_English.pdf (accessed 23 October 2010).

Yescombe, E.R. (2002) *Principles of Project Finance*. San Diego and London: Academic Press, pp 8–15.

Yescombe, E.R. (2007) *Public–Private Partnerships: Principles of Policy and Finance*. Oxford: Elsevier.

3 Promoting social opportunities in the procurement of construction projects

John Hawkins
Institution of Civil Engineers, United Kingdom

Introduction

The infrastructure challenge

Millions of poor people in low-income countries lack access to basic infrastructure. Infrastructure not only contributes to economic growth, but it is also an important input to human development (Foster and Briceño-Garmendia, 2010: 2). This is encapsulated by the Millennium Development Goals (MDGs) which address infrastructure issues both directly and indirectly in six of the eight goals. In 2000, 189 states adopted the United Nations Millennium Declaration which set quantified, time-bound targets for addressing extreme poverty, hunger and disease, and for promoting gender equality, education and environmental sustainability. The MDGs are also an expression of basic human rights: the rights of everyone to good health, education and shelter (United Nations, 2010). Both developed and developing countries committed themselves to deliver these goals for international development by 2015. However, progress towards achieving the MDGs has been mixed; the United Nations (UN) (2010: 3) reports that, without additional efforts, several of the targets are likely to be missed in many countries.

Many of the goals are interdependent, particularly those targets relating to infrastructure. For example, Target 10 aims to halve, by 2015, the proportion of people without sustainable access to safe drinking water and basic sanitation, whilst Target 11 aims to achieve a significant improvement in the lives of at least 100 million slum dwellers by 2020.

The infrastructure challenge to achieve the MDGs is enormous, particularly in Africa where it is estimated that US$93 billion is required per annum, about one-third of which is for the maintenance of the existing stock. This is despite strong economic growth of 4 percent a year between

2001 and 2005, of which about half is due to investment in infrastructure. To attain the MDGs in Africa, annual growth rates of an average of 7 percent would be required, along with huge improvements in access to power, the transport network and the sustainable provision of water (Foster and Briceño-Garmendia, 2010: 2). To achieve this growth and development, it will be necessary for planners to find the right balance between the social opportunities derived from the following: the service provided by the infrastructure asset; and the process of design, construction, operation and maintenance.

The opportunity

Despite the challenges highlighted above, this represents a major opportunity for the construction sector.

The greatest benefit to society from the investment in infrastructure lies in the services derived from the constructed asset. The extent to which the social development opportunities are maximised will depend upon the impact, performance and quality of an infrastructure asset, and the service it delivers to consumers and society. Less well understood, and a matter of some conjecture, is the potential to derive benefits from incorporating social development opportunities during the *process* of design, procurement, implementation and operation. Embracing both concepts has the potential to increase the contribution which investment in infrastructure can make towards economic growth, poverty reduction and attaining the MDGs (Hawkins et al., 2006: 5).

The social development opportunities that are incorporated into the procurement of infrastructure fall into three categories:

- *Industrial:* provides opportunities to increase employment, the input of local consultants, contractors and suppliers, training, and improved on-site working conditions;
- *Environmental:* provides the opportunity to look for sustainable procurement solutions as well as to manage the environmental impact of the construction process and the constructed item;
- *Society:* provides opportunities to empower community groups, particularly women, as well as to manage the social impact such as resettling people affected by the project.

Balancing competing objectives

Procurement professionals are often concerned that the pursuit of social development opportunities has the potential to compromise the ultimate purpose of delivering the infrastructure asset (Watermeyer, 2005). Typically, the concerns raised revolve around the risk of the following occurring (Watermeyer, 2005: 14): loss of economy and inefficiency in procurement;

the exclusion of certain eligible tenderers from competing for contracts; the reduction in competition; unfair and inequitable treatment of contractors; lack of integrity or fairness; lack of transparency in procurement procedures; and failure to achieve the social opportunity through the procurement itself.

Socio-economic policy is invariably set at a macro level and then implemented at a micro level with planners who are tasked with balancing the primary project objectives and the secondary social opportunities. To achieve both the project objectives and the social opportunities, it is important that the procurement process is integrated into the design and planning phase. Watermeyer (2005: 14) states:

> Procurement is a process which creates, manages and fulfils contracts. If procurement is a process, it can be documented as a succession of logically related actions occurring and performed in a definite manner which culminates in the completion of a major deliverable or the attainment of a milestone. Processes in turn are underpinned by methods and procedures, which are informed and shaped by the policy of an organisation.

By following a sequence of actions from project planning through to contract implementation and into operations and maintenance, the social opportunities can be attained. Thus, in this chapter, the widest possible definition of procurement is adopted, and it includes: project identification, planning and design; procurement strategy; tender and selection; contract implementation; and operations and maintenance.

This chapter aims to demonstrate that the inclusion of social development opportunities through the planning, design, construction and operation process can contribute towards sustainable economic growth, poverty reduction and attaining the MDGs. The chapter is divided into two sections. The first section focuses on the importance of identifying social development opportunities as a policy objective and the opportunities that can be pursued. The second section identifies the actions to achieve the social development opportunities required at each stage of the procurement cycle that will enable the delivery of the objective.

Promoting social development opportunities

In this section, the promotion of social development opportunities is discussed under the three categories highlighted above: industrial, environmental and society.

Industrial

Increasing the local input of people, goods and services in the delivery of infrastructure projects can make a major contribution to economic growth.

Commonly known as "local content", the measures taken to ensure such increases also open up opportunities for the poor to participate in the growth process through employment creation and development of locally-owned enterprises in the construction and supply industries (Wells and Hawkins, 2008: 6–7). Local content is a recognised term in the oil and gas sector. It can be defined as the involvement of local enterprises in planning, design and construction services as well as locally added value in transactions occurring throughout the contractor's supply chain. In the context of local content, "local" means promoting national economic interest against foreign competitors or promoting an economically disadvantaged community within a country. These policies may include (Wells and Hawkins, 2008: 6–7):

- creating opportunities for local consultants;
- increasing work opportunities for local contractors;
- increasing employment throughout the supply chain;
- creating market openings for locally produced materials and components;
- providing training opportunities for local employees.

The immediate benefits of investing in local content lie in increased opportunities for local people to earn income through employment on the construction site or in the supply industries; better opportunities for local businesses as contractors, sub-contractors, producers, or suppliers of materials and equipment; and increased local knowledge and skills. The longer-term development outcomes are poverty reduction, inclusive growth and more sustainable infrastructure. The latter benefit is particularly important for African countries, considering the high costs of capital investment and of rehabilitating and maintaining infrastructure. However, few countries have clear policies for the promotion of local content in infrastructure, and those countries that have such policies have difficulty implementing them.

The history of governments using procurement to achieve industrial goals goes back to the 19th century with both the US and Great Britain linking fair wages and labour conditions to procurement (McCrudden, 2004: 2). Both countries also used the procurement of construction works to address sudden rises in unemployment. This set the tone for the type of social opportunities pursued by countries during the majority of the 20th century. The best known example is the "Buy America" policy of the US. Under the Buy America Act (1933) and subsequent legislation, the basic legal requirement on federal purchasers is that materials, supplies, articles or, since 1990, services that are acquired for public use should be substantially American (McCrudden, 2004: 26).

Examples of using procurement to advance social goals in the developing world have usually related to the promotion of local content. India has pursued this policy, resulting in high local content in the procurement of infrastructure. It is estimated that 100 percent of the inputs to water and rural road projects and 75 percent of the inputs to power projects in that country are local (Wells and Hawkins, 2008: 6–7).

In Malaysia, the procurement policy supports the National Development Policy which seeks to improve the economic participation of the *Bumiputera* (the native Malay population) and to make them equal partners of development in the country (Watermeyer, 2005: 6). The government has achieved these goals through a complex arrangement of set-asides – effectively contracts are 'set aside' for the targeted community – and price preferences. The Malaysian government created these goals to overcome the belief amongst native Malays that they were economically disadvantaged against Chinese Malaysians.

Similar arrangements have operated across Asia. For example, Indonesia has used procurement to enhance the capacity of small- and medium-sized enterprises. In 2000, Indonesia introduced a construction law that established a procurement system that had the specific purpose of providing work with the highest possible local content to the exclusion of foreign competition. In its Country Procurement Assessment Report, the World Bank (2001: 2) stated that these regulations had the effect of limiting competition, thus violating the principle of one country, one market, and implying that Indonesia was forgoing the benefits that arise from increased competition. Similar criticisms have also been made by the World Trade Organisation (WTO) (1997: para 122) concerning the Malaysian procurement system:

> These preferences not only restrict competition among suppliers, thereby impairing economic efficiency, but also raised the cost to the Government and state owned enterprises of procuring goods and services.

The WTO and World Bank believe that the inclusion of business opportunities for local firms, such as those in Indonesia and Malaysia, can distort the market, and are at odds with fundamental elements of global public procurement rules. Malaysia believes that the application of the WTO and World Bank rules would damage the pursuit of its macro-policy objectives of improving the economic participation of the *Bumiputera*. Thus, it has been forceful in resisting being bound by international public procurement rules (McCrudden and Gross, 2006).

The Multilateral Development Banks (MDBs) play a key role in funding infrastructure projects in low-income countries. Despite recognising the need to balance local development with procurement objectives, they insist on the application of international competitive bidding rules for the appointment of contractors, with selection being based on the lowest evaluated price. For example, the World Bank specifically does not allow any evaluation criteria other than price and resources and the capability of the bidders to undertake the contract. With the tender process led by price, the technical component is often weak, and objectives such as developing the local economy are overlooked (Wells and Hawkins, 2008: 11). In contrast, consultants are appointed principally on the basis of quality.

The World Bank system has been incorporated into many developing country procurement regulations as part of the recent reform programmes (World Bank, 2001). In the late 1990s and early 2000s, the World Bank embarked on a series of reviews of national procurement systems, in partnership with regional development banks and country governments. The World Bank was concerned that many countries, such as Indonesia, had procurement systems that were prone to corruption and collusion, lacked transparency, and failed in their principal objectives: to procure goods and services for the government with due consideration to maximising economy and efficiency, and promoting competition and fair and equitable treatment of all suppliers, contractors and consulting firms (World Bank, 2001).

The outcome from each review was a Country Procurement Assessment Report that typically would recommend the establishment of a new procurement law based on the United Nations Commission on International Trade Law (UNCITRAL) Model Procurement Law, and a government-led regulatory authority to aid and monitor the implementation of the law. While recognising the desire to introduce strict rules where few previously existed, the insistence on countries fully opening their markets restricts their ability to take advantage of opportunities to use procurement as a tool for achieving industrial policy objectives. The African Development Bank (2006) reported that less than 40 percent by value of new works contracts in 2006 were awarded to companies from the continent. This figure includes contracts awarded to international firms that have established offices in African countries.

The WTO's Government Procurement Agreement incorporates a basic non-discrimination provision which requires parties to accord the products, services and suppliers of any other party treatment 'no less favourable' than they give to their domestic products, services and suppliers (WTO, 1994). It also requires that parties should not discriminate among goods, services and suppliers of other parties. However, it recognises the need for developing domestic industries, including the nurturing of small-scale and cottage industries as well as the development of other sectors of the economy. This apparent contradiction illustrates the difficulty that countries have in balancing competing objectives, and may explain why so few developing countries have signed the agreement.

South Africa has developed a procurement system that attempts to balance its macro-policy objectives with a fair, equitable, transparent, competitive and cost-effective procurement system (Watermeyer, 2005: 6). The construction industry in South Africa is being used to develop black-owned small- and medium-sized enterprises and to generate employment under Black Economic Empowerment laws. Known as "Targeted Procurement", the system provides for two methods for achieving the local content objectives:

- A preference is provided to the targeted group of black-owned businesses in the tender evaluation criteria;
- The social opportunities are defined as contract participation goals, and companies compete on how they would achieve the goals. Points are allocated within the tender evaluation criteria for the methodology to achieve the goals.

These methods meet the requirements of the South African government's macro-policy objectives and the requirements of the WTO and World Bank for a procurement system based on competition. A procurement system which allows companies to compete on achieving specified industrial opportunities can also be applied to achieve environmental and society opportunities.

Environmental opportunities

Over the last 30 years, the breadth of social opportunities that planners and designers of construction projects have been asked to consider has been widened to include environmental opportunities. Initially, the planners and designers had to consider the environmental impact of the infrastructure asset and the construction process on the natural environment. More recently, the concepts of sustainable procurement and green procurement are leading planners and designers to consider the environmental opportunities, such as minimising carbon emissions and using sustainable materials. For example, in Brazil, the law requires proof of the legality of the source of wood used in public construction and infrastructure projects (UN Department of Economic and Social Affairs, 2008: 5).

Environmental Impact Assessments (EIAs) have become mainstream policy, as most donors require them as a condition of providing finance. Most countries have introduced them as a standard requirement for large infrastructure projects. Traditionally, an EIA addressed the adverse consequences of a specific project on the natural environment. Project modifications and remedial actions were then included in an environmental plan to reduce the identified impact on the environment. EIAs have since evolved to consider the impact that the construction and operation of the infrastructure asset would have on the local community.

A World Bank (2006: ix) report highlighted weaknesses across Asia in the implementation of EIAs, including weak enforcement and late implementation of assessments. The effectiveness of these assessments is also undermined by poor coordination between central and local-level government bodies, low levels of public consultation and information disclosure, and under-funding. The Ugandan National Environment Agency reports that there is inadequate follow up after project approval, and the quality of some of the EIA reports is poor (World Conservation Union, 2007: 19). The allocation of funds for the implementation of mitigation measures is often overlooked or the funding provided is not sufficient, and methods of soliciting views from the public to ensure effective public participation require significant improvement.

EIAs are dealt with at the micro-project level, with mitigation measures managed through the procurement and construction processes. To achieve these measures, it is essential that they are clearly defined in the project design and contract specifications. Moreover, the financing of the mitigation measures identified in the EIA should be considered an integral part of the project.

There is considerable pressure on developing countries to incorporate environmental opportunities into infrastructure procurement based on strategic environmental assessment. The Organisation for Economic Co-operation and Development (OECD, 2006) is encouraging each developing countries to put in place a Strategic Environmental Assessment as a response to the donor community changing their approach from providing financial support for individual projects to giving support for such broad policies and strategies. The scale, nature and regional or sectoral reach of major infrastructure investments requires more than traditional EIAs to account for induced downstream economic and social changes which may cause significant environmental and social effects. It also creates strategic choices to identify social opportunities.

These pressures are often seen as demands by Western countries. The call to care more for the environment can be seen by developing countries as a kind of neo-colonial paternalism, especially when it is combined with conditions attached to the obtaining of loans (Schubert, 2007: 11). The cost of including environmental objectives may be perceived to be prohibitive when balanced with the cost of delivering the principal objectives of the project. However, environmental incidents in developing countries cause substantial damage to society, the economy and the environment (Schubert, 2007: 7). The consequences of climate change are likely to be most severe on the poorest people of the world. Therefore, the importance of environmental objectives is likely to increase, leading to the governments of developing countries having to find the balance between development and sustainability. To achieve this, they will have to develop environmental policies that allow planners and designers to identify the environmental opportunities. These opportunities may include low-carbon design solutions, improved energy and water efficiency, support for the reuse and recycling of materials, and the use of sustainable materials (UN Department of Economic and Social Affairs, 2008: 1).

A number of governments are considering innovative non-carbon emitting energy projects. For example, the government of Rwanda is exploiting the methane gas trapped beneath Lake Kivu. The gas concession alone will more than double the amount of electricity currently generated in the country, and could enable Rwanda to sell electricity to neighbouring countries as well as making possible the extension of electricity supply to many more Rwandan homes (Goodsir, 2009: 9).

The government of Sri Lanka is removing barriers to, and reducing implementation costs of, renewable energy. It is also taking action to promote and improve energy efficiency. In addition, the private sector and local communities are involved in order to foster rural economic development. The targets of the project include (Goodsir, 2009:9): provision of electricity access to 100,000 rural households through solar home systems and off-grid electricity connections via independent mini-grids powered by micro-hydro, wind and biomass generators; and electrification of 1000 small- and medium-sized rural enterprises through renewable energy resources. The off-grid projects completed between 2002 and 2007 include the electrification of 66,267 rural homes through solar home systems, and 3036 homes through 72 village hydro projects.

There is also innovation in the water sector where water capture and storage provide clean drinking water as well as water for livestock watering and farmland irrigation. For example, a primary objective of a water supply programme in Zimbabwe is to establish alternative sources of clean water for domestic use and livestock watering, and to improve systems of household food security and income generation through effective utilisation of sand abstraction systems. The Dabane Trust in Zimbabwe has been successfully engaged in the development of basic, low-cost abstraction equipment that avoids the need, both to dig open sand-wells and to deplete dryland forest areas. The Dabane programme utilises perforated pipes known as wellpoints that are driven into the water-bearing river sand. The water is then drawn out by simple hand pumps that are straightforward and easy to maintain. The training of people in the project area in the lucrative cultivation and irrigation of the fields is important for the sustainable success of the project.

The examples presented above demonstrate how identifying the environmental opportunities can lead to substantial benefits to society.

Society

The greatest benefit to society from the procurement of infrastructure is from the service derived from the constructed asset. As has been shown earlier in this chapter, infrastructure has a direct impact on the delivery of the MDGs. For example, the third MDG is "to promote gender equality and to empower women", measured against the targets of achieving gender parity in education and levels of female literacy and representation in government. There is no particular target among the MDGs for infrastructure provision, but providing clean drinking water and basic sanitation facilities – target 10 – can have an immediate impact on the quality of life of women and lead to an empowering effect.

This not only demonstrates the inter-dependency of the MDGs but it also shows how the process of planning and design offers social opportunities which can improve the service provided to the local community and the end user. In recent years, Social Impact Assessments (SIAs) have become a legal requirement in many countries as part of the EIA. Traditionally, the SIA would set out to identify and ameliorate the negative or unintended outcomes from, and impact of, the constructed asset and the construction process on the local community, such as the potential resettlement of people affected by the project (Vanclay, 2003: 6). Applying SIAs in this manner potentially misses social opportunities and the attainment of better social outcomes. Consultants carrying out SIAs can work with the communities and other stakeholders to identify the additional social opportunities that could be incorporated as part of the objectives of the project.

For example, if a woman can get a sufficient amount of safe water for her family from a nearby pump, she will have more time and opportunity to improve her life and the lives of the members of her family through education, paid work and better health (Fisher, 2008: 223). However, as some development experts point out, just building a new pump does not necessarily mean that it will be used when the long walk to fetch and carry the water enables women to socialise away from the village. Thus, the social benefit of that pump would also be greater if the women were involved in making the decisions on where that pump needs to be placed and what needs to happen to ensure that it is well used.

The same women can also benefit from the procurement process by being employed and trained to plan, install, operate and maintain the pump. The UN (2010) reports that, whilst the participation of women in the labour force has increased since 2000, there are still significant gender gaps in participation rates, occupational levels and wages. Paid employment for women has expanded slowly and women continue to assume the largest share of unpaid work. Close to two thirds of all employed women in developing countries work as contributing family workers or as workers on their own account, typically in forms of employment that are vulnerable and lack job security and benefits.

There are many examples of women participating in the design, construction and operation of basic infrastructure facilities. Not only does this provide women with training, skills and an income but it challenges traditional perceptions about women's status, skills and capabilities held by the community, the family and even the women themselves. A demonstration of the change in women's status due to their involvement in a community water committee can be found in the words of Nakwetikya, of Ndedo village in Tanzania (Fisher, 2008: 223):

> Three years ago, before we formed a committee and prepared ourselves as a community for the water source, men just saw women as animals.

> I think they thought of us as bats flapping around them. They had no respect and no one would allow you to speak or listen to what you had to say. When I stand up now in a group meeting I am not an animal. I am a woman with a valid opinion. We have been encouraged and trained and the whole community has learnt to understand us. I was treated like a donkey only fit to carry baggage all the time. I can assure you though, that if you come back in a few years you will see that women will be leaders of this village. That will bring so many benefits to everyone!

These examples can apply to any community group, whether in a rural or an urban location. The key is to develop holistic, community-led programmes that, as the UN (2010: 14) reports, are more successful than standalone programmes.

Key actions to be taken to realise the opportunities

In the first section it has been argued that it is important to promote social opportunities in infrastructure projects. The potential opportunities that planners and designers can consider in order to achieve specific policy objectives were discussed. In the present section, the actions that can be taken at each stage of the procurement cycle to achieve the social opportunities are set out. The industrial, environmental and society opportunities may be quite different but the processes that can be followed to deliver them are similar. The proposed actions are at the following stages of the project cycle: project identification and planning; detailed design; procurement strategy; tender and selection; contract implementation; and operations and maintenance.

Project identification and planning

The project identification and planning stage is the most important part of the project when considering how the project objectives and social opportunities will be realised. To maximise the socio-economic benefits, the client needs to consider a number of strategic issues to be able to define the project objectives and social opportunities. Key to this decision-making process is what type of facility and related service will best meet the need and how to ensure that the facility provided will be maintained.

The first action for the client is to decide if the need is to be met through a series of individual projects or a programme of projects. The series of individual projects or the programme of projects that is decided upon should be in line with national, regional and sector development plans. Following this decision, the actions to define the social opportunities as project objectives can include those which are now outlined.

- A feasibility study can assess the benefits of the project and the various social opportunities, and find the right balance between competing objectives;
- Consultation with users and affected communities can help to identify social opportunities and define the objectives of the project to meet local needs. Well designed surveys of intended beneficiaries should reveal what kind of asset or service is sustainable. Involvement of the beneficiaries at the planning stage can also promote ownership, and build cohesion and accountability among the members of the community (Wells and Hawkins, 2008: 13);
- The SIAs should be expanded to consider social opportunities such as engaging community groups, and industrial opportunities including employment generation, training, and the use of local enterprises and products. A relevant question is: "Can the asset have a use beyond the immediate purpose?";
- The EIAs should reflect national environmental policy. The assessments should be expanded to consider environmental opportunities such as low-carbon design solutions, improved energy and water efficiency, support for reuse and recycling of materials, and the use of sustainable materials;
- A delivery plan should be prepared, to ensure that the results from the SIA and EIA are followed through into the detailed design and procurement strategy. The results should be incorporated into the tender and contract documents as project objects. These are then well defined in the contract specification with the necessary finance allocated;
- A strategy for operation and maintenance should also be formulated, as this will ensure that the asset has a long life, saving considerable sums of money. It is estimated that spending US$1 on road maintenance provides savings of about US$4 to the economy (Foster and Briceño-Garmendia, 2010: 10). It is also likely to result in greater use of local products, enterprises and skills.

The social opportunities should then be included in the business case for the procurement of the project.

Detailed design and specification

The detailed design and the specification can have a huge impact on the likelihood of realising the social development opportunities. For example, over-design can have an impact on the cost of the capital infrastructure as well as on realising the social-economic benefits over the lifecycle of the constructed asset. Many countries do not have the finance or the capacity and capability to operate and maintain high-technology design solutions.

This effectively excludes local suppliers from participating in such projects, and can lead to poor maintenance of the asset.

Project designs should consider the whole-life value that includes the operation and maintenance of the asset. By adopting lower-cost technologies such as those for roads, or lower-end solutions such as those for water standposts, one can reduce capital costs by one-third. Alternatively, capital-intensive, low-maintenance technologies may represent a higher investment cost in the short run but overall lifecycle costs may be lower if reconstruction can be avoided or postponed (Foster and Briceño-Garmendia, 2010: 12).

The design can also have an impact on the ability to build up the capacity and capability of local contractors. Over-design can potentially exclude local suppliers from participating in the project, whilst low-cost technologies can provide more opportunities for them to win contracts. Competition will increase as more companies will be technically qualified to bid (Wells and Hawkins, 2008: 13). Clients should also standardise designs; this will allow local contractors to build up their skills and increase their efficiency, thus enabling them to win further contracts in the future

Local sourcing of materials, components and equipment can create opportunities for local suppliers and generate employment. The MDB rules on the procurement of goods allow local enterprises to be given up to 15 percent preference. However, clients are often unable to exercise this preference because materials, components and equipment are usually purchased by the contractor.

The social opportunities identified in the EIA and SIA should be incorporated into the design. The design should consider the physical and psychological demands of the users to ensure that the asset is used. It should also consider whether the asset can have more than a single purpose. For example, a water storage unit can be used for agriculture and utilised to provide clean drinking water.

Procurement strategy

The procurement strategy identifies the best way of achieving the project objectives, including the social opportunities and value for money, taking into account risks and constraints. The client or its agent will need to package the works, and decide on the contracting strategy and procurement arrangements to finalise the procurement strategy. Formulating the procurement strategy is the key to maximising the opportunities for local firms to participate in the project, and also the chance to generate employment opportunities. Key to this decision-making process will be engaging the market to identify the available external capability and capacity, including: design; cost consultancy; construction; project and programme management; and transaction advisors (Construction Industry Development Board, 2010).

Market engagement is achieved though a two-way communication process with potential consultants, contractors, subcontractors and suppliers. Clients should provide the market with a clear understanding of the project objectives, including the social opportunities, and then expect suppliers to ensure that they understand this and take steps to match their bids to the clients' requirements. Potential suppliers benefit from an early insight into the intentions of the client, whilst the client benefits from the incorporation of the expertise of the market into the development project.

Market engagement can identify the potential hurdles that restrict local firms from gaining access to public procurement markets. The biggest hurdle for local firms is having access to secure finance, particularly working capital. Even if banks are willing to lend funds to these firms, interest rates can be prohibitive (Wells and Hawkins, 2008). To overcome these hurdles, the client can: introduce a prompt payment regime to ensure cash flow; provide advance payments; and/or waive or lower the level of performance bonds.

Local firms can also have difficulties in gaining access to tender markets; pre-qualification is often costly and time consuming. To overcome these problems, clients can: hold regular supplier conferences; simplify tender procedures; have project documents written in the local language; use SMS texting to alert firms of tender opportunities; offer training through pre-tendering workshops; lower the cost of gaining access to tender information by either charging only for the cost of reproducing the documents or supplying the documents free of charge; waive capacity building or training requirements for local firms on the grounds that it is local firms that such levies are designed to assist; and maintain a robust register of contractors graded according to ability (Wells and Hawkins, 2008: 16).

Having identified the capability and capacity of the client and the market place, the client has a number of options to maximise the local content without compromising the delivery of the project.

Contractors are often unable to bid for work because the contracts are too large for them. There are strong pressures from donors and businesses to combine requirements into larger and fewer contracts in order to derive benefits from economies of scale and lower administration costs. Letting projects in smaller contracts (unbundled) can increase competition and give lower prices, as it allows a greater number of local contractors to bid for the work. Unbundling is allowed by the World Bank and many country procurement regulations as long as it can be shown that the objective is to increase local content (Wells and Hawkins, 2008: 13).

As shown earlier in the chapter, engaging with the local community as participants in the delivery process can create opportunities for local firms and generate employment for local people. Not only is the asset delivered but it generates "ownership" and increases the likelihood of long-term maintenance. However, both unbundling and community contracting have difficulties in delivering at scale. Thus, projects should only be broken down into smaller contracts when:

- there is administrative capacity to administer the increased number of contracts that result from the unbundling of the project;
- the unbundling does not result in an inappropriate division of responsibilities, increased contractual risk, duplication of establishment charges and under-utilisation of resources (Watermeyer, 2005).

In management contracting, a consultant is engaged to manage the project on behalf of the client. The consultant then sub-contracts the works to an appropriate number of companies. This approach may be used where the client lacks the administrative capacity to manage the work, particularly in the case of large-scale projects. It also provides the opportunity to unbundle the work into a number of contract packages. The key is to define measurable contract-specific goals for local enterprises, employment generation and training opportunities for inclusion in the tender documents and contract specification. This would also apply where the client is engaging a single contractor on a design–build, design–bid–build or turnkey project and wishes to include similar goals at the sub-contract level.

South Africa has embedded a measurable system for the inclusion of contract participation goals within its National Standards for Construction Procurement. The purpose of the goals is to provide standardised options to meeting the government's Black Economic Empowerment policy objectives. These standards formed the basis of the recently published International Standard for Construction Procurement. The standard (International Standards Organisation, 2010) consists of eight parts, including:

- Part 5: Participation of targeted enterprises in contracts;
- Part 6: Participation of targeted partners in joint ventures in contracts;
- Part 7: Participation of targeted enterprises and targeted labour (local resources) in contracts;
- Part 8: Participation of targeted labour in contracts.

The objective of this series is to provide a generic and standard set of "process, procedures and methods for the establishment within an organisation of a procurement system that is fair, equitable, transparent, competitive and cost effective" (International Standards Organisation, 2010: vi) and which may be used to promote social opportunities as objectives in addition to those associated with the immediate objective of the procurement (International Standards Organisation, 2010: vi). The standard establishes standards for core processes including those for defining industrial opportunities as contract-specific goals; these can also be applied in relation to other social opportunities. These will be explored in the following sections of this chapter.

Framework agreements offer particular advantages to both the client and the local construction industry. They are used where there is a programme of similar projects to meet an identified need. Intense competition amongst the many small contractors, and selection on the basis of price, make it difficult for a firm to obtain continuous work. Not only do framework agreements offer the potential of continuous work, but also contractors benefit from not having to repeatedly tender for projects of the same client. Contractors can also develop their relationships with sub-contractors and suppliers, as well as purchase plant and equipment, and write off the costs against on-going programmes. Framework agreements also offer the possibility of more stable employment and skills development. Clients achieve efficiencies in administrative costs and professional fees as tenders need to be invited for fewer contracts, and a smaller number of projects managed. They may also benefit from lower tender prices as contractors do not have to tender repeatedly for projects (Wells and Hawkins, 2008: 19–20).

Reservations and set-asides have been used extensively in Asia, for example in Malaysia and Indonesia. This provision of the procurement method restricts access to tender lists in order to favour local enterprises. As an alternative to offering preferences on the basis of price, the whole contract value – or a proportion of it – may be set aside for implementation by local contractors. The attraction of a policy of set-asides is that it can produce the quickest, most visible gains in the entry of local firms into public procurement markets. Concern is often expressed that this method distorts the market, but the entry of new firms into the market can also boost competition and result in lower prices.

Tender documents and process

Key to the tendering stage is to ensure that the social opportunities identified in the project planning and design stage are defined in the tender documents. Points can be awarded in the tender evaluation process in response to the social opportunities such as an environmental plan and a community plan. The International Standard for Procurement (International Standards Organisation, 2010) provides for tender-evaluation points to be allocated for the achievement of social opportunities. Social opportunities are defined as contract-specific goals, and tender evaluation points are granted using one of the following methods (International Standards Organisation, 2010: 3):

1 award a fixed number of points for attaining a specific goal;
2 award a variable number of points in proportion to the degree to which a tenderer responds to a particular goal (i.e. in proportion to the quantum of the goal offered); and
3 award points on a comparative basis in terms of which
   - the best offer received scores the maximum number of allotted points,
   - the worst offer scores no points, and
   - the remaining offers are scored between the limits of 1and 2 above.

These methods provide a competitive and flexible process to motivate a potential supplier to maximise the use of, for example, recycled materials or the input of the local community.

The tender process adopted will partly depend upon the procurement strategy and the roles and responsibilities of each member of the project team.

Consultants are often appointed at an early stage of the planning and design of the project and thus can exert a strong influence on the social opportunities included in the project. Appointing a consultant who is familiar with the local context will aid the development of the project to maximise the social opportunities (Wells and Hawkins, 2008). The tender documents for consultants should specify the social opportunities as contract-specific goals with the pre-qualification criteria framed around track record and local knowledge with respect to the social opportunities.

Consultants should demonstrate in their tender submission how they will achieve the contract-specific goals, such as the extent to which they plan to use local enterprises, labour and community groups. This can also include a written plan for the transfer of technology and development of skills. Evaluation of quality can also include points for the percentage of nationals in the total key staff time proposed, and for the transfer of technology to local enterprises and staff, as permitted in World Bank guidelines (Wells and Hawkins, 2008: 14).

As was explained earlier in this section, contractors can be motivated to deliver the social opportunities through the tender evaluation process. The social opportunities that are included in each tendering exercise will depend on the capacity and capability of the market that has been targeted to undertake the project; a standard procurement process that suits that market is adopted.

Providing a preference on price to local firms in the tender evaluation process is the most common method for promoting local contractors. On World Bank funded contracts, a preference of 7.5 percent may be provided on works contracts; many country procurement systems also offer a similar preference. This means that if the tender of a local contractor is less than 7.5 percent above the lowest tender, the contractor can be awarded the contract.

The International Standards Organisation (2010: 93) goes beyond granting a financial preference by either:

(a) granting a preference by the employer in the evaluation of tender offers, in return for the tendering of a contract participation goal or an undertaking to attain a specified contract participation goal at the time that tenders are evaluated; or
(b) requiring a contractor to achieve a minimum specified goal in the performance of a contract; or
(c) involving both (a) and (b).

For example, the client could set a contract-specific goal of 10 percent for the employment of local labour for each contract. No tender-evaluation points are granted in respect of contract-participation goals that are less than 10 percent. Tenderers that meet or exceed 10 percent are granted a preference in proportion to their tendered contract-specific goal; the maximum points are awarded to the tenderer with the highest tendered contract-participation goal (International Standards Organisation, 2010: 93). Incentives to enable firms to meet this type of social opportunity can also be included within the contract agreement.

Contract implementation

Key to the contract implementation is to ensure that the social opportunities identified in the previous stages are defined and measurable in the contract specification with a delivery plan including the incentives to realise them.

As social development opportunities have gained strong currency, there has been pressure to include environmental and health and safety opportunities as clauses within the conditions of the contract. The International Labour Organisation (ILO) successfully lobbied the MDBs to include the decent work agenda as a number of clauses within the MDB Harmonised Edition of the FIDIC[1] Conditions of Contract for works. It was thought that conditions of contract on issues such as forced labour, child labour, wages, and supply of food and water would reinforce local legislation or provide a minimum standard for donor funded projects where local legislation was absent (International Federation of Consulting Engineers, 2010). However, the conditions of contract are generally statements of intent or tests of reasonableness that require further definition in the specification or other contract document if they are to be realised.

Conditions of contract define the relationship between the employer and contractor by formulising the risk allocation and the contract management procedures such as payment. Thus, it becomes a challenge to draft a condition of contract that is a measurable objective. Social development opportunities are more likely to be realised if they are defined alongside the project objectives within the contract specification. The specification allows the opportunity to be translated into a clearly defined objective that can be measured at the conclusion of the contract. For example, the FIDIC MDB Harmonised Edition includes a clause that requires the contractor to implement an HIV-AIDS awareness programme (International Federation of Consulting Engineers, 2010: 25). The standard HIV-AIDS specification of the Construction Industry Development Board of South Africa (2003: 2) has a measurable requirement that the programme reaches a minimum of 90 percent of the workforce.

1 FIDIC stands for the International Federation of Consulting Engineers.

This approach can also be used to translate the recommendations from the EIA into clear actions to be achieved by the contractor that the employer or the employer's agent can evaluate to measure success. A pertinent example would be a target that 20 percent of the materials used for the project should be recycled. When social development opportunities are included in the specification, the contractor should consider them as a constraint on doing the work with a methodology to deliver the objective and price accordingly.

The specification can also include either the payment of a financial incentive to a contractor for the attainment of a specified contract participation, or require the contractor to record and report on the quantum of work generated for the targeted labour. If one considers the earlier example in the tender section, the client can include incentive payments of up to 3 percent of the contract value for the attainment of key performance indicators relating to contract-participation goals in excess of 35 percent (International Standards Organisation, 2010: 93).

Sanctions and contractual remedies for failing to achieve the social opportunity are an equally important incentive. These can include:

- withholding a completion certificate until the contractor has satisfactorily demonstrated that the contract-specific goals have been achieved;
- imposing a financial penalty more severe than the financial preference calculated at the time when tenders were evaluated or more severe than the cost of complying with contractual obligations; or
- restricting the contractor and its shareholders and directors from being awarded further contracts by the procuring entity for a specified period of time (International Standards Organisation, 2009: 16).

Sanctions should be sufficiently punitive to encourage compliance. Care should be taken when offering preferences that the sanction is not more attractive than compliance: for example, the quantum of the penalty for non-compliance should not be less than that of the preference offered.

Conclusion

Scaling-up investment in infrastructure is fundamental to achieving the MDGs, as it underpins so many of the goals and targets. With only four years at the time of writing to realise these ambitious goals and targets, gaining access to the finance required for this programme during a period of severe economic recession will be challenging. This provides an opportunity for clients, planners and designers to be creative by utilising the procurement process to promote social opportunities as a means to maximising the potential social and economic benefits derived from the investment.

There has been a long history of the procurement process being utilised to promote industrial opportunities and, more recently, environmental and society opportunities. This experience demonstrates that, when correctly identified, planned, designed, financed and implemented, promoting social opportunities should not compromise the core project objectives. Instead, the opportunities become core to the project. Critical to their success is ensuring that the opportunities are defined with realistic and measurable goals at each stage of the project cycle. The actions to take to identify and then deliver the social opportunities are presented in summary form in Table 3.1.

Table 3.1 Actions to realise the social opportunities

	Actions	*Society*	*Industrial*	*Environment*
Planning	Feasibility study	✓	✓	✓
	Social impact assessment	✗	✓	✗
	Environmental impact assessment	✓	✗	✓
	Delivery plan	✓	✓	✓
	Operations and maintenance plan		✓	✓
	Include the social opportunities in the project business case		✓	✓
Design	Local sourcing of materials, components and equipment	✗	✓	✗
	Low-cost technologies	✗	✓	✗
	Capital-intensive	✓	✗	✗
	Incorporate recommendations from EIA	✓	✗	✓
	Incorporate recommendations from SIA	✓	✓	✗
	Consider physical and psychological health of user	✓	✗	✗
Procurement	Market engagement	✓	✓	✓
	Let smaller contracts	✗	✓	✗
	Define contract-specific goals	✓	✓	✓
	Management contracting	✗	✓	✗
	Framework agreements	✗	✓	✗
	Reservation and set-asides	✗	✓	✗
Tender selection	Social opportunities identified in the project planning and design stage are defined in the tender documents	✓	✓	✓
	Tender evaluation points for social opportunities	✓	✓	✓
	Appointing a consultant who is familiar with the local context	✓	✓	✓
	Consultants should demonstrate in their tender submission how they will achieve contract specific goals	✓	✓	✓
	Provide a preference base for meeting a contract specific goal	✓	✓	✓

	Actions	*Society*	*Industrial*	*Environment*
Contract implementation	Define the social opportunities as contract specific goals in the specification	✓	✓	✓
	Include incentive payments benchmarked against including the contract specific goals	✓	✓	✓
	Include penalties and contractual remedies for failure to meet the contract specific goals	✓	✓	✓

References

African Development Bank (2006) *Annual Procurement Report 2006*. Abidjan: ADB.

Construction Industry Development Board (2003) *Specification for HIV/AIDS Awareness*. Pretoria: CIDB.

Construction Industry Development Board (2010) *Delivery Management Guidelines: Practice Guide 2 – Construction Procurement Strategy*. Pretoria: CIDB.

Fisher, J. (2008) Women in water, sanitation and hygiene programmes. *Proceedings of the Institution of Civil Engineers – Municipal Engineer*, 161(4): 223–9.

Foster, V. and Briceño-Garmendia, C. (2010) *Africa Infrastructure: A Time for Transformation*. Washington D.C.: Agence Française de Développement and World Bank.

Goodsir, S. (2009) Engineering capability in Rwanda. SISTech, http://www.unesco.org.uk/uploads/SISTech%20-%20Engineering%20Capability%20in%20Rwanda.pdf/

Hawkins, J., Wells, J. and Herd, C. (2006) *Modifying Infrastructure Procurement to Enhance Social Development*. London: Institution of Civil Engineers and Engineers Against Poverty.

International Federation of Consulting Engineers (2010) *Conditions of Contract for Building and Civil Works*, Harmonised Edition. Paris: FIDIC.

International Standards Organisation (2009) *Draft International Standard for Construction Procurement – Part 7: Participation of targeted enterprises and targeted labour (local resources) in contracts*. Geneva: ISO.

International Standards Organisation (2010) *International Standard for Construction Procurement – Part 1: Processes, Methods and Procedures*. Geneva: ISO.

Jones, B. (2010) *Beyond the state: Community development in rural Uganda*. Presentation at the *Engineering a Better World* Conference, Institution of Civil Engineers, 11 March, London.

McCrudden, C. (2004) Using public procurement to achieve social outcomes. *Natural Resources Forum*, 28: 257–67.

McCrudden, C. (2007) *Buying Social Justice: Equality, government procurement and legal change*. Oxford: Oxford University Press.

McCrudden, C. and Gross, S. (2006) WTO government procurement rules and the local dynamics of procurement policies: A Malaysian case study. *European Journal of International Law*, 17(1): 151–85.

Ministry of the Environment of Finland and Marrakech Task Force on Sustainable Buildings and Construction (2010) *Best Policy Practice, Buildings for a Better*

Future, http://esa.un.org/marrakechprocess/pdf/Buildings_for_a_Better_Future_SBC_MTF_Finland.pdf (accessed on 7 February 2011).

Organisation for Economic Cooperation and Development (2006) *Applying Strategic Environmental Assessment: Good practice guidance for development co-operation*. Paris: DAC Guidelines and Reference OECD.

Schubert, M. (ed.) (2007) *Final Report of the Marrakech Task Force on the Project "Cooperation with Africa": Best Practice in African Countries*. Weimar: Federal Ministry for the Environment, Nature Conservation and Nuclear Safety.

United Nations Department of Economic and Social Affairs, Division for Sustainable Development (2008) *Sustainable Development Innovation Briefs*, Issue 5, August, p. 5.

United Nations (2010) *Keeping the Promise: a forward-looking review to promote an agreed action agenda to achieve the Millennium Development Goals by 2015*. Report of the Secretary-General. New York, United Nations.

Vanclay, F. (2003) International principles for social impact assessment. *Impact Assessment and Project Appraisal*, 21(1): 5–11.

Watermeyer, R. (2005) Project Synthesis Report: Unpacking Transparency in Government Procurement – Rethinking WTO Government Procurement Agreements, CUTS International, http://cuts-international.org/pdf/synthesis-report.pdf (accessed on 7 February 2010).

Wells, J. and Hawkins, J. (2008) *Increasing Local Content in the Procurement of Infrastructure Projects in Low Income Countries*. London: Institution of Civil Engineers and Engineers Against Poverty.

World Bank (2001) *Indonesia Country Procurement Assessment Report: Reforming the Public Procurement System*. Washington, D.C.: World Bank.

World Bank (2006) *Environmental Impact Assessment Regulations and Strategic Environmental Assessment Requirements Practices and Lessons Learned in East and Southeast Asia*, Safeguard Dissemination Note No. 2. Washington, D.C.: World Bank.

World Conservation Union (2007) African Experts Workshop on Review of Effectiveness of EIA Systems – Report. Addis Ababa, 12–13 April, http://www.uneca.org/sdd/documents/Report-AfricanExpertsWorkshop-on-Effectiveness-of-EIA.pdf (accessed on 7 February 2010).

WTO (World Trade Organisation) (1994) *Government Procurement Agreement*. Geneva: WTO.

WTO (World Trade Organisation) (1997) *Policy Review: Malaysia*, WT/TPR/S/31. Geneva: WTO.

4 Transparency in construction

George Ofori
National University of Singapore

Introduction

"Ethics", "integrity", "transparency" and "good governance" have become words in currency over the past few decades. Political campaigns have been fought on them. In most countries, statutory agencies have been formed to pursue them. Most countries have dedicated anti-corruption agencies. South Africa has the Office of the Public Protector. Several governments seek to improve the ranking of their countries on international benchmarks such as the Corruption Perception Index (published annually by Transparency International). Some non-governmental organisations pursue aspects of the subjects. National strategies and campaigns have been launched in many countries. For example, Bolivia marked its first National Day Against Corruption on 8 February 2011. Malaysia has a national integrity programme. Many companies, in all sectors in that country, have integrity or compliance officers. In the construction industry, ethics, professionalism and "best practice" are also words one hears frequently these days.

The UNDP (2008b) notes that, in the Asia-Pacific region, people are becoming increasingly concerned about corruption and malpractice, especially in public life, and that politicians are starting to respond; most want to be associated with fighting corruption. Civil society groups, too, are making greater efforts to hold public- and private-sector organisations to account, and the media are also focusing on the issue and finding new ways to expose it.

International agreements on corruption include the UN Convention Against Corruption that was adopted by the UN General Assembly in October 2003, and came into force in December 2005. Others are the Inter-American Convention Against Corruption, adopted in 1996, OECD Anti-Bribery Convention, Paris Declaration, and Accra Agenda for Action and African Union Convention on Preventing and Combating Corruption, adopted in 2003. The G20 Summit in Seoul in 2010 also issued a statement on the subject. The World Bank's Integrity Vice-Presidency was established in 2001 to investigate fraud and corruption in World Bank-financed projects as well as possible staff misconduct (World Bank Office in Bangkok, 2010).

This movement to engender ethical behaviour has generally been in response to the many recent cases of high-level malpractice and corruption in both developed and developing countries. The construction sector has been found to be particularly prone to such practices. Mason (2008) cites several studies that found that construction practitioners in many countries (including Australia, South Africa, the UK and the US) consider there to be a significant amount of corruption in the industry. Doree (2004: 146) observes that many investigations by parliament, the cabinet, justice and antitrust authorities had found "widespread use of cartels and structural bid rigging within the Dutch construction industry". Pearl et al. (2005) found that construction professionals in South Africa, especially contractors, appeared to have a reputation for unethical behaviour. The range of ethical problems that they found included collusion, bribery, negligence, fraud, dishonesty and various unfair practices. Most construction professionals felt that the industry suffers from a high incidence of unfair tendering practices, over-claiming and/or withholding payment for service delivered. Sohail and Cavill (2008) note that corruption in construction includes bribery, fraud, embezzlement and kickbacks. Shakantu (2006) reports that the following forms of corruption have been found in construction industries: extortion, bribery, theft, fraud, collusive bidding or bid rigging. Those found in South Africa include: submission of fictitious invoices for materials and labour; payment of bribes for work, or for prompt payment; fraud; writing off of loans; and fictitious payments.

Many authors have observed that, although it is widely known that unethical behaviour is rife in the construction industry, there is little mention of it, and thus not much is being done to address it (see, for example, Shakantu, 2006). Flyvberg (2009: 155) notes that:

> ... we need to honestly acknowledge that infrastructure investment is no easy fix but is fraught with problems. It is a great help that Barack Obama openly identified "the costly overruns, the fraud and abuse, the endless excuses" in public procurement as key problems. The *Washington Post* calls this "a dramatic new form of discourse" and so it is. Before Mr Obama, it was not *comme-il-faut* to talk about overruns, deception, and abuse and the few who did were ostracised. However, we cannot solve problems we cannot talk about. So talking is the first step.

Ethics

Ethics are considered to be conceptions of which actions are right or wrong, and what individuals in the society or organisation should seek to do or avoid. It is suggested that ethical problems do not have easily defined solutions but are usually surrounded by ambiguities, complexity and ill-defined boundaries (see, for example, Lynch, 1998). Crane and

Matten (2007: 6) note that business ethics "is primarily concerned with those issues not covered by the law, or where there is no definite consensus on whether something is right or wrong". Royal Institution of Chartered Surveyors Working Committee (2000) suggests that professional ethics involves "giving of one's best to ensure that clients' interests are properly cared for, but in doing so the wider public interest is also recognised and respected".

It appears that, in many countries, the standard of ethics in construction is not satisfactory (see, for example, Richter and Burke, 2007). Vee and Skitmore (2003) found that most unethical behaviour in the construction industry in Australia is experienced in the form of unfair conduct, negligence, conflict of interest, collusive tendering, fraud and bribery. The instances of unfair conduct included biased tendering evaluation systems, re-tendering and shopping for prices after tenders have closed, and non-payment of consultants' fees by clients. Examples of negligence could be lack of proper care, and the omission of such duty of care for the interests of others as the law may require. The study by Vee and Skitmore (2003) found a high incidence of conflict of interest in the Australian construction industry. Collusion could lead to increased construction prices, possible quality compromises, company failures through unfair competition, a negative industry image, and decreased employee productivity through moral dissatisfaction. Fraud involves deceit, or breach of confidence. Finally, bribery could be categorised into: paying someone to do what the person should ordinarily do to expedite matters; paying someone to use the person's power or influence over others to get something done by those others; and compelling a person to pay through threats of what will occur if the payment is not made.

The rest of this chapter focuses on the forms of unethical behaviour that may be grouped under the collective word "corruption" in the context of the construction industry, with special consideration of the issue in the developing countries.

What is corruption?

There are several definitions of "corruption". The definition also appears to change over time, as the issue is dynamic. For example, the UNDP (2008a) indicates that, in 1998, it had defined corruption as "the misuse of public power, office or authority for private benefit – through bribery, extortion, influence peddling, nepotism, fraud, speed money or embezzlement" (p. 6). The International Federation of Consulting Engineers (FIDIC) defines it as "the misuse of public power for private profit" (FIDIC, undated a). However, corruption also exists in the private sector. Thus, UNDP (2008a) notes that corruption is now defined more broadly as: "misuse of entrusted power for private gain" (p. 6).

Authors such as Stansbury (2005) and UNDP (2008a) distinguish between "grand" and "petty". Grand corruption typically involves the payment of relatively large sums of money by companies, and is generally associated with high-level politicians or officials. Petty corruption involves smaller amounts but more frequent transactions. For example, junior officials might demand money to provide places in schools or hospitals or connections to public utilities.

UNDP (2008b) notes that corruption has plagued the world since the emergence of organised forms of government, and, until the nineteenth and early twentieth centuries, many of the now developed countries were riddled with corruption. It took many decades to bring the levels down. Thus, it believes that the developing countries are likely to shed many corrupt practices as a natural by-product of modernisation, but they need not simply wait for development to take its course. They can devise programmes specifically to tackle corruption. It is pertinent to observe that corruption remains a universal problem. As Box 4.1 shows, continuous vigilance is necessary in all countries.

Box 4.1 Principles of public life

In an effort to hold its public servants to high standards, the UK introduced a broad code of behavior in 1994 for those in public life. The Organisation for Security and Cooperation in Europe (OSCE, 2003) suggests that The Seven Principles of Pubic Life "can be applied universally, regardless of differences in politics, history or culture" (p. 6):

Selflessness – Holders of public office should take decisions solely in terms of the public interest. They should not do so in order to gain financial or other material benefits for themselves, their family, or their friends.

Integrity – Holders of public office should not place themselves under any financial or other obligation to outside individuals or organizations that might influence them in the performance of their official duties.

Objectivity – In carrying out public business, including making public appointments, awarding contracts, or recommending individuals for rewards and benefits, holders of public office should make choices on merits.

Accountability – Holders of public office are accountable for their decisions and actions to the public and must submit themselves to whatever scrutiny is appropriate to their office.

Openness – Holders of public office should be as open as possible about all the decisions and actions that they take. They should give reasons for their decisions and restrict information only when the wider public interest clearly demands.

Honesty – Holders of public office have a duty to declare any private interests relating to their public duties and to take steps to resolve any conflicts arising in a way that protects the public interests.

Leadership – Holders of public office should promote and support these principles by leadership and example.

Source: Committee on Standards in Public Life (undated)

The OSCE (2003) notes that codes of conduct – for ministers, legislators, civil and foreign service officers, the judiciary, and local government – can help countries to put the seven principles highlighted in Box 4.1 to work. Similarly, citizens' charters can compel government agencies to provide certain levels of service to citizens and to solicit complaints and sanctions if these levels are not met.

Corruption has wide ramifications. UNDP (2008a) notes that corruption weakens national institutions, and leads to inequitable social services, economic inefficiency and unchecked environmental exploitation; and it hits hardest at the poor who depend on public services and the natural environment and are least able to pay bribes for essential services that should be theirs by right. Moreover, corruption undermines human development. It is likely to corrode ideals of public service, so that administrations will tend to be less interested in investing in health and education. It is also likely to undermine efforts at poverty reduction – by diverting goods and services targeted for the poor to well-off and well-connected households who can afford to bribe officials.

Corruption and mismanagement in construction and their effects

Writing in *Fortune* magazine, Kimes (2009) declares that "the construction industry globally remains rife with corruption". She relates the often repeated observation that Transparency International had consistently found construction the most corrupt of all the sectors in its annual Bribepayers Index. Stansbury (2005) highlights two international surveys undertaken in 2002 which showed that, in many countries, corruption was higher in construction than in any other sector of the economy. Ameh and Odusami (2010) note that there is growing consensus within and outside the Nigerian construction industry that corruption and other unethical practices are endemic in the industry.

FIDIC (undated a) notes that corruption lowers the quality of public infrastructure by diverting public resources to private uses, and allowing the avoidance of codes and standards. It considers corruption:

> … morally and economically damaging. Firstly, it jeopardizes the procurement process, is always unfair, and often criminal. It saps money from required development projects and adversely affects their quality. Secondly … corruption is more basically wrong because it undermines values of society, breeds cynicism, and demeans the individuals involved.

Transparency International (undated) observes that:

> Corruption in the construction sector not only plunders economies; it actually shapes them. Corrupt government officials steer social and economic development towards large capital-intensive infrastructure projects that provide fertile ground for corruption, and in doing so neglect health and education programmes. The opportunity costs are tremendous, and they hit the poor hardest.

It has also been found that corruption has a negative impact on investment, on innovation and on entrepreneurship as the processing of licences and permits gives opportunities for rent seeking, thus creating uncertainty, additional cost and inefficiency. For these reasons, corruption discourages investment, both foreign and local; thus, it has a negative impact on economic growth. Moreover, it reduces the efficiency and productivity of investment and increases operating costs. It encourages risky and often dangerous practices. Corruption distorts markets as the playing field becomes no longer level where bribery is involved. It increases inequality, and the poor has even more reduced access to goods and services. Corruption impedes sustainable development as it can facilitate the avoidance of environmental regulations.

The link between corruption and the attainment of the Millennium Development Goals (MDGs) in developing countries is often made. For example, the German agency, *Deutsche Gesellschaft für Technische Zusammenarbeit* (GTZ) (2010) observed:

> Corruption is increasingly understood as an obstacle to sustainable development. It is one of the major stumbling blocks on our road to meet the Millennium Development Goals: scarce resources are diverted, and services become unaffordable to poor households due to corruption (p. 1).

Kenny (2007) reviews a number of studies and concludes that failed governance can lead to the construction of the wrong infrastructure, poor construction and quality of provision, insufficient maintenance and high levels of theft and loss. This can reduce economic returns to projects as well as the entire infrastructure stock. It is suggested that corruption in public procurement affects economic growth by[1] skewing public investment and spending to areas offering the highest opportunities for personal gain, not where the needs are greatest; affecting the investment climate as international companies bidding in good faith for contracts find the system behaving in a way they did not expect and spread the word that the investment climate is hostile; increasing business costs; and hindering the development of competent and genuine local contracting and consulting enterprises.

Some authors put forward the concept of "efficient corruption". They argue that bribery may enable firms to get things done in economies that face bureaucracy and bad, rigid laws (see a discussion in Svensson, 2005). Moreover, where the system for allocating licences, giving permits and awarding public-sector contracts is built on bribery, this may result in the most efficient firms being able to afford to pay the highest bribes. On the other hand, in such a system, public officials may extend or delay approval procedures to attract more bribes. As the UNDP (2008a) notes, the argument that, by "greasing the wheels", corruption speeds things up is hardly supported by evidence from the aggregate standpoint. It suggests that it seems more likely that if businesses readily pay bribes they will simply encourage bureaucrats to introduce more steps in the administrative processes.

Studies indicate that the extent of "leakage" on public-sector construction projects is substantial. The losses take place in all countries, regardless of their levels of development. They also occur on all types of projects, including those involving foreign financial assistance. Kenny (2007) gave detailed coverage of various implications of corruption and malpractice on infrastructure projects around the world. He notes that corrupt acts raise the price of infrastructure and can also reduce its quality and economic returns to investment in it. Kenny (2007) found a relationship between the average unit cost for upgrading a road with a bitumen surface across countries and the country's Corruption Perceptions Index. The *Global Corruption Report 2008* (Transparency International, 2008) notes that estimates by the World Bank indicate that between 20 and 40 percent of investments in the water sector are lost as a result of corruption. Also, research by the World Bank has found that, by tackling corruption and improving the rule of law, countries can increase their national income by as much as four times in the long term, and child mortality can fall as much as 75 percent.

Collier and Hoeffler (2005) reviewed a number of studies and found the costs of corruption to be substantial in the infrastructure sector. For example, corruption delays and reduces expenditure on infrastructure investment; corruption reduces the growth generated by a given expenditure

1 www.worldbank.org.bd

on infrastructure investment; and corruption raises the operating cost of providing a given level of infrastructure services.

Brosshard (2005) gave many examples of "Monuments of corruption" which do not make any economic sense and had major environmental and social impacts, but went ahead owing to corruption and cronyism. They include the US$2 billion Bataan nuclear power plant in the Philippines which sits on an active fault line, and was completed in the 1980s but has never been used; the Yacyretá Dam on the border of Argentina and Paraguay which is flooding the Ibera Marshes, a unique ecosystem – due to cost overruns partly attributable to corruption, the power generated by Yacyretá needs to be subsidised by the government; and Enron's Dabhol power plant which threatens to destroy a fragile coastal area in India – it was mothballed in 2001 because its electricity was prohibitively expensive.

Corruption also has a negative impact on post-disaster and post-conflict reconstruction. For example, it is estimated that 30–40 percent of the more than US$7 billion which was pledged for Aceh after the tsunami in 2004 was stolen (UNDP, 2008a). Also, about one quarter of the 50,000 housing units built for the victims are in a bad state and will have to be rebuilt because much of the timber used did not meet the requirements in the building codes (U4 Anti-Corruption Resource Centre, 2007).

Reasons for corruption in construction

There are many reasons for the relatively high level of mismanagement and corruption in the construction industry, and some of them may be considered. Stansbury (2005) discusses "the foundations of corruption in construction" and observes:

> The construction sector is complex, diverse and fragmented, all of which contribute to a lack of effective control and the absence of uniform integrity standards. ... The lack of transparency surrounding projects and the contentious environment both tend toward bribery and deception (p. 49).

The first reason why the level of corruption in the construction industry is high is that construction projects are relatively large. The large sums involved increase the temptation towards rent seeking behaviour. Administrators do not appreciate the financial issues on projects as they deal with the flows of large sums of money, and they have an inflated perception of the contractor's "profit". Thus, they tend to press the contractor for "rewards for their services". Second, construction projects have long periods of gestation. The passage of time increases the financial risks of the construction firm in the form of increases in interest rates and inflation, and changes in currency exchange rates. This provides incentives for the firm to seek to expedite the process.

Third, constructed items must be built in the locations where they are to be placed. Thus, in each country, not only are a large number of construction projects of many different sizes underway at any one time, but also these projects are geographically dispersed. This makes the control of practice and procedure on more than a few projects difficult. It also increases the number of "stakeholders" who might be interested in seeking opportunities to extract rent. Fourth, construction projects are technically complex. Thus, in many cases, only experts can detect malpractice or mismanagement, for example, in the form of poor quality of materials or workmanship. Moreover, there are many different ways of solving the same problem on a construction project, making it difficult to determine whether an approach that has been adopted is the correct one. It is pertinent to note that Transparency International (2010) disputes the notion that public procurement is a complex subject. It suggests that "everyone, from individual citizens to high-level government officials, can play a role in ensuring that tax payers' money spent on procurement delivers good quality services at a fair economic cost for all" (p. 1).

Fifth, the administrative procedures involved in a construction project can be extensive and complicated. The building team requires approval from many agencies at various stages of the planning, design and construction process. These include the planning and building control authorities, as well as those in charge of fire protection, safety and health, and the water, sewerage and power agencies. After the project is completed, the commissioning and inspection by public officials before occupation requires professional expertise and administrative formalities. Finally, considering the physical nature of constructed items, in which much of the completed work is not visible, it is possible and relatively easy to conceal substandard quality of materials or workmanship, with or without approval from officials who might have been unduly influenced.

Similarly, Stansbury (2005) discusses the characteristics of construction which make it corruption-prone under the following headings: size, complexity, uniqueness and lack of frequency of projects; government involvement; number of contractual links and phases; the fact that work is concealed; a culture of secrecy; entrenched national interests; the fact that no single organisation governs the industry; lack of "due diligence"; and the cost of integrity to firms. The *Global Corruption Report 2008* observes (Transparency International, 2008) that the water sector is especially vulnerable to corruption because: water governance spills across many agencies, regulations and systems dispersed across countries, with many loopholes to exploit; and is viewed as a largely technical issue so consideration for its political and social dimensions, including corruption, is limited; and large water management, irrigation and dam projects are complex and difficult to standardise, making manipulation difficult to detect.

Some cases of malpractice on construction projects

In many countries, the law expressly forbids many of the practices in construction including some that appear to be normal to the practitioners. The Dutch Parliamentary Inquiry and Audit Commission (see, for example, Doree, 2004), which was set up to investigate fraud in the construction industry in 2002, identified several shortcomings, although it did not find actual corruption. The traditional market structures that involved short-term relationships provided no opportunity to optimise the price–quality relationship and inhibited competition based on performance. It had also led to a fragmented industry, with conflicting objectives. Other factors affecting the industry's performance were: in some parts of the country and on some types of construction, the incidence of mergers, cost of preparing bids for projects, and stringent pre-qualification criteria reduces the size of the market; and "collaboration" among some firms in tenders had further reduced competition.

Darroch (2005) presents an account of a number of cases brought by the government of Lesotho against international contractors. The trials tested many of the legal aspects of corruption, including the definition of bribery, and resulted in the first debarment of a major international company by the World Bank. Darroch (2005) suggests that international financial institutions should coordinate anti-corruption strategies so that the debarment of a company by one of them automatically results in debarment by *all* of them. In October 2007, the European Union Anti-Fraud Office (OLAF) revealed that companies from France, Italy and Germany convicted of paying bribes to win contracts in a dam project in Lesotho were fined €4.4 million (FIDIC, undated b). The chief executive of the project was jailed for 15 years. The involvement of OLAF and the use of bank records obtained from the Swiss authorities and the World Bank underline concerted efforts to track, trace and prosecute corruption. The European Commission unearthed a cartel amongst lift manufacturers discovered to be operating in Germany, Belgium, Luxembourg and the Netherlands that had been in operation between 1995 and 2004. The fines imposed totalled €992 million, and the companies involved included some of the leading names in the sector.

In 2008, Halliburton, a US construction firm, agreed to pay fines of about US$559 million to the US government to settle charges concerning bribes alleged to have been paid by a subsidiary, Kellogg, Brown and Root, on the construction of a liquefied natural gas plant in Nigeria, from 1996 to the mid-2000s (Iriekpen, 2009). That would be the largest amount paid by a US firm in a bribery investigation. In 2008, Siemens AG of Germany agreed to pay US$800 million in fines in the US to settle investigations involving alleged payments to government officials around the world to win infrastructure contracts. As part of the settlement, Siemens did not admit to bribery allegations but it admitted to having had inadequate controls and keeping improper accounts. In 2009, the UK Office of Fair Trading (OFT) announced its decision to fine 103 construction firms £129 million for engaging in bid rigging, largely in the form of cover pricing, on 199 tenders between 2000 and 2006 (Power and Jordan, 2009). Prior to this, the OFT had issued a Statement of Objections in April 2008 against 112 construction firms for alleged bid rigging. The decision followed a four-year investigation. The results of a technical and management audit of a series of projects in Uganda, which showed many instances of malpractice, are presented in Box 4.2.

Box 4.2 Results of technical audit of projects in Uganda

The technical audit on the provision of facilities for the Commonwealth Heads of Government Meeting (CHOGM) in Uganda, which included the design and construction of several venues, found many irregularities. Some examples are now presented.

- The procurement process for most projects followed Public Procurement and Disposal of Public Assets (PPDA) Regulations contained in the Act of 2003;
  - 31 jobs were procured using Restricted Domestic Tendering, 8 using Open Domestic Bidding, 4 were procured by Direct Procurement, and one as a variation;
  - information on 4 jobs transferred from Ministry of Works and Transport (MoWT) to Ministry of Local Government (MoLG) was not given for verification; but the process was flawed by the starting and in some cases completion of works prior to concluding the contracts;
- In 4 cases selection of firms was based on "Pseudo Estimates" which is not allowed by PPDA. MoWT Contracts Committees were not specific in recommending and awarding contracts, leaving it to the Accounting Officer to determine the final contract price;
- 35% of jobs lacked Engineer's estimates to guide the Evaluation Committee in assessing bids; where available, they were far below the bids received from bidders;
- 33 projects had addenda or variations issued, 7 were signed after implementation, 24 were still waiting for signature, and were not yet approved by contracts committee;

- In 7 cases, payments to contractors were effected prior to signing of contracts;
- There was inefficient retrieval and management of financial documents in client agencies;
- Payments to contractors had inadequate attachments, such as interim certificates and measurement sheets. Many contractors did not issue receipts for payments received;
- There were many delays in processing and effecting payments to contractors due to bureaucracy and poor documentation. This could have affected the timely performance of services providers by limiting their cash flow.

Source: COWI Uganda (2008)

Research findings on malpractice in construction

As noted above, mismanagement, fraud and other forms of corruption manifest themselves on construction projects in many ways. Some examples are now discussed.

OSCE (2003) observes that, on a contract, both the purchaser and seller have many ways of corrupting the procurement process, and at any stage of the process. Before contracts are awarded, the purchaser can: tailor specifications to favour particular suppliers; restrict information about contracting opportunities; claim urgency as an excuse to award a contract to a single contractor without competition; breach the confidentiality of suppliers' offers; disqualify potential suppliers through improper pre-qualification requirements; and take bribes. Suppliers can also: collude to fix bid prices; promote discriminatory technical standards; interfere improperly in the work of evaluators; and offer bribes. OSCE (2003) further states that the most serious and costly forms of corruption may take place after the contract has been awarded, during the performance phase. The purchaser may: fail to enforce quality standards, quantities or other performance standards; divert delivered goods for resale or for private use; and demand other private benefits. For their part, the contractor or supplier may: falsify qualities or standards certificates; over- or under-invoice; and pay bribes to contract supervisors. The list of 47 possible examples of corruption in infrastructure projects outlined by the Global Infrastructure Anti-Corruption Centre (GIACC) (2008) is presented in Box 4.3.

Box 4.3 Examples of corruption in infrastructure projects

Pre-qualification and tender

1. Loser's fee
2. Price fixing
3. Manipulation of pre-qualification
4. Bribery to obtain main contract award
5. Bribery during sub-contract procurement
6. Corruptly negotiated contract
7. Manipulation of design
8. Specification of overly sophisticated design
9. Inflation of resources and time requirements
10. Obtaining a quotation only for price comparison
11. Concealment of financial status
12. Intention to withhold payment
13. Submission of false quotation
14. Falsely obtaining export credit insurance

Project execution

15. False invoicing: supply of inferior materials
16. False invoicing: supply of less equipment
17. False work certificates
18. Excessive repair work
19. Overstating man-day requirements
20. Inflated claim for variation (1) – contractor inflating its claim
21. Inflated claim for variation (2) – contractor paying a bribe to the architect to approve an inflated claim
22. False variation claim
23. Issue of false delay certificate
24. False extension of time application
25. False assurance that payment will be made
26. Delayed issue of payment certificates
27. Concealing defects (1) – contractor concealing a defect in the work
28. Concealing defects (2) – sub-contractor paying a bribe to the supervisor to accept a defective material or work
29. Set-off of false rectification costs
30. Refusal to issue final certificate
31. Requirement to accept lower payment than is due
32. Extortion by client's representative
33. Facilitation payment

34. Overstating of profits
35. False job application

Dispute resolution

36. Submission of incorrect contract claims
37. Concealment of documents
38. Submission of false supporting documents
39. Supply of false witness evidence
40. Supply of false expert evidence
41. Bribery of witness
42. Blackmail of witness
43. False information as to financial status
44. False statement as to settlement sum
45. Over-manning by law firm
46. Excessive billing by lawyer
47. Complicity by lawyer

Source: Global Infrastructure Anti-Corruption Centre (2008)

Inception and feasibility stage

Often, in the physical planning of a project, there is no attempt to relate it to the longer-term national or district physical and socio-economic plans, or the programme for the sector. Thus, projects are built as discrete facilities with weak interconnections, and the local infrastructure does not operate as a system and is thus ineffective and inefficient. It is suggested, also, that projects are sometimes launched when there are no funds to support them. In some cases, their costs are deliberately underestimated to present the government with an ongoing project for which it would then be obliged to provide the funds needed. This leads to delays and difficulties in meeting interim payment obligations, and results in sub-optimal project performance in terms including time, productivity and quality.

Studies have also found that the requests for proposals from consultants are often poorly formulated and not in compliance with statutory requirements. Moreover, the bases for the selection of consultants, which are mostly qualitative and subjective, are often unclear and opaque. Another aspect of construction projects on which there are often problems is the feasibility study. The planned project's cost, time and return on investment estimates are often falsified. As many of the items to be estimated cannot be quantified, and the interest rate applied in the analysis is subjective, the results of feasibility studies are subject to misinterpretation. It is suggested that, if consultants assess projects solely on their merits, they risk obtaining no future contracts (Darroch, 2005). This causes what is termed "excessive appraisal optimism", a form of corruption that distorts the planning process to the benefit of projects with large budgets, contracts and prestige. Kenny (2007) reports that, in Indonesia in the early 1990s, road maintenance accounted for about 47 percent of the central government grants to districts for roads. By the end of the decade this figure was 15 percent.

Finally, the environmental impact assessment, which is a mandatory requirement for all projects in many countries, is also often inaccurate or false, or contains presumed mitigation methods that both the planners and the proposers know will not be undertaken.

Planning and design

At the planning stage, the possible deficiencies include inappropriate location. The siting of buildings or routing of roads is often influenced by factors other than the technical or professional ones that determine appropriateness of sites for particular construction projects. Also, at the planning stage there can be inflation of the available resources, and of the time required to complete the project.

The design of the whole building or some of its aspects, such as the selection of the structural frame, or the mechanical or electrical installations, is often not the most beneficial. For example, it is suggested that, in some projects, there is over-design of structural elements, or complex design and specification of overly sophisticated materials, components or installations and those that are not suited to the contexts of the projects. Another aspect is where designers specify particular brands of products owing to various influences.

Procurement

The procurement process is often focused upon in the discussion on malpractice and corruption in the construction industry. Thus, major procurement reform programmes have taken place in many countries, developed and developing, in the past decade (this is discussed later). At the pre-qualification and tender stages there are many possible inappropriate practices. Some aspects of the procurement of goods or services might be restricted to particular business enterprises. For example, the projects might be advertised only in certain newspapers; or the contractors' insurance has to be obtained from certain companies. In many cases, the pre-selection criteria, or even the contract itself, will be designed in such a way as to favour a particular preferred tenderer. A common occurrence is that the information on the assets, resources and track record submitted by contractors at both the pre-selection and tendering stages is false.

There can be collusion in the bidding process. There are many instances of *price fixing*, when businesses conspire to set prices higher than they would be if normal competition were at play. In *bid rigging*, companies work together to have one company in the group submit a "lower" price (while the others put up *cover prices*), but this would actually be higher than the companies would have priced the project if they had not been working together. Contractors might also be involved in bid rotation schemes that can limit the market in certain regards, including type of work and geographical location. *Low balling* is a sales practice whereby the company offers an unrealistically low price for a good or service that will not be honoured when the client actually expresses the demand.

Shakantu (2006) and Mason (2008) are among authors who have suggested that, in construction, the line between ethical and unethical behaviour is not always clear. Similarly, Vee and Skitmore (2003) suggest that it is necessary to distinguish illegal acts from those that are unfair but not illegal. However, some practices, such as *bid shopping*, whereby clients or main contractors disclose the prices of competing bidders as they seek further discounts from other contractors or subcontractors, were thought to be illegal by almost all survey respondents in their study. The putting up of *cover prices*, where some of the bidders submit deliberately inflated prices in order not to win projects, is a common occurrence in the construction industry.

There are other instances where the integrity of the bidding process is compromised. An example is the payment of bribes to obtain the contract award. In many countries there is anecdotal evidence that there is a norm of the proportion of contract sum that must be paid to secure the contract. Kenny (2006) observed that, in Eastern Europe and Central Asia, construction firms reported paying an average of 7 percent of the value of public-sector projects in bribes to win bids or alter the terms. The UNDP (2008b) notes that, across the Asia-Pacific region, the standard kickback for infrastructure projects is often quoted as 10 percent of the value of the contract. Officials regularly demand bribes to accept bids or to approve the completion of the work. Often, construction companies also take the initiative, colluding with officials or other bidders. In many cases, the bases for the selection of the contractor are unclear and opaque. The tender evaluation report is seldom disclosed. Another common problem on public-sector projects is the leakage of key information such as the tender assessment procedure, or even the "engineer's estimate".

Implementation and execution stage

Where the contract provides for the contractor to be reimbursed for expenses, the padding of expense accounts also occurs. The timesheets, daywork sheets or invoices that should accompany claims for reimbursement might be fraudulent (Sohail and Cavill, 2008). There are many instances where the materials used on the project, and the methods that are applied, do not comply with the specifications. Examples include fewer reinforcement bars, smaller thicknesses of floor slabs, and smaller widths of the carriageways of roads. These situations often occur where the contractor has to make up for a low bid, or recover a bribe. Lewis (2005) notes that, while earthquakes may not be preventable, it is possible to avoid the disasters they cause. In the previous 15 years, there had been more than 400 recorded earthquakes in 75 countries, making some 9 million people homeless, injuring 584,000 and causing 156,000 deaths. Many of the deaths and injuries were the result of buildings that collapsed because concrete was diluted, steel bars were excised, or otherwise substandard building practices were employed, often (as in Italy and Turkey) while corruption among contractors and public officials resulted in ignored building codes, lax enforcement and the absence of on-site inspection.

Making arrangements for the issuing of false test and work certificates also occurs. This can lead to substandard construction. In India, the Central Vigilance Committee inspected 16 venues for the Delhi Commonwealth Games and found that every quality certificate scrutinised was either suspect or "forged" (Ganapathy, 2010). The team reported "major failures of concrete samples and preparation of forged testing records". It found lower standards of steel and aluminium and inferior quality anti-corrosion coating on steel.

There are other instances where the integrity of the process of administering the construction process during physical activity on site is compromised. These include unauthorised sub-letting of all or parts of the contract that might be further sublet, in a situation of multi-layered subcontracting. This makes it difficult to determine the parties responsible for particular aspects of the project, and leads to poor performance in many regards. For example, as the main contractor does not have control of the workforce that actually undertakes the work, performance in terms of quality and safety is often compromised (Debrah and Ofori, 1997).

Contractors often do not safeguard the health and safety of their workers or that of the residents in the environs of their projects from the lack of attention to training, lack of provision of appropriate tools and equipment, failure to provide and enforce the use of personal protection equipment, and lack of attention to the development of a safety culture in the workplace.

Contractors try to recover potential losses by submitting claims for various reasons, such as errors in the contract documents, and results of the actions or omissions of the client or consultants, or putting in inflated claims for work done owing to variations to the design requested by the client or by the consultants. Also, claims for extension of time are often made, or granted, on bases that are not legitimate.

The construction contract provides for the contractor to be paid at certain intervals within specified periods, and upon the issue of relevant certificates. A common occurrence on construction projects, especially in developing countries, is delays in the issuing of payment certificates, and in the honouring of certificates. The bureaucracy is a contributory factor.

Another cause is the lack of fit between the volume of projects launched each year and the amount in the government's budget to pay for the work.

Collusion among the suppliers of key construction materials such as cement, steel or reinforced concrete in order to keep their prices high has been a common issue that the competition authorities in several countries (such as South Africa: Competition Commission, undated) have had to deal with.

Closing and operating phases

At the end of the construction project, the procedures that might be subject to mismanagement and inappropriate behaviour include the issuing of the certificate of completion, and preparation of the final certificate, preparation and agreeing of the final account for the project. The inspection of the completed building before granting occupation is another activity that is vulnerable to malpractice.

The situation in developing countries

The Constitution of South Africa (Act 108 of 1996) requires the public-sector procurement system to be fair, equitable, transparent, competitive and cost effective. Various relevant legislation has drawn on these broad principles. For example, the Local Government: Municipal Systems Act (Act 32 of 2000) also states that the selection process for service providers must: be fair, equitable, competitive and cost effective and allow equal and simultaneous access to tender information; minimise the possibility of fraud and corruption; and make the municipality accountable to the local community regarding decisions in the selection of the service provider. Whereas such declarations might be found in the laws of many countries, the procurement arrangements in developing countries often fall short of the guiding principles and best practices. Thus, in the run-up to the building of the World Cup 2010 venues, the Construction Industry Development Board (CIDB) of South Africa (2006: 4.7 to 4.8) highlighted the risks in the nation's public procurement system that must be managed.

Nkinga (2003) notes that, in Tanzania, although since 1966 many laws had been passed and initiatives launched to combat corruption, unethical behaviour and abuse of power, reports of the Prevention of Corruption Bureau show that it remains a problem. He outlines the weaknesses in the procurement system in Tanzania before the reform process started in 2001 as: the laws on procurement were fragmented with several loopholes and no enforceable penalties; the laws were inadequate for the procurement of works and selection and engagement of consultants; there was no statutory body to enforce the legal framework and procedures; regulations were outmoded, and originated in stores management rather than competitive procurement; the authority of the Central Tender Board was eroded by ad hoc arrangements in other sectors; many exceptions were made, and rules were often ignored; management information, especially for enforcing accountability, was lacking; too much was left to the discretion of individual public officers who were given no clear policy or guidance; and there was a shortage of adequately trained and experienced procurement officers.

Zou (2006) reviews the prevailing corruption prevention practices in China's construction industry and finds that corruption happens in different forms during all stages of the construction project, and the current anti-corruption practices are reactive rather than proactive. He proposes a corruption prevention strategy including random and regular checks, severe punishment and prosecution of corrupt personnel, and the promotion of a healthy and clean construction culture.

The case of Pakistan (as outlined in the Country Report in the *Global Corruption Report 2009*) (Transparency International, 2009) further illustrates the gap between the laws and attainment in developing countries. Pakistan endorsed the ADB-OECD Anti-Corruption Action Plan for the Asia-Pacific region in November 2001, and ratified the UN Convention Against Corruption in August 2007. The Public Procurement Rules came into force in 2004. In 2007, the government "claimed that transparency was the 'hallmark' of government policy and that the government was promoting e-governance as a tool for more openness and in order to make processes more efficient". Moreover, a senior official indicated that "government had made it mandatory that integrity pacts are signed for all government contracts over Rs10 million". In October 2007, a new Competition Commission was set up in Pakistan. However, it is suggested that "corruption is a serious problem in Pakistan, and this position is corroborated by a number of recent studies and reports". In September 2007, the newly constructed Sher Shah Bridge in Karachi collapsed only a few weeks after it had been completed, killing 14 people and injuring several others. Shoddy construction and corruption were suspected to have been the cause. In a 2009 court case in Pakistan which illustrates the difficulty of implementing the laws in many countries, and presents a way to realise possible improvement in the situation, the Sindh High Court ordered the Karachi Building Control Authority to post all documents relating to building and development projects on its website within one week, according to *Business Recorder* (2009). The Trustees of Transparency International Pakistan, after considering the violation of laws (such as the Karachi Building and Town Planning Regulations, 2002, and the Karachi Building and Town Planning Regulations (Amendment) 2005) which prescribed procedures for transparent dealing in real estate by builders and developers, had filed a constitutional petition in the court.

Shakantu and Chiocha (2009) found that there is "deep-rooted corruption in the construction process" in Malawi (p. 1575) which is engaged in by most project participants at every level and at every phase of the process. The corrupt practices result mainly in project delivery delays. The practices include bribery, fraud, collusion and negligence. The local conditions in the industry and procurement systems appear to shape the form and extent of corruption. Collusion, price differentiation, bid rigging and tampering with claims and payment certificates are among the major corrupt practices. Shakantu and Chiocha suggest measures that can be taken to improve the situation. Phiri and Smallwood (2010) made similar findings in a study that focused on the impact of corruption in the procurement process on the quality of built items and the balance between new work and maintenance.

Osei-Tutu et al. (2010) attempt to identify the corruption-related challenges in the procurement of public infrastructure projects in Ghana that must be addressed to enable the expected economic gains of infrastructural projects to be realised. They found that conflict of interest, bribery, embezzlement, kickbacks, tender manipulation and fraud were observed on the projects. They suggest that, to control corruption, there should be a sound procurement system and pro-social equity policies that would foster good governance, corporate social responsibility, transparency, accountability, judicious public expenditure and national progress. They found that there were challenges in using the Public Procurement Act 2003 (Act 663) to proffer solutions for these underlying constructs.

Ameh and Odusami (2010) assess the perceptions of construction professionals regarding ethical issues in the Nigerian construction industry. The results of their study indicated that there was a decline in unethical practices within the industry compared to the pre-1999 era following the steps taken by the Budget Monitoring and Price Intelligence Unit to close loopholes in the procedures involved in contract awards and execution at the federal level. However, bribery is endemic in the construction industry. Among the professionals in the industry, the quantity surveyor was perceived as the most susceptible to bribery; and the construction manager as facing the greatest pressure to act unethically. Ameh and Odusami recommend that professional institutions should give priority to enhancing awareness of ethical issues. Clients should ensure adequate and prompt remuneration for professional services, and limit the discretionary powers of quantity surveyors. There should also be strict independent monitoring, supervision and auditing of contract progress and performance in conjunction with government anti-corruption agencies. The government should establish an independent "National Council for the Built Environment" to police standards, receive and investigate petitions on professional misconduct, and sanction individuals and organisations which breach ethical principles and rules.

Irumba and Mwakali (2007) study ethics in the construction industry in Uganda. They focus on the spate of fatal accidents on construction sites and cases of designs that fail to preserve the environment, which they consider to be the result of professional negligence and poor construction practices. They also note that the industry has recorded cases of corruption, especially in the tendering process. The proposals they make for addressing the issues include strengthening codes of professional ethics, and enforcing the safety and environmental regulations.

Sichombo et al. (2009) study the need for technical auditing in the construction industry in Zambia to address the unethical practices that were common in the industry. They establish that the pre-contract stage of projects is more susceptible to unethical acts than the post-contract stage; this makes the appointment of technical auditors at the planning stage more appropriate. They find that the benefits of technical auditing are ranked as client confidence, followed by enhanced accountability, reduced project costs and fewer disputes. They suggest prevention strategies and necessary policies and guidelines for the introduction of mandatory technical auditing on publicly funded construction projects in Zambia and elsewhere.

The issues of context and culture are relevant to this discussion. For example, in developing countries, the existing construction project management systems that were inherited from the former colonial administrators bear no relation to the local culture, administrative systems, and authority structures. This deprives them of the contextual foundations necessary for their effective implementation. In a study on the countries in southern Africa, Rwelamila et al. (1999) suggest that lack of cultural appropriateness of procurement approaches accounts in large part for the poor performance on construction projects in those countries. Sohail and Cavill (2008) show how accountability initiatives in construction projects can reduce corruption in infrastructure services. To attain this, they suggest that it is necessary to pay attention to "cultural considerations" (Sohail and Cavill, 2008: 729). Similarly, Transparency International (2008) states that a particular country's dynamics determines the right mix and sequence of anti-corruption reforms.

What is to be done?

Much of the efforts to reduce malpractice, corruption and mismanagement have focused on the procurement stage. Transparency International (2006) suggests that procurement systems should incorporate several key ingredients. First, contracting opportunities should be widely publicised. Second, the awards that are made should go to the companies or individual professionals who meet the contract requirements and make the best offers. Third, the rules governing procurement should be clear and fair. Fourth, the process should be transparent, with predictable results. Finally, public officials should be accountable. The key objectives of the public-sector procurement reform initiatives (which occurred in many countries, developed and developing, in the early 2000s) that are often cited in the literature include: value for money; predictability; contestability; efficiency; transparency; accountability; and effective implementation. An international study found that the main concerns of the review programmes were similar, but that the objectives and aspirations were different from one country to another (Harland et al., 2005). Harland et al identified the following top ranking priorities: accountability; competition; transparency; probity; value for money; efficiency; legal compliance; cost effectiveness; and education of public procurement personnel.

There have been some wider-ranging proposals. Transparency International (2008) notes that the case studies and experiences presented in the *Global Corruption Report 2008* yield lessons for fighting corruption in the water sector, including: understand the local water context, otherwise reforms will fail, as one size never fits all in fighting corruption; and build pressure for water reform from above *and* from below because, whereas leadership from the top is necessary to create political will and drive institutional reform, bottom-up approaches add checks and balances that include the monitoring of money flows or benchmarks of utility performance.

The German agency GTZ (2010) suggests that there are four principles which help to effectively fight corruption on projects: transparency, accountability, voice and participation, and integrity. GTZ (2010) further recommends that anti-corruption measures should be integrated into sector programmes. It suggests four steps: (i) identify risks – draw up a risk map identifying corruption-prone areas, by adopting a value-chain approach; (ii) set priorities – consider the gravity of corruption and its detrimental impact on the services and outcomes, in the context of the political economy of the sector; (iii) devise anti-corruption measures – these should incorporate transparency, accountability, voice and participation, and integrity, which are the principles of anti-corruption, and extend to broader governance issues which might involve reform; and (iv) measure progress – using actionable indicators. OLAF notes that, when operating in high-risk regions, companies should also map out corruption risks in advance and take preventative steps to mitigate problems rather than reacting to a crisis (FIDIC, undated b).

The UK Anti Corruption Forum (2006) recommends several measures. First, there should be efforts to increase awareness of corruption and its consequences through publicity and training. Second, there should be improved international cooperation to counter incidences of malpractice. Third, appropriate codes of conduct and management systems should be adopted and enforced. Fourth, there should be fair, reasonable, objective and transparent procurement and project management procedures. Fifth, effective anti-corruption monitoring and reporting procedures should be introduced. Finally, there should be fair and effective prosecution and blacklisting procedures. Sohail and Cavill (2008) suggest that successful reforms to combat corruption in the construction sector must combine top-down government monitoring with private-sector motivation and grassroots oversight. In a report on post-disaster reconstruction, World Bank (2010: 294) recommends: "Funding sources should work to establish common transparency standards. They should require … that the use of their funds be widely disclosed to the public".

The UNDP (2008b) notes that, for the countries in the Asia-Pacific region, there is no one route or single answer; action should come from many directions – from governments, civil society organisations, the media and private business working together. It proposes the following common initiatives:

1 *Join with international efforts* – such as the United Nations Convention Against Corruption;
2 *Establish benchmarks of quality* – such as for anti-corruption agencies and the media;
3 *Strengthen the civil service* – for example by raising salaries, ensuring merit-based recruitment and promotion, and rigorous systems of control;
4 *Encourage codes of conduct in the private sector*;
5 *Establish the right to information* – by enacting laws and encouraging their application;
6 *Exploit new technology* – such as information technology and e-governance, to break the information monopoly of officials;
7 *Support citizen action* – by publishing basic information on contracts to facilitate citizen auditing.

The *Global Corruption Report 2008* (Transparency International, 2008) presents the following strategies and tools to tackle corruption in the water sector: scale up and refine the diagnoses of corruption in water, and adapt them to specific local contexts; strengthen regulatory oversight, build capacity, provide adequate resources, create a clear institutional mandate, implement transparent operating principles and introduce public consultation and appeals processes; ensure fair competition for, and accountable implementation of, contracts – governments and contractors can enter into integrity pacts and local and international financial institutions should expand their due diligence requirements to include anti-bribery provisions; and adopt and implement transparency and participation as guiding principles for water governance, including opening up project budgets, disclosure of performance indicators, community involvement in project selection and civil society participation in auditing, water pollution mapping and performance monitoring of utilities. The report notes that implementing these recommendations requires a strategic vision. There should be a global response, local expertise and adaptation and buy-in from stakeholders.

Anti-corruption initiatives in construction

Many sectoral, national and international initiatives aim to address corruption in construction. Some examples are considered in this section.

Non-governmental organisations

The UK-based Global Infrastructure Anti-Corruption Centre (GIACC) seeks to provide resources and services for the purpose of preventing corruption in the infrastructure, construction and engineering sectors (http://www.worldbank.org.bd). Its web-based resource centre provides information, advice and tools to help stakeholders to understand and try to prevent corruption, and to identify it when it occurs. The resources include: analysis and examples of corruption, why and how it occurs, liability for and cost of corruption; anti-corruption programmes for governments, funders, project owners, companies and associations; the Project Anti-Corruption System (PACS); anti-corruption tools including claims code, contract terms, corporate code, due diligence, employment terms, gifts and hospitality policy, procurement, reporting, rules, training, transparency; advice on how to deal with corrupt situations; and information on anti-corruption conventions, indices, surveys, forums and initiatives.

Transparency International is the world's largest non-governmental anti-corruption organisation. It has chapters in over 90 countries. Among its broad range of activities, Transparency International (undated a) promotes anti-corruption measures by working in close collaboration with construction participants worldwide. In 2005, the organisation's global corruption report focused on construction and post-disaster reconstruction. The report in 2008 also put the spotlight on the water sector. In a bid to prevent corruption on construction projects, Transparency International aims to raise awareness of corruption in the sector, to develop anti-corruption tools, and to promote the implementation of anti-corruption actions. Transparency International also hosts the secretariat of the Water Integrity Network (see Box 4.4).

Many organisations are collaborating to prevent corruption in the construction industry. Some of the initiatives are shown in Box 4.4. Some of the initiatives are at the national level, whereas others are more global. The latter include those peculiar to the whole construction industry or part of it, and those which are multi-sectoral.

Box 4.4 Some transparency initiatives in construction

The *Global Anti-Corruption Education and Training Project* (ACET) is spearheaded by the American Society of Civil Engineers (ASCE) and involves many international organisations including the World Federation of Engineering Organisations (WFEO), International Federation of Consulting Engineers (FIDIC), World Economic

Forum Partnership Against Corruption Initiative (WEF PACI), Pan-American Union of Engineering Associations (UPADI), World Bank, Inter-American Development Bank (IADB) and Transparency International. It provides an anti-corruption training programme for the construction and engineering industry. ACET develops and distributes education and training programmes devoted to the importance of individual integrity among all the participants in the performance of construction and engineering projects (http://email.asce.org/international/February07.html).

The *Construction Industry Ethics and Compliance Initiative* is a non-profit-making private association of US construction companies committed to the highest level of ethics and conduct and compliance with the law. This association seeks to enable its members to share best practices for addressing the ethics and compliance risks in construction work, and approaches to current and emerging ethics and compliance issues (http://www.ciecinitiative.org/).

The *Ethical Edinburgh* initiative is based on the observation that bribery and fraud are thriving and undermining the effectiveness of aid in many aspects of international development. It was set up to prepare the ground for the establishment of an international centre for transparency in construction which would provide or facilitate the specific services that would help curb corruption. The services could include: support to and external verification of companies' anti-corruption procedures; accreditation of independent assessors for projects on which integrity pacts are used; and the maintenance of related databases. The initiative suggests that, with time, it might be replaced by the GIACC (http://www.ethicaledinburgh.org).

The *UK Anti-Corruption Forum* is an alliance of British business associations, professional institutions, civil society organisations and companies with interests in the domestic and international infrastructure, construction and engineering sectors, to promote industry-led actions that can help to eliminate corruption domestically and internationally. The members of the forum believe that corruption can only be eliminated if governments, banks, business and professional associations, and companies working in these sectors cooperate in the development and implementation of effective anti-corruption actions. The forum represents the interests of over 1000 UK companies and 300,000 construction and engineering professionals.

The *UN Global Compact* seeks to promote responsible corporate citizenship. It is promoting 10 principles on human rights, labour, the environment and anti-corruption. Over 2000 companies have signed up to these principles. The tenth principle states that "Businesses should work against corruption in all its forms, including extortion and bribery."

The mission of *U4 Anti-Corruption Resource Centre* is to make development aid more efficient by promoting an informed approach to anti-corruption. Funded by bilateral development agencies, U4 provides resources, helpdesk and training for donors who address corruption challenges through development support. U4 is operated by the Chr. Michelsen Institute based in Norway.

The *Water Integrity Network* (WIN) aims to fight corruption in water worldwide in order to reduce poverty. WIN is an open and inclusive global network that promotes anti-corruption activities and coalition-building at the local, regional and global levels, and between actors from civil society, private and public sectors, media and governments. The WIN is committed to accountability, transparency, integrity, honesty and mutual support and knowledge exchange among its members. These characteristics, to which any member should adhere, unite the network.

Under the *World Economic Forum – Partnering Against Corruption Initiative* (WEF PACI), over 100 companies in the construction and engineering, mining and mineral, and oil and gas sectors, with a combined annual turnover of over US$500 billion, have committed themselves to zero tolerance of corruption by, or within, their organisations. They have adopted the PACI Principles for Countering Bribery.

The *World Federation of Engineering Organisations* (WFEO) is an international umbrella for associations representing individual professional engineers from over 90 countries, which represent 8 million engineers. It has formed an Anti-Corruption Standing Committee that is tasked with promoting anti-corruption actions internationally.

Source: Based on information from Global Infrastructure Anti-corruption Centre at http://www.giaccentre.org/initiatives.php (accessed 17 May 2011).

Some legal developments

Two relatively recent laws which have the aim of fighting corruption have attracted much attention in the construction industry.

The Sarbanes Oxley Act was passed in 2002 (107th Congress, 2002). It establishes ethical standards and financial accountability for all publicly-traded US (and foreign) corporations doing business in US. The primary goal is to restore public trust in US corporate and financial reporting systems. Section 302 provides internal guidelines mandated to be developed and implemented, ensuring proper and accurate financial disclosure of all corporate activity. After these internal protocols are set up, responsible, signing officers must acknowledge their accountability for so doing, creating a powerful benchmark of culpability.

The UK Bribery Act came into force on 01 July 2011 (legislation.gov.uk, 2010). Under the Act, the company must show compliance with six principles if they wish to rely on statutory defence in its section 7(2). Otherwise, the company will be strictly liable for the acts of employees, agents, intermediaries and joint venture partners, and subject to unlimited fines and potential debarment from EU and other publicly funded contracts. The principles are:

1 Risk assessments – the firm must routinely and comprehensively assess its bribery risks;
2 Top-level commitment – most senior levels of management must commit themselves to preventing bribery, and communicate this throughout their organisations;
3 Due diligence – firms should execute due diligence procedures which cover all parties with whom they do business;
4 Effective policies and procedures – clear, practicable, accessible and enforceable procedures must be in place to prevent bribery;
5 Effective implementation – the implementation of the procedures must be effective and embedded;
6 Monitoring and review – there should be mechanisms to monitor and review programmes.

These recent developments and the Lesotho case (FIDIC, undated b) underline the risks of involvement in corruption overseas. The case also illustrates that individuals can be punished, whereas previously fines were imposed only on legal entities. The case involved the use of intermediaries, and the defence claimed that the company did not know about the bribes paid by its agent. This argument was dismissed. Indeed, the OECD anti-bribery convention and the laws of most countries apply to bribes paid 'directly or indirectly'.

Professional institutions

Apart from their codes of conduct, some professional institutions and trade associations have specific anti-corruption programmes. Mason (2008) examines the benefit of promoting a single ethical code for the construction industry using the Society of Construction Law's Statement of Ethical Principles, drawn up in 2003. The statement was intended to apply to the work of all construction professionals whatever their original qualification or affiliation. It states that ethical conduct is compliance with the following ethical principles: honesty; fairness; fair reward – avoid acts which are likely to result in another party being deprived of a fair reward for their work; reliability – maintain up-to-date skills and provide services only within one's area of competence; integrity – have regard for the interests of the public, particularly people who will make use of or obtain an interest in the project in the future; objectivity – identify any potential conflicts of interest and disclose the conflict to any person who would be adversely affected by it; and accountability – provide information and warning of matters within one's knowledge which are of potential detriment to others. The Royal Institution of Chartered Surveyors (RICS) first published the "Rule Book" in 2004. In the 2006 edition (RICS, 2007), the nine RICS Core Values were: (i) Act with integrity; (ii) Always be honest; (iii) Be open and transparent in your dealings; (iv) Be accountable for all your actions; (v) Know and act within your limitations; (vi) Be objective at all times; (vii) Always treat others with respect; (viii) Set a good example; and (ix) Have the courage to make a stand.

FIDIC (undated a) highlights the need for transparency, noting that the preparation of a shortlist of engineering consultants is an important part of the process, and must be carried out openly. It has developed the Business Integrity Management System (BIMS) for the supply side of the procurement of projects, and the Government Procurement Integrity Management System (GPIMS) for the public, demand-side of the procurement process. It has also prepared policy statements on corruption, and on integrity. FIDIC (undated a) recommends that:

1. Firms and individuals should internally develop and maintain systems to protect their high ethical standards and codes of conduct;
2. Firms should have a commitment to integrity through the implementation of a dedicated system involving all levels of management and every employee;
3. Associations of firms should assist member firms to develop a BIMS, by providing guides, training and general support;
4. Qualification-based selection procedures and competitive tendering, should be used in procurement;

5 Firms should notify funding agencies of any irregularities, in order that cancellation or other remedies may be exercised, in accordance with the loan agreement;
6 Firms should be aware of local law regarding corruption and should promptly report criminal behaviour;
7 Associations of firms should take prompt disciplinary actions against any member firms found to have violated the FIDIC Code of Ethics. This could include expulsion, and notification to public agencies;
8 Associations should foster and support the enactment of legislation in their own countries, aimed at curbing and penalising corrupt practices.

Companies

Construction companies have also set up their own codes and systems (see, for example, those of the Bechtel Group and Skanska AB). Skanska has declared that its Five Zero's Vision includes: "Zero ethical breaches, meaning that we take a zero tolerance approach to any form of bribery or corruption". Many US construction firms, including the Bechtel Group and Fluor Corporation, have heads of compliance, and provide all their employees with anti-corruption training. *Fortune* magazine reports that the executives of the US construction firm Fluor (US$17 billion annual revenue in 2008, more than half of it overseas) "promote an open-door policy and a hotline for reporting crime – as well as tough penalties for violators, who receive zero tolerance for infractions. This dedication to transparency has helped make Fluor the World's Most Admired engineering firm and fourth among all businesses in global competitiveness" (Kimes, 2009).

In the next section, an initiative that addresses the issue of transparency in construction from a broad perspective is presented.

What is CoST?

The Construction Sector Transparency Initiative (CoST) is a multi-stakeholder initiative that is designed to increase transparency and accountability in public-sector construction projects (CoST, 2009).[2] The initiative is funded by the Department for International Development (DfID) of the UK government, and sponsored by the World Bank.

CoST was inaugurated in Dar es Salaam, Tanzania, in June 2008 after a series of high-level consultations among key players in the construction industries in many countries. It seeks to attain transparency in the management of the entire process of public-sector construction projects involving the planning for, design, construction and operation of, the built

2 This part of the chapter is based on various publications by the CoST initiative. More information on CoST can be found at: http//www.constructiontransparency.org

assets that are so critical to the national socio-economic development of all countries.

CoST was piloted in seven countries. These were: Ethiopia, Malawi, Tanzania and Zambia in Africa; the Philippines and Vietnam in Asia; and the UK. Guatemala became a CoST associate country during the period. The two-year pilot phase ended on 31 December 2010. The main aim was that if, during pilot period, it could be demonstrated after evaluation that CoST was useful, then it would continue, and might be rolled out in other countries, becoming an international standard for the management of construction projects.

CoST follows the principles and architecture of similar international transparency initiatives for other sectors. Two of these programmes also involved the DfID. The Extractive Industries Transparency Initiative (EITI) is aimed at attaining full publication of each company's payments, and of government revenues from oil, gas and the mining sector (http://eiti.org). The Medicines Transparency Alliance (MeTA) aims to have publicly available information on the price, quality, availability and promotion of medicines (http://www.medicinestransparency.org).

The CoST principles

The CoST initiative is based on a belief that public-sector infrastructure projects should support economic growth that contributes to sustainable development and poverty reduction (CoST, 2009). However, the mismanagement and malpractice that have been evident on such projects in many countries can undermine potential social and economic benefits and value for money.

The second principle of CoST is that the governments of all countries should be accountable to their citizens for public expenditure on projects. The governments should also be committed to encouraging high standards of transparency and accountability in both the public and private sectors.

The third principle of CoST is that the disclosure of basic project information throughout the project cycle could be an effective and efficient way to improve the possibility that value for money would be attained on projects. There would also be greater transparency during the implementation of the project.

The fourth CoST principle is that transparency in the administration and delivery of public-sector projects would provide the catalyst for its wider adoption in the industry as a whole, and lead to the entrenchment of the norms of professionalism and ethics in the construction industry. The impact of, and benefits from, this situation would go beyond attaining good value for the money spent by society and individual private clients on their construction projects. It would promote competition within the industry, and also enhance the environment for both domestic and foreign direct investment.

The final principle of CoST is that a collaborative multi-stakeholder group can play an oversight and interpretative role in ensuring greater transparency and public understanding of the information disclosed on projects.

Nature of CoST

The CoST architecture

CoST relies on cooperation among stakeholder groups to monitor the key stages of public-sector construction projects and highlight differences between specification and delivery (CoST, 2009).

In each country, the key stakeholders in public-sector construction projects are organised in a Multi-Stakeholder Group (MSG). The group includes: the procuring entities of public sector projects, such as the ministries of education and health, and the roads agency; the public financial management bodies; design and construction companies and their professional institutions and trade associations; and civil society. Other institutions that might have an interest in the MSG include the financial institutions and the local representatives of the bilateral and multilateral aid agencies. Figure 4.1 presents examples of the stakeholders that might participate in an MSG, and their relationships and interests. Each MSG is supported at a high level by the national CoST champion, who is appointed by the government.

The CoST International Secretariat, based in London, provides the MSGs and the entire CoST process with co-ordination, advice, templates for various procedures and actions, and information and research support.

The CoST International Advisory Group (IAG) provides guidance for the CoST pilot programme, and offers a forum for lesson-sharing. The IAG comprises a representative from the DfID, one from the World Bank, and one from each of the pilot countries. The other members appointed onto the IAG include someone chosen to represent the views of industry, and two persons from academic institutions. It was intended that the IAG would meet at least twice each year. In reality, it has met more frequently; some of the meetings have been via video-conferencing.

The lesson sharing and knowledge transfer in the CoST process takes place in many ways. For example, a meeting of all the CoST champions was held in Addis Ababa, Ethiopia, in April 2010. The MSG Coordinators of the pilot countries have been invited to attend many IAG meetings; and to hold their own meetings in conjunction with those events.

Each pilot country received funds towards the cost of implementing CoST. The funds, drawn from the financial provision made by the DfID, were administered by the International Secretariat.

The CoST criteria

The main criteria of the CoST process are now outlined. The first criterion of CoST is that, in each country, for public-sector construction projects, a threshold (in terms of project value) would be set. On projects in this category, there would be regular disclosure of material project information in a specified format to a wide audience in a comprehensive and comprehensible manner.

The CoST process is not an end in itself; it is intended to supplement the existing framework for ensuring good practice on public-sector projects. For example, it is presumed that the main public-sector procuring entities of construction projects would be subject to a credible audit process of their projects, and their projects would go through independent financial and technical audits.

The adequacy of the material project information disclosed by the procurement entities and the audits are assessed by an independent Assurance Team comprising experienced construction professionals appointed by the MSG.

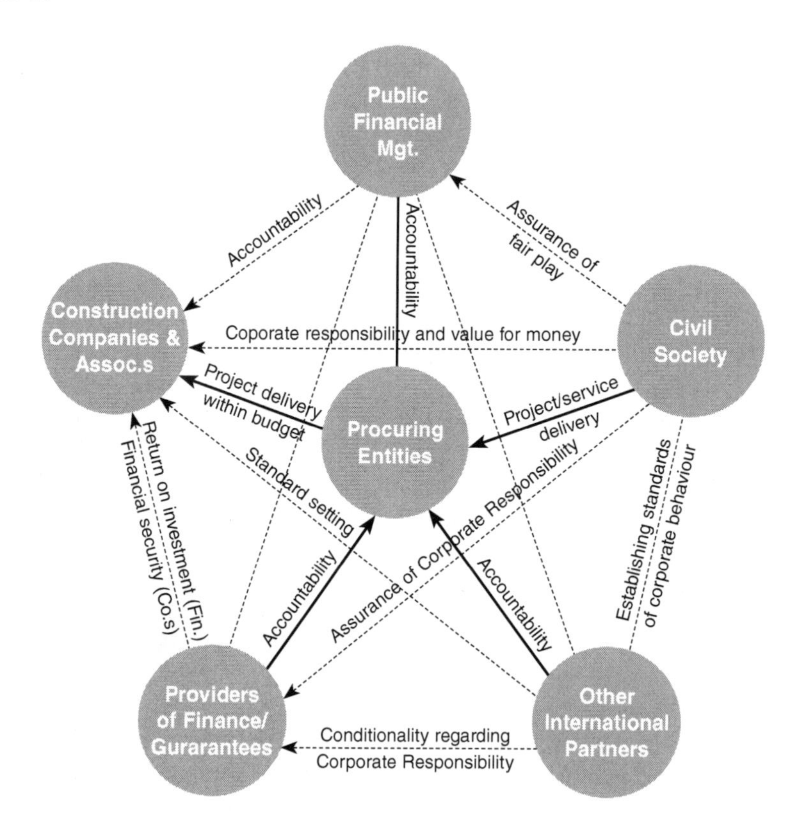

Figure 4.1 CoST stakeholders and their interests (source: CoST)

The MSG has oversight of the CoST process. The MSG releases the findings of the Assurance Team to the stakeholders. The information disclosed includes comments by the Assurance Team on relevant aspects of the project addressing the question: "Are you getting what you paid for?" An important feature of CoST is that civil society is actively engaged in the design, monitoring, evaluation and normal administration of the construction project transparency process within the country concerned. In these regards, another criterion of CoST is that the MSG undertakes capacity building to prepare the public to understand and to use the disclosed information.

It is pertinent to note that it is not the role of CoST to make accusations of corruption. Instead, the stakeholders are empowered under CoST to receive accurate information, to understand what it means, and to act on it (as and when necessary) to support the country's own procedures for uncovering and prosecuting corrupt practices and instances of malpractice, incompetence and negligence on public-sector construction projects.

Information to be disclosed

Table 4.1 presents an outline of the information to be disclosed by the procuring entities at various stages of public-sector projects under the CoST process: it is termed "Material Project Information" (MPI).

Benefits of CoST

The main general benefits of the CoST initiative could be outlined (CoST, 2009). First, the accountability of procuring entities and contractors for the cost and quality, as well as other critical performance aspects of public-sector projects, will be enhanced through improved transparency in the process of administering the planning, design, construction and operation of these projects. Second, there will be improved management of public finances, and the governance of public-sector construction projects will be strengthened. Third, there will be greater efficiency in the procurement and delivery of public projects. The effect of this will be higher quality infrastructure at lower capital and lifecycle costs.

The specific benefits to the government are that there will be greater confidence of the citizenry in the government, and in the process for procuring and delivering public-sector projects. As mentioned above, this is a sector which has not had a good record in terms of governance, accountability and both technical and managerial performance. Moreover, the financial transparency which the CoST process involves will lead to improved confidence and trust of business enterprises, both local and foreign, which will increase the prospects for both domestic and foreign investment. Furthermore, the process will result in a higher quality of projects that will lead to better social and economic outcomes as well as returns.

Table 4.1 Material project information to be disclosed

Stage of project	*Information to be disclosed*
Initial planning stage	Project purpose Specification Location Intended beneficiaries Feasibility study
Project funding	Financing agreement
Tender process	Tender evaluation report
At contract award (main contract for works)	Contractor's name Contract price Contract scope of work Contract programme
During contract implementation	Significant changes to the contract which affect the price (threshold to be determined) and reasons for those changes Changes to the contract which affect the programme and duration (threshold to be determined) and reasons for those changes Details of any re-award of the main contract
At contract completion	Contractor's name Total contract cost Final contract payment Actual scope of work Actual contract programme Project evaluation report (upon completion and on-going)

Source: CoST, 2010.

The CoST initiative also offers benefits to the construction industry. First, the increased transparency will increase confidence that there is a "level playing field" on construction projects in the particular country. It will give the assurance that the pre-qualification and tendering processes for both consultants and contractors, as well as contract award and project administration, will be fair. This will enhance competition for projects, and contribute to giving society value for money. Second, the reduced exposure to financial risk and potential technical difficulties will enable the construction enterprises to have better access to credit, loans and insurance on more favourable terms. Third, the implementation of CoST should lead to improved professional standards in the construction industry. Finally, the enhanced transparency and better standards will lead to a better image for the construction industry. Thus, a major player in the UK construction industry, Mr Peter Gammie, Chief Executive of the Halcrow Group, expressed strong support for CoST (CoST, undated):

> As a values driven business, the way we behave individually and as a company underpins those values. Transparency, honesty and integrity

> go right to the heart of the way we work in Halcrow, and our corporate purpose – sustaining and improving the quality of people's lives, is wholly aligned to the objectives of the CoST initiative. We are therefore proud to be able to support CoST, as well as a number of other programmes that target corruption in the construction sector.

There are also benefits to civil society and to citizens. First, civil society, especially potential beneficiaries of the project, will have increased access to information on the project and the asset which it will create, and this will give them the opportunity to participate in the governance of the construction industry, as it relates to projects intended to benefit them. Second, with the information disclosed to them, the citizens and civil society can hold the procurement agencies, the members of the design team and contractors to account. This will encourage good quality performance and better returns on public investment in projects. Finally, the appropriate occupational health and safety (OHS), environmental and social safeguards will be implemented, and this will ensure that hazardous materials are not used and health and safety laws are observed.

The CoST initiative has received much support and interest. Several professional institutions have covered it in their publications (see, for example, Smith, 2011). The following clause was included in the Bangkok Declaration that followed the 14th International Anti-Corruption Conference in Bangkok in November 2010 (IACC, 2010: 4):

> We also closely assessed the construction sector, exploring the application of the CoST initiative, and concluding that there is a fine balance in degrees of detail for efficient reporting, while fighting systemic corruption does not necessarily require witch-hunts and investigations.

Results of CoST pilot

Requirements to disclose information

The Baseline Studies in the CoST pilot countries found that each of the pilot countries had several procuring entities for public-sector construction projects (they ranged from 155 in Malawi to nearly 10,000 in the Philippines, and more than this number in the UK). The rest of the section on CoST is based on material drawn from the IAG report, CoST, 2010. As part of the wave of public-sector procurement reform discussed above, each of the pilot countries has undertaken a review of its procurement statute, and produced a new one in the period between 2003 and 2008. Each of the countries has also formed an oversight agency for public-sector procurement; an example is the Public Procurement Regulatory Authority in Tanzania.

The Baseline Studies also revealed that some of the items of information on public-sector construction projects are already required by law to be disclosed on a routine basis in all the pilot countries. In Tanzania, nearly 60 percent of the CoST information is already required by law to be disclosed. In the UK, this figure is just under 50 percent. Even in Ethiopia, the country where, among the pilot nations, the smallest proportion of the information is required to be disclosed by law, the figure is a significant 22 percent. They include: the invitation to tender (which also provides details of the project); and the name of the contractor/consultant which wins the tender or commission. The items which are required to be disclosed in some of the countries include: the engineer's estimate (in UK and the Philippines); the actual contract price and programmes, after completion (Tanzania); evaluation and audit reports (Vietnam); and the price of the winning tender (all the countries except Ethiopia and Vietnam).

The items that are *not* required to be disclosed in all the pilot countries include: the feasibility study; and the tender evaluation report. The disclosure of the tender evaluation report is prohibited in Ethiopia, Tanzania, Zambia and Malawi. In none of the pilot countries was material project information during the project implementation stage required to be disclosed.

Actual disclosure

In terms of pro-active disclosure, the Baseline Studies showed that the information that was actually published fell short of what the law required. For example, whereas the Baseline Studies found that the invitation to tender was disclosed in newspapers and websites, it was not always disclosed in Ethiopia and Zambia. The name of the winner of the contract is also not always disclosed in Tanzania. The contract price is not always disclosed in Ethiopia, the UK and Zambia. Finally, the engineer's estimate is not disclosed in the Philippines and the UK.

Some of the information is made available on request in some countries. For example, unsuccessful bidders can request information about tender evaluation. In the Philippines and the UK, information is only released at the discretion of the procuring entity. In the UK, such requests may be refused on the grounds of confidentiality.

Barriers to disclosure

The barriers to the disclosure of information on public-sector construction projects include the legal restriction that has been discussed above. There are several administrative barriers, and they include: poor record-keeping and management of information on projects in the procuring entities; the maintenance of information on projects only on documents available only in hard copy, and kept in different places; and the electronic storage and data release occurred only in some procuring entities in the UK. However, this did not mean that there was instant access. These observations indicate that there is inadequate capacity in the procuring entities to gather and process the required information on projects.

Other barriers included the situation that the information is usually not in a suitable format for disclosure, and needs to be reprocessed. The final one is the issue of the cost of disclosure, in comparison with the perceived benefits.

Lessons from CoST implementation

A summary of the lessons that were evident during the implementation of the CoST pilot is now presented:

1. CoST is a viable and effective way to ensure transparency on public projects;
2. The support of the government at the highest level is critical;
3. The role of the CoST Champion is very important;
4. A legal mandate on information disclosure seems to be key;
5. The multi-stakeholder approach is effective and beneficial, but effort is required to maintain it;
6. In each country, the CoST process should be systematically planned and managed;
7. The nature of the information that is disclosed and the medium of disclosure should be most appropriate in the context of the country;
8. The nature of CoST will differ from one country to another;
9. International coordination, exchange of lessons, and transfer of knowledge on CoST are important.

Problems with CoST implementation

A summary of the problems that would be encountered in the implementation of the CoST initiative is now presented:

1. It is necessary to persuade procuring entities to collect and provide material project information;
2. Much work is involved in collating the required CoST material project information. Thus, the procuring entities require some capacity to undertake this task;
3. The information on projects should be presented in a format and in a medium that the stakeholders, especially the beneficiaries of the projects, can easily access, and also understand;
4. There is a need to build capacity in the beneficiaries to understand the CoST project information;
5. There is some cost involved in many aspects of the programme for disclosing material project information as required under the CoST initiative;

6 The collection and processing of information would have a time implication on projects. Thus, it is necessary to consider these aspects when planning the projects;
7 There might be legal issues with respect to liability when assuring, or disclosing, the information;
8 The mainstreaming of CoST will be a great challenge.

Establishing an anti-corruption system

The review of the literature and implementation of anti-corruption and transparency programmes throughout the world shows that the key components of an anti-corruption programme could be as now discussed.

1 Government's declaration of intent and legislation – championing at the highest level; political will to act against all wrongdoers, applying equality under the law; comprehensive legislation (based on global models);
2 Multi-stakeholder action and interest – involving government (at various levels, such as procurement, audit, law enforcement); professional institutions, trade associations, unions; companies – consultancy, contracting, supply; civil society (embracing and representing beneficiaries and society);
3 Transparency as a national norm and practice -- providing information for action, if necessary, including legal provision, routine application on projects;
4 Effective enforcement – anti-corruption agency (with high-level reporting channels); close collaboration among related agencies "to close the loop"; predictability of detection and prosecution (routine audit; transparency); staffing with qualified personnel; adequate compensation for officers; strict code of behaviour;
5 Sanctions and penalties – suspension or blacklisting of firms; punishment being effective deterrent;
6 Education and awareness building – coverage in syllabi of educational institutions; continuing professional development of transparency, ethics, anti-corruption;
7 Ethical business dealing – integrity pacts for sectors containing enforceable sanctions; integrity pacts on projects; codes of conduct for professions and trades; integrity and ethics codes for companies; development of national ethical standards firms can aspire to;
8 International cooperation and action – global conventions and protocols; action at home on corrupt action abroad; mutual learning and sharing of experience.

Conclusion

What would a construction industry with effective strategies and programmes be like? In such an industry, professionalism, ethical behaviour and practice, and transparency would be part of all construction projects. Appropriate practice and behaviour of construction practitioners and administrators would be inculcated in them as an integral part of their education and professional development. The disclosure of material project information will be the usual practice in the construction industry. Individuals and firms that provide services as transparency specialists would emerge. Most large organisations in the construction industry will have compliance officers. The construction industry will shed its poor public image. Such a transformation would impress other sectors, and these would seek to emulate it. Transparency would become a culture throughout the construction industry.

References

107th Congress (2002) Sarbanes-Oxley Act of 2002, Public Law 107–204 – July 30. http://www.sec.gov/about/laws/soa2002.pdf (accessed on 2 December 2010).

Ameh, O.J. and Odusami, K.T. (2010) Professionals' ambivalence toward ethics in the Nigerian construction industry. *Journal of Professional Issues in Engineering Education and Practice*, 136(1): 9–16.

Bosshard, P. (2005) The environment at risk from monuments of corruption. In Transparency International (ed.) *Global Corruption Report 2005 – Corruption in construction and post-conflict reconstruction*. Berlin: Transparency International: 19–23.

Business Recorder (2009) KBCA asked to post building/development projects documents on website within week. September 11. http://www.alacpakistan.com/news.php (accessed on 30 December 2010).

Collier, P. and Hoeffler, A. (2005) The economic costs of corruption in infrastructure. In Transparency International (ed.) *Global Corruption Report 2005 – Corruption in construction and post-conflict reconstruction*. Berlin: Transparency International: 12–19.

Committee on Standards in Public Life (undated) *The Seven Principles of Public Life*. http://www.public-standards.gov.uk/About/The_7_Principles.html (accessed on 3 January 2011).

Competition Commission (undated) *Infrastructure and construction*. http://www.compcom.co.za/infrastructure-and-construction (accessed on 12 February 2011).

Construction Industry Development Board (2006) *Procure 2010: Manual to Guide Infrastructure Development for the Soccer World Cup – Manual 4: Best value procurement*. Pretoria: CIDB.

CoST (Construction Sector Transparency Initiative) (2009) *CoST Factsheet*. London. http://www.constructiontransparency.org/ResourceCentre/Communications2/detail.shtml?id=942147648 (accessed on 13 March 2011).

CoST (Construction Sector Transparency Initiative) (2010) *Report of the CoST International Advisory Group*. London. http://www.constructiontransparency.org/News/detail.shtml?id=1463404087 (accessed on 13 March 2011).

CoST (Construction Sector Transparency Initiative) (undated) *Companies and associations: Private sector endorsement of CoST*. http://www.constructiontransparency.org/CountriesSupporters/CompaniesAssociations (accessed on 19 March 2011).

COWI Uganda (2008) *Engineering Audit of CHOGM Activities*, Final Report. Office of the Auditor General, Kampala. http://www.oag.go.ug (accessed on 26 December 2010).

Crane, A. and Matten, D. (2007) *Business Ethics: Managing corporate citizenship and sustainability in the age of globalization*, 2nd edn. Oxford: Oxford University Press.

Darroch, F. (2005) Case study: Lesotho puts international business in the dock. In Transparency International (ed.) *Global Corruption Report 2005 – Corruption in construction and post-conflict reconstruction*. Berlin: Transparency International: 31–6.

Debrah, Y.A. and Ofori, G. (1997) Flexibility, labour subcontracting and HRM in the construction industry in Singapore: can the system be refined? *International Journal of Human Resource Management*, 8(5): 690–709.

Deutsche Gesellschaft für Technische Zusammenarbeit (GTZ) (2010) *Anti-Corruption in Practice*. Eschborn: GTZ.

Doree, A.G. (2004) Collusion in the Dutch construction industry: an industrial organisation perspective. *Building Research and Information*, 32(2): 146–56.

FIDIC (undated a) *FIDIC Policy Statement: Corruption*. Geneva. http://www1.fidic.org/about/statement16.asp (accessed on 2 December 2010).

FIDIC (undated b) *European Union underlines risks of corruption*. http://www1.fidic.org/news/content.asp?ArticleCode=25Et (accessed on 2 December 2010).

Flyvberg, V. (2009) Comment: Evidence suggests that infrastructural stimulus initiatives often detract from the economy in the long term. *The Banker*, October: 154–5.

Ganapathy, N. (2010) Another scandal hits C'wealth Games. *The Sunday Times* (Singapore), August 1: 20.

Global Infrastructure Anti-Corruption Centre (2008) *Examples of Corruption in Infrastructure*. London, http://www.giaccentre.org/documents/GIACC.CORRUPTIONEXAMPLES.pdf (accessed on 2 December 2010).

Harland, C., Bakker, E., Caldwell, N.D., Phillips, W.E. and Walker, H.L. et al. (2005) *The Changing Role of Public Procurement*. Executive report from the Second Workshop, 17–19 March, Geneva.

IACC (International Anti-Corruption Conference) (2010) The Bangkok Declaration: Restoring Trust – Outcome of the 14th International Anti-Corruption Conference, 10–13 November. http://14iacc.org/about/declaration, accessed on 3 June 2011.

Iriekpen, D. (2009) Nigeria bribe – Halliburton to pay US$559 million settlement. *World News Journal*, 28 January, http://africannewsanalysis.blogspot.com/2009/01/nigeria-bribe-halliburton-to-pay-us-559.html (accessed on 19 March 2011).

Irumba, R. and Mwakali, A. (2007) Ethics in construction: Examples from Uganda. Presented at the CIB World Building Congress *Construction for Development*, 14-17 May, Cape Town, South Africa.

Kenny, C. (2006) *Measuring and Reducing the Impact of Corruption in Infrastructure*, World Bank Policy Research Working Paper 4099, Washington, D.C.: World Bank.

Kenny, C. (2007) *Infrastructure Governance and Corruption: Where Next?* World Bank Policy Research Working Paper 4331, Washington, D.C.: World Bank.

Kimes, M. (2009) Fluor's corporate crime-fighter. *Fortune*, February 9. http://money.cnn.com/2009/02/05/news/companies/Fluors_crime_fighter.fortune/index.htm (accessed on 02 December 2010).

legislation.gov.uk (2010) The Bribery Act 2010. http://www.legislation.gov.uk/ukpga/2010/2/3/cross-heading/general-bribery-offences (accessed on 2 December 2010)."Transparency International (2006) *Handbook for Curbing Corruption in Public Procurement*. Berlin.

Lewis, J. (2005) Earthquake destruction: corruption on the fault line. In Transparency International (ed.) *Global Corruption Report 2005 – Corruption in construction and post-conflict reconstruction*. Berlin: Transparency International: 23–6.

Lynch, P. (1998) Professionalism and ethics. In P. Sadler (ed.) *Management Consultancy: A handbook of best practice*. London: Kogan Page.

Mason, J. (2008) Promoting ethical improvement in the construction industry – a single professional code? Presented at RICS Construction and Building Research Conference, 4–5 September, Dublin.

Nkinga, N.S.D. (2003) Public procurement reform – the Tanzanian experience. Paper presented at Joint WTO – World Bank Regional Workshop on Procurement Reforms and Public Procurement for the English-Speaking African Countries. 14–17 January, Dar es Salaam.

OSCE (Organisation for Security and Cooperation in Europe) (2003) *Best Practice for Combating Corruption*. Vienna. http://www.osce.org/files/documents/9/a/13738.pdf (accessed on 2 December 2010).

Osei-Tutu, E. Badu and Owusu-Manu, D. (2010) Exploring corruption practices in public procurement of infrastructural projects in Ghana. *International Journal of Managing Projects in Business*, 3(2): 236–56.

Pearl, R., Bowen, P., Makanjee, N., Akintoye, A. and Evans, K. (2005) Professional ethics in the South African construction industry – a pilot study. Presented at The Queensland University of Technology Research Week International Conference, 4–8 July, Brisbane.

Phiri, M.A. and Smallwood, J.A. (2010) The impact of corruption on the Malawian construction industry. *Acta Structilia*, 17(2): 107–25.

Power, H. and Jordan, D. (2009) OFT fines construction sector £129m for bid-rigging. *The Sunday Times*, 22 September, http://business.timesonline.co.uk/tol/business/industry_sectors/construction_and_property/article6843798.ece, (accessed on 3 June 2011).

Richter, W.L. and Burke, F. (eds) (2007) *Combating Corruption, Encouraging Ethics: A Practical Guide to Management Ethics*. 2nd edn. American Society for Public Administration. Lanham, MA and Plymouth: Rowman and Littlefield.

RICS (Royal Institute of Chartered Surveyors) (2000) *Guidance Notes on Professional Ethics*. London: RICS.

RICS (Royal Institute of Chartered Surveyors) (2007) *Rules of Conduct for Members*. London: RICS.

Rwelamila, D., Talukhaba, A.A. and Ngowi, A.B. (1999) Tracing the African project failure syndrome: The significance of *ubuntu*. *Engineering, Construction and Architectural Management*, 6(4): 335–46.

Shakantu, W. (2006) Corruption in the construction industry. *Civil Engineering*, July: 43–7.

Shakantu, W. and Chiocha, C. (2009) Corruption in the construction industry: the

case of Malawi. Proceedings, RICS COBRA Research Conference, University of Cape Town, 10–11 September: 1568–76.
Sichombo, B., Muya, M., Shakantu, W. and Kaliba, C. (2009) The need for technical auditing in the Zambian construction industry. *International Journal of Project Management*, 27(8): 821–32.
Smith, K. (2011) Transparency in action. *International Construction Review*, 01 Quarter: 26–7.
Sohail, M. and Cavill, S. (2008) Accountability to prevent corruption in construction projects. *Journal of Construction Engineering and Management*, 134(9): 729–38.
Stansbury, N. (2005) Exposing the foundations of corruption in construction. In Transparency International (ed.) *Global Corruption Report 2005 – Corruption in construction and post-conflict reconstruction*. Berlin: Transparency International: 36–40.
Svensson, J. (2005) Eight questions about corruption. *Journal of Economic Perspectives*, 19(3): 19–42.
Transparency International (2008) *Global Corruption Report 2008: Corruption in the Water Sector*. Water Integrity Network. Cambridge: Cambridge University Press.
Transparency International (2009) Country reports: Asia and the Pacific. In *Global Corruption Report 2009*: 282–4. http://www.scribd.com/doc/22743947/Corruption-Reports-Pakisan-2009 (accessed on 13 March 2011).
Transparency International (2010) *Corruption and Public Procurement*. Working Paper No. 5/2010. Berlin: Transparency International.
Transparency International (undated) Construction, engineering and post-disaster reconstruction. http://www.transparency.org/global_priorities/public_contracting/key_sectors/construction_and_engineering (accessed on 17 May 2011).
UK Anti-Corruption Forum (2006) *Anti-corruption Statement*. London. http://www.anticorruptionforum.org.uk/acf/forum_publications/statement/acas.pdf (accessed on 26 February 2006) http://www.u4.no/helpdesk/helpdesk/query.cfm?id=138.
U4 Anti-Corruption Resource Centre (2007) *Summaries of Literature on Costs of Corruption*. U4 Brief, 8 June.
UNDP (United Nations Development Programme) (2008a) *Corruption and Development: Anti-corruption interventions for poverty reduction, realization of the MDGs and promoting sustainable development*. New York: UNDP.
UNDP (United Nations Development Programme) (2008b) *Tackling Corruption, Transforming Lives: Accelerating Human Development in Asia and the Pacific*. New Delhi: Macmillan.
Vee, C. and Skitmore, M. (2003) Professional ethics in the construction industry. *Journal of Engineering, Construction and Architectural Management*, 10(2): 117–27.
World Bank (2010) *Safer Homes, Stronger Communities: a handbook for reconstructing after natural disasters*. Washington, D.C.: World Bank.
World Bank Office in Bangkok (2010) *The World Bank's Strengthened Engagement on Governance and Anti-Corruption in East Asia and the Pacific*. Bangkok: World Bank.
Zou, X.W. (2006) Strategies for minimizing corruption in the construction industry in China. *Journal of Construction in Developing Countries*, 11(2): 15–29.

Part II

Industry performance on projects

5 Information communication technology usage by construction firms in developing countries

Koshy Varghese
Indian Institute of Technology Madras

Evolution of ICT and applications in construction

The evolution of application of Information and Communication Technology (ICT) in most domains has been driven by technological innovation in the ICT sector. During the early days of commercial computing, numerical processing was the primary application. Eventually, the capabilities of computing expanded to include data processing, visualisation, communication and collaboration.

General applications in construction have shown a similar technological growth pattern. The initial application of computers in construction was for computationally intensive and repetitive tasks such as financial accounts, analysis and design of construction works and network analysis. With the availability of the personal computer (PC), supported by spreadsheets and database systems, the applications expanded to include some site-based activities such as materials management and monthly progress reporting. During the late 1980s and early 1990s, computer-aided design and drafting (CADD) technologies became increasingly available on the PC platform and this enabled digital drawings to be viewed at the construction site. The steady progress of computer-aided design (CAD) technologies to 3D, nD and parametric models were enabled by the availability of faster processors, enhanced graphics capabilities, and larger hard disks and memory storage. However, the computing applications and usage during this stage were largely focused on the Architecture, Engineering and Scheduling functions of the construction process.

Electronic mail and the internet added a new dimension to ICT applications in construction. Instant global communications enabled collaboration of engineers distributed across the globe, and immediate access to information and data required applications that could process the data and enable decision support. Moreover, the internet is an apt medium for construction organisations, as sites are geographically distributed and data access between the sites and the home office enable better decision-making. This also enabled applications such as enterprise resource planning (ERP) to be applied effectively on construction projects.

The recent developments in wireless communication have given a further fillip to ICT applications in construction. This has enabled pervasive and ubiquitous computing, to the extent that data can be retrieved and uploaded even at the construction workface. Much of today's research and development (R&D) focuses on harnessing wireless technologies to support various construction tasks. With the availability of powerful, rugged, wearable computers with high-speed wireless communications, the hardware technology to link all levels of the project team is nearly complete. Appropriate software to utilise the hardware is required to enhance the efficiency of the various processes. Moreover, efforts in ensuring interoperability among the different software used along the construction value chain are also an evolving area.

The gap between the developed and developing countries in using these computer technologies and applications was large in the 1980s and early 1990s. However, with the availability of affordable hardware and global communications, the developing countries are rapidly bridging the technology divide. Today, any person or organisation with internet access in a developing country has equal access to the information and software available on the web. Figure 5.1 shows the growth of internet usage in the developed and developing countries. It can be seen that, although there is still a gap between the two groups of countries, the developing world is also getting connected at a rapid rate. However, there are still many non-technological barriers that limit the widespread application of these technologies in developing countries.

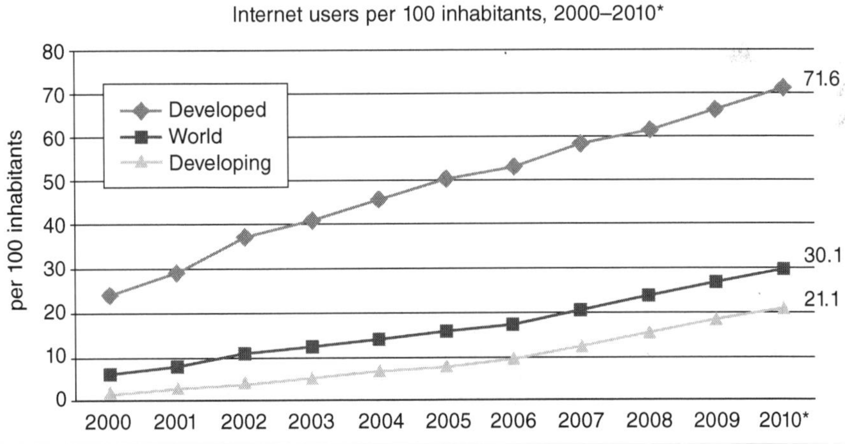

*Estimates
The developed/developing country classifications are based on the UN M09, see: http://www.itu.net/ITU.html
Source: ITU World Telecommunication/ICT indicators database

Figure 5.1 Global growth of internet users

ICT adoption in construction

Globally, the construction industry has been considered to be a late adopter of changing technology and management paradigms. Figure 5.2 shows a popular representation of the technology adoption lifecycle which has been used by several authors to show how new ideas and technologies spread in different cultures (see, for example, Rogers 1962; Savage 1985). The innovators represent a small number of users who are willing to experiment with the technology even before it is commercially viable. "The Chasm" represents the stage that a technology should cross to be widely accepted and used with little risk on direct return on investment. The users at the right-hand end of the curve represent the laggards who adopt the technology when they have no other choice.

In comparison with other industries, the innovators of the construction industry are considered to be late adopters in a global context. Thus, there is a significant lag between the early users of the technology and the widespread usage in the global construction industry. Moreover, even the most progressive (innovators) construction organisations in emerging countries are significantly behind the early majority in advanced economies.

To illustrate the scenario, consider a technology such as CAD. Software for CAD has been available since the 1980s. The manufacturing industry has been steadily implementing CAD, and today, in industrialised countries, even the small and medium-sized enterprises (SMEs) have applied sophisticated CAD concepts to integrated design with the supply chain and manufacturing process. This has made it possible for them to target high levels of efficiency. The manufacturing practice in emerging nations is not far behind, and, in some situations, some multi-national firms have implemented the latest technologies in their new production plants in emerging nations before applying them in their existing plants in the industrialised countries. In contrast, the innovators in construction have partially integrated the project delivery systems with the parametric design approach. Lower in the technology chain, 3-D CAD is largely adopted in

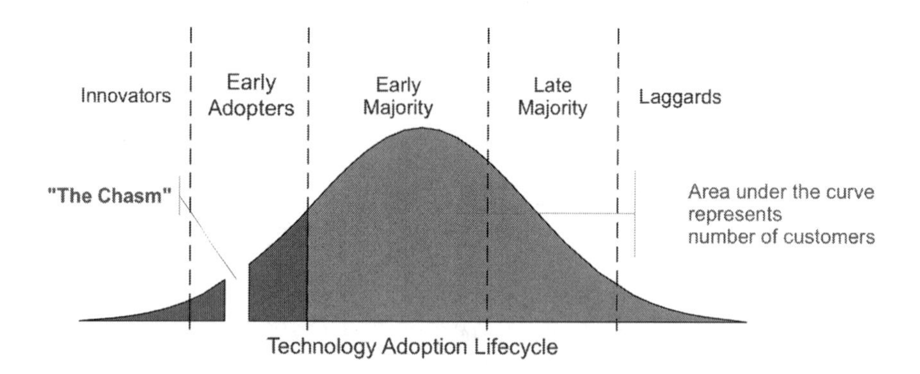

Figure 5.2 Technology adoption lifecycle (source: http://upload.wikimedia.org/wikipedia/commons/d/d3/Technology-Adoption-Lifecycle.png)

the industrialised countries, while most of the organisations in the emerging nations have only recently moved from manual drafting to 2-D CAD.

There are a number of reasons for the late adoption of technology in the construction industry. One of the key reasons is that construction is project-oriented and the process results in the production of unique products at differing locations. As a result of this situation, the clients, organisational structure, design and construction teams, applicable statutory regulations and a myriad of other factors vary from project to project, making it challenging to standardise the practices. Moreover, the construction industry is highly fragmented; the largest constructors are responsible for only about 1 percent of the market whereas, in sectors such as manufacturing, large organisations can control up to 40 percent of the market (International Labour Organisation, 2001). The more significant proportion of the market which the large firms control enables the firms to implement progressive technological ideas and mandate that other firms in the supply chain also align to adopting the same technologies.

The construction industry in emerging nations faces further challenges. In many of these countries, construction was a traditional practice and not considered as a full-fledged industry. As a result, the firms in the sector were not included in the policy-based incentives that were provided to firms in other industries to encourage them to adopt new technologies and implement innovation. Traditional practices continued to ignore the fast pace of change in other industries. Moreover, in emerging economies, most large projects were government-owned and the bureaucracy typically resisted changes in practice, primarily owing to the related uncertainties and the large initial cost of implementation that are usually associated with the new technologies. Although the cost of hardware has dropped rapidly over the years, the cost of specialised software is still beyond economic justification for most firms in developing countries. Another factor that caused resistance to change was that organisations had to undergo a culture shift, as ICT tends to bring in higher levels of transparency and accountability. Transparency was never an attribute that was encouraged by the governments of developing countries, especially those with a colonial past.

These factors combined to dampen the demand for ICT in developing countries. As there was no demand from the construction industry for new technologies and practices, graduates from construction-related programmes in the universities who had been exposed to ICT concepts and technologies were not given the opportunity to explore and implement the technologies. For this reason, in many of these countries, the construction industry lost a large number of bright graduates to the organisations in the other industries which were recruiting graduates to implement their IT systems. For example, a survey of the Nigerian construction industry reports that 98 percent of the construction organisations are computer enabled, but most computer-based tasks in these firms are routine office work such as word processing and communications, not engineering or planning (Oladapo, 2007).

The recent spurt in infrastructure construction in emerging nations, coupled with the growth of privately funded projects, has given an impetus to efforts by the professional organisations to improve the efficiency of construction processes. This demand, coupled with maturing technology suitable for construction application, makes this an appropriate period for the construction industry to infuse ICT-driven change into its processes. The expectation of this trend is supported by the fact that the developing countries have been the focus of many recent publications (Serpell and Barai, 2007).

Roadmaps for ICT in construction

Change in construction industry and application of ICT

Efforts to improve the general practices and productivity of the construction industry have been a focus of numerous forums and committees. One of the earliest efforts in this regard was a study on improving the effectiveness of the construction industry by the Business Round Table in the United States (Roberts, 1987). A report with similar objectives developed in the United Kingdom is the Egan (1998) Report. Both reports were change agents on the respective construction industries and have been widely referred to by other countries, and emulated by some nations such as Hong Kong (Construction Industry Review Committee, 2001) and Singapore (Construction 21 Steering Committee, 1999).

Similarly, to enable research organisations and the construction industry to plan ICT adaptation, adoption and usage, roadmaps outlining industry-level strategies have been formulated. Two widely referred-to roadmaps are those developed by the European Construction Technology Platform (ECTP) (2008) and FIATECH (2010). Some aspects of the roadmaps are futuristic in the current global context. However, they provide an accepted benchmark in the construction ICT domain.

Table 5.1 summarises the common themes of the roadmaps as identified by Hannus (2007). In this study, the FIATECH areas were mapped to the ECTP themes. The business drivers and research areas are as specified by the ECTP report. It can be seen from the roadmaps that most of the research areas focus on harnessing and integrating available technologies to re-engineer the processes rather than suggesting further basic technological development.

Status of areas in developed vis-à-vis developing countries

Each of the areas covered by the roadmaps is relevant to both developed and developing countries. While the roadmaps are primarily based on opinions and trends in the developed countries, the global pattern has been that the developing nations will eventually adapt and apply the technology.

Table 5.1 Comparison of ECTP and FIATEC roadmaps (source: Hannus, 2007)

European Const. Tech. Platform	*FIATECH*	*Business Drivers*	*Research Areas*
Value Driven Business	Intergrated automated procurement and supply network	Performance-driven process, value to customer, total life-cycle support, product and service customization	ICT for customer centric product and service definition requirement management and compliace assessment
Industrialized Production	Intelligent and Automated Construction Job Site	Supply network management; Open market; Effective manufacturing and construction.	ICT support for modular provision of customized constructions; Logistics, on-site production and assembly; Integration of construction site in the process.
Digital Models	Automated Design	Semantics and interoperability user and lifecycle orientation real-time adaptive models.	nD models, access to life time information for all stakeholders anywhere anytime; ICT for design, configuration, analysis, simulation, and visualization
Intelligent Constructions	Intelligent Self-maintaining and Repairing Operational Facility	Integrated automation and control, remote diagnostics and control, context-aware seamless configurability	Smart embedded systems and devices for monitoring and control, embedded learning and user support
Interoperability	Lifecycle Data Management and Information Integration	Data/file exchange, Data sharing, Flexible interoperability	Model servers; Distributed adaptive components; Ontologies and open ICT standards for semantic communication; ICT infrastructures
Collaboration Support	Real-time Project and Facility management coordination and control	Rapid and easy connectivity, robust team interaction, seamless enterprise integration	ICT tools for information sharing, project steering, negotiations, decision support, risk mitigation, etc.
Knowledge Sharing	Technology and Knowledge Enabled Workforce	Improving efficiency of project delivery process through knowledge availability	ICT for transforming project experiences into corporate assets. Object repositories, IPR protection of complex shared data, context aware applications

This section presents each area in more detail and discusses the status of each area in the developed and developing countries.

Although the roadmap is presented as distinct areas, these areas are not independent and there are significant overlaps and inter-linkages among some of them. Moreover, in the comparison of the areas, only developments in selected aspects of each area are considered.

Value-driven business

Over the years, constructed facilities have become increasingly more complex and, in tandem with this, the owners' requirements with respect to cost, quality, safety, health and the environment have become more demanding. Defining, specifying and communicating details of user needs at early stages of the project have a significant influence on delivering value to the owner and ensuring the success of the project. Conventional tools such as drawings and models require significant effort to prepare, and are limited in their ability to communicate details of design intent and simulate the impact of various scenarios through what-if analysis.

Computer-based modelling and simulation offer the potential for the project team to virtually synthesise, test and visualise alternative project scenarios in a collaborative environment. Such virtual modelling will assist the design team in enabling a more holistic definition of client needs and a better understanding of the needs by the participants in various phases of the project. The development of environments for integrated virtual project modelling is in its infancy. The state of the art in advanced countries is currently limited to modelling a few aspects of the project.

An illustration of the utilisation of ICT technology (in a developed country) for adding value for the customer is the Presidio Parkway Project in San Francisco, California. The scope of this project involves replacing the existing link between the Golden Gate Bridge and the city of San Francisco with a new roadway meeting the expectations of today's citizens. As stated in the project's web page (Arellano, 2009):

> The project team strived to create a roadway that reduces impacts to biological, cultural and natural resources; respects the project setting within a national park, the National Historic Landmark District and surrounding neighborhoods; meets community needs; and provides a safer roadway.

State-of-the art modelling and visualisation technologies were used to illustrate the route and the impact of the proposed road to citizen groups. Through these visualisations and simulations, the project team and citizens eventually agreed on an acceptable design that best met the requirements (Autodesk, 2010). Following this acceptance of the basic design in the preliminary stages of the project, other stages of the project could be carried out with better understanding from the users and citizens.

The use of such technologies is viable in developing countries. Countries such as Brazil, India and China have large infrastructure projects on which such technologies could be utilised. These countries also have informed policy-makers to mandate the use of the technologies, and skilled engineers and technicians who can develop and deliver the models. However, one critical aspect is that there is very little public interest and participation in projects, especially during the preliminary phases. As a result, there is no user-driven requirement to develop such models and most projects are not modelled in detail in order to finalise client and user requirements. However, in some developing countries, a few owners of private projects require the development of visualisation and decision-support models for early-stage decision-making. One such example is the airport projects in India that are being built using the Public–Private Partnership mode of procurement. On these projects, the private owners and their consultants require the development of detailed models to explore, and obtain an understanding of, how the passenger experience will be influenced by alternative design schemes. As the trend to utilize private funding for infrastructure projects is increasing in developing countries, it is likely that the barriers for utilising such technologies will reduce.

Industrialised production

The concept of transforming the construction process to bring about manufacturing-type efficiencies has been a mission of the construction industry in many countries such as those in the US and UK. The construction industry has constantly adapted industrial engineering methods in an attempt to improve cost-effectiveness. Key areas identified by the roadmaps are standardisation and modularisation of components, and effective supply chain management.

In a fragmented industry such as construction, best practices of the modularisation and supply chain management are implemented in pockets. Modularisation of homes is one area that has been practised for decades and is slowly gaining popularity in the mainstream in both the developed and the developing countries. Complex building components such as ducting which had to be custom fabricated can now be specified in standardised form and delivered to site ready for final installation. Although software technology enables efficient customisation, it also encourages standardisation for improved productivity and cost-effectiveness. For example, the standard families available in building information modelling (BIM) software are ultimately linked to specific suppliers, and if the designer chooses these components from the existing options, the geometry and other specifications will be accurately modelled in the software, enabling better design, estimating and procurement. This will result in better delivery of the final product along the same lines as industrial production.

Many parts of the supply chain management process in the construction industry are similar to those of manufacturing. However, one of the challenges faced in construction is effective tracking before and after delivery. Unlike manufacturing, the locations of construction projects change, and this results in a more dynamic supply chain. The ability to track the delivery of components and efficient warehousing of the components on site is an issue that is being addressed with Geographical Positioning Systems (GPS) and Radio-Frequency Identification (RFID) technology. Both technologies were expensive until a few years ago. However, today, due to the large volume of demand, the cost has dropped sharply and technology developments such as metal-friendly RFID have enabled wider application in construction. A few industrial sites in developing countries are also using similar technologies (CORE, 2010). As location technologies are implementable using off-the shelf components and the basic information provided is of significantly high value, these technologies are likely to become cost-effective for implementation in the developing countries.

Digital models

The application of ICT in design has been popular worldwide. Today, most design functions are assisted by corresponding software for computational and drafting functions. The use of software in design offices worldwide has become common and, in many cases, the leading design offices develop innovative design tools to meet their requirements. In developing countries also, design offices are using software for standard analysis and drafting functions. In countries where it is not economically viable for a single firm to acquire the software, specialised firms offering software-based services have emerged to meet the needs in a cost-effective manner.

The frontier technologies in the area of digital design are BIM and nD CAD. BIM technologies allow a multi-disciplinary design team to work with a common database to visualise the facility at different stages of design. BIM also enables coordination between design teams, as well as suppliers and contractors. Thus, BIM is widely considered to be an integral part of the Virtual Design and Construction process and an enabler of lean construction.

In industrialised countries, design organisations are steadily shifting to the use of BIM. One of the key drivers is that many government and other owner organisations have contractually mandated the use of BIM. Similar policy steps are also suggested in emerging nations (Wong et al., 2011). Although it is ideal that the initiative to implement BIM should be owner-driven, architects and even contractors on engineering, procurement and construction (EPC) projects have taken the initiative to explore the utility of the technology as it has been found that significant cost and time savings can be achieved through a well managed BIM execution plan. There are numerous cases of how BIM technologies have enabled better coordination between the engineering, procurement, construction and operations on projects.

In the developing countries, the use of BIM is limited mainly due to lack of owner-driven demand and the fact that the cost of the required hardware and software is still relatively high when compared to that of manpower. Moreover, there are technological limitations in providing the seamless integration of software and models through the various phases of a project. As technology adopters in developing countries prefer to invest in fully proven systems, there are delays in adopting the technology. Although there are many barriers to the widespread use of the technology today, it is likely that BIM will follow a similar technology adoption path as CAD. During the 1980s and 1990s CAD technology made steady progress to become the ubiquitous platform for drafting. It is only a matter of time before BIM becomes widely used globally as the platform for design and a tool for integrated project delivery.

Intelligent construction

The current common name of "Intelligent Building" is "Smart Building", and it enables a building to have the intelligence to sense the dynamic needs of its inhabitants and adjust its responses to match their needs. Examples are the response of the structure to dynamic loads, functional response to change of weather, or special functions to monitor the status of elderly inhabitants. The technology required to implement such systems includes an intelligent network of sensors, communications and control systems.

Today, the hardware and software technology required to implement basic levels of intelligent construction exist. Structural health monitoring is a widespread application, and although the technology is largely implemented in developed countries, its use in developing countries is also increasing. Many important structures, such as the stadia of the 2008 Olympic Games in Beijing, China, and key bridges in the earthquake-prone zones in China and India, are appropriately instrumented using sensors; the status of the sensors is transmitted to centralised locations where the response of the structure to internal and external conditions is monitored to ensure functionality and safe service to its users (McDonald, undated; Anderson and Shivam, undated).

Another popular application of the intelligent construction concept is the monitoring and control of buildings for sustainable resource usage. Sensors are used to monitor energy and water usage based on the needs of the occupants, and controls are activated to ensure that the needs are met by optimising the utilisation of energy. The popularity of the Leadership in Energy and Environmental Design (LEED) and similar rating systems for assessing the environmental performance of buildings around the world has motivated owners and architects to specify the use of intelligent building systems. Although such systems are not common, and are generally expensive to implement in a developing country, the publicity of sustainable practice and the potential of generating savings in the lifecycle cost of the building and its operation encourage owners to adopt them.

A third application area where the potential of intelligent construction is being explored is in assisted living for disabled and elderly people. In such applications, the building or home is instrumented to specifically recognise the needs of its inhabitants and take action to make the environment more liveable for them. Technology for assisted living is an area of intense research in the developed world where a significant proportion of the population are senior citizens and this proportion will increase in the future. In developing nations, the population age is more evenly distributed and the culture of extended families, with senior citizens living with their children, is common. However, due to rapid urbanisation, nuclear families, smaller living spaces and frequent job relocations, the culture is gradually changing and it is likely that in the future a large proportion of senior citizens in developing nations will also need the benefits of ICT-assisted living. As the cost of sensors and ICT in general drops and the need for the technology grows, the cost–benefit relationship of implementing such systems will make them more viable.

Interoperability

Interoperability is the seamless integration between different data and data models required in the various phases of a building's lifecycle and is critical for other areas of the ICT development and implementation roadmap to be functional. In its basic form, interoperability requires a unified data representation of the building that enables diverse software packages to receive their inputs from the representation and then update the representation with the outputs after performing their functions. A large number of software packages have been developed for the construction industry applications and the development of each has evolved based on different influences and standards. For example, in the CAD space, AutoCAD and Bentley are popular platforms. However, files developed on one platform require transformation in order to be read by another platform. For example, a quantity take-off software package, using a CAD file as an input, needs to be able to read different file formats for versatile operation. Interoperability suggests a single common data representation enabling seamless operation of the software without the requirement to read multiple formats.

The lack of interoperability between software platforms in the construction industry has resulted in the loss of process efficiency and productivity. Lack of interoperability is also one of the key barriers to ICT adoption in the construction industry. Jacoski and Lamberts (2007) have quantified the losses due to interoperability among small firms in Brazil. They found that there was a 22.5 percent loss in design productivity due to interruption in the flow of design information and digital documents. The study proposed a

Portuguese XML-based[1] software tool to enable interoperability. Although this is a reasonable solution from a research perspective, industry-wide adoption requires a broader framework that is applicable to both small as well as large companies globally.

Industry Foundation Classes (IFC) is widely considered to be the interoperability solution for the construction industry. IFC has represented the data for today's buildings using an object-oriented format and has been widely accepted by software developers and end users. Although for most users IFC is an "under the hood" aspect of interoperability, participation in, and contribution to, the development of IFC is a critical element in the design of the IFC structure. The forums that contribute to IFC development, the International Alliance for Interoperability (IAI) and BuildingSmart, have extensive representation from developed countries and almost no representation from developing countries. As a result, the data and objects represented by IFC are primarily based on architecture, building components and building functionality required in the represented developed nations. In the long term, buildings modelled and constructed using the IFC representation will be more cost-effective and efficient to construct, and it is likely that components not available in the standard representation will become defunct in construction practice.

Collaboration support

The scale and demands of today's projects require a collaborative effort for effective project delivery. Collaboration can be facilitated through a number of technologies such as facsimilie, electronic mail, conference-calls and video conferencing. ICT has been the prime channel for promoting collaboration, and with developments in ICT, the quality of the collaboration environment has improved dramatically. As the capabilities of collaboration technologies have improved, it has become possible to source expertise and services from people and organisations located in different geographical areas. As a result, the project teams of today are globally distributed and rely on collaboration environments to ensure appropriate coordination of their contributions to the projects. The telephone is the most commonly used tool for collaboration, and the rapid development of wireless technologies and web-accessible hand-held computing has enabled people to be reachable and to participate in projects from almost any part of the world. Other forms of collaboration, such as video conferences, are richer than the telephone environment and are being integrated into the mobile environment.

In the developed countries, the reliable existing ICT infrastructure enables collaboration to be productive. As a result, virtual meetings and

1 XML stands for Extensible Markup Language. It is a set of rules for encoding documents in machine-readable form.

conference calls have become commonplace in many business practices. While the platforms for collaboration are maturing rapidly, it has often been lamented that people seem to spend more time communicating than doing actual work. A large proportion of such activities occur because of the unplanned nature of collaboration and the urgency with which issues have to be resolved. Tools for planning and enabling collaboration are being integrated, and constitute the next generation of project planning tools (Eppinger, 2001). One of the key areas in which planned collaboration is critical is the design phase of a concurrently engineered project. Design collaboration on concurrent projects has been identified to be a critical area for success of downstream activities and sophisticated tools to plan information exchange and communication are under development.

The availability of communication and collaborative technologies has been a key driver in the outsourcing of tasks to developing countries where the task can be done more cost-effectively. Developing nations, especially India, Brazil and the emerging countries of Eastern Europe, have been supporting outsourced engineering and design for a number of global construction projects. The provision of these services has required the establishment of advanced design centres in these developing nations, and dedicated infrastructure for collaboration and communication from these centres. The distribution of these centres around the world has also enabled companies to do round-the-clock work on engineering projects. On these projects, work from one centre is handed over to another as the day/night progresses in different parts of the globe. Although outsourcing has established state-of-the art collaboration technologies and practices in some developing countries, the work done in these facilities is largely for projects executed in the developed countries. Recent studies have shown that the sophistication of the collaboration environments has limited influence on effective collaboration, and the knowledge of the culture and working practices of the collaborating teams has a significant influence on the quality and outcome of the collaboration (DiMarco et al., 2010).

For projects within the developed countries, collaboration support can improve project performance. However, the infrastructure for ubiquitous collaboration is limited. The personnel of leading construction organisations are still limited to collaborating through mobile telephone discussions or video conferences when available. As the wireless infrastructure for data transfer in the developing countries improves, it can be expected that more collaboration options will become available to the construction industry.

Knowledge sharing

Knowledge sharing and knowledge management enable an entity to become a learning organisation. This is particularly important to construction organisations, as construction projects are geographically distributed; capturing the lessons learned on each project and making it accessible to others on a project is a valuable decision support asset. ICT-based knowledge-sharing platforms can be of different levels of sophistication. Tools such as web sites, videos and case studies are simple media for sharing unstructured knowledge. Sophisticated knowledge databases with indexed representations and categorised searches can be more effective for focused decision support.

However, capturing, documenting and publishing lessons learned in a formal framework is a challenging task due to the project-type nature of the work, which results in the team being under intense pressure during the project, only to be disbanded and sent to new projects once the project is completed. To overcome this limitation, a few leading global construction groups have invested in forming specialised knowledge management groups within the organisations, thus giving them a better opportunity to capture and code the knowledge and lessons learned on a project. Work on developing ontology and query mechanisms for project-based knowledge is also essential for appropriately structuring and sharing such knowledge.

Data mining is another area of development in knowledge representation. Data mining technologies have the potential to generate the knowledge gained on a project by analysing the data documented for a project as a part of the regular processes. However, this technology is still in its preliminary stages of development.

As the features of specific knowledge sharing technologies are limited, they are not widely used in construction organisations. Some of the leading organisations in developed nations have created their own proprietary platforms for knowledge sharing and have been using them for several years (Thomas and Keithley, 2002). In the developing countries, attempts are also being made by construction organisations to design and develop knowledge portals. Chan and Liu (2007) have surveyed the Hong Kong and Chinese construction industries and found the urgent need for industry-wide knowledge sharing through region-specific custom portals. However, in addition to the technology availability and organisational mandate, a culture of sharing information and learning from mistakes has to exist for the efforts to be successful.

Looking ahead

A majority of the volume of construction in the coming decades will be in the developing world. The developing countries need significant transport, energy and urban infrastructure to cater to the basic needs of their people. Adopting appropriate ICT-based practices will enable the project teams to achieve the construction and operational utility of the project. Ugwu and Kumaraswamy (2007) have proposed a number of critical success factors for the adoption and diffusion of ICT in the construction industries of emerging economies.

Overall, the key factors influencing the adoption of ICT are governance, cost, process, appropriateness of technology, people skills and culture. Governance refers to the laws and policies of a country or region that promote the use of ICT. It can also be in the form of contractual conditions specified by the owner on the use of specific ICT tools and methodologies for all projects undertaken for the owner. For example, if a city decides that all its tender documents will only be available in an electronic form and must be submitted electronically, it will compel the interested bidders to adopt the new method. In Singapore, CORENET, the government's construction and real estate web site,[2] has brought about changes and a level of standardisation in the industry practice. However, the government and owners in developing countries are generally conservative, and hesitate to bring about such changes due to the lack of in-depth knowledge about the technology and uncertainties concerning the impact of its implementation. It should be emphasised that they stand to gain the most as ICT implementation eventually improves national level productivity.

An owner organisation will ultimately gain the savings or bear the losses due to ICT implementation. Thus, a good understanding of both the costs and the benefits is essential. The relatively high cost of implementation has been one of the strongest barriers in developing countries. Moreover, the rapid rate of obsolescence of hardware and the need for investments in new versions of software with uncertain compatibility, bugs and expenses for training, have made the investment decision bewildering to most owners and contractors. Typically, introduction of the software to a process will result in an initial decline in productivity. Few decision-makers want to be responsible for this type of change, and it is usually less controversial to maintain the status quo. Organisations have to systematically build capabilities to deal with rapidly changing technologies if they are to remain competitive in future markets (Sreepuram and Rao, 2006). Furthermore, as the cost of ICT implementation drops and the value of people skills in developing countries grows, the risk of the ICT investment will become much more palatable to bear, and decision-makers will favour the exploration of ICT implementation.

Existing paper-based processes will have to be re-engineered to get the benefit from ICT implementation. Organisationally, re-engineering a process requires changes at many levels, and it is only natural that they will be resisted. Understanding and managing the changes through appropriate organisational directives, training and incentives is a key requirement for effective adoption of ICT-based processes. Managers need clarity of goals, and they must be aware of the change processes and be trained to lead their subordinates through these processes. Surveys of construction industry ICT

2 CORENET: http://www.corenet.gov.sg (accessed 18 May 2011).

usage in Jordan have shown that ICT-based changes in procedures are a challenge but, once management commitment is firm, they result in fewer mistakes in documentation, improved document quality, faster rate of work, ability to handle greater complexity of work, and, eventually, reduced costs of doing business (El-Mashaleh, 2007).

Software has become sophisticated, with numerous options and features. Expertise is essential to determine if a particular technology can be adapted to the processes required by the organisation. Often, organisations are left with unusable software as a result of enthusiastic marketing by the software vendors and a minimal assessment of its adaptability to the requirements of the organisation. Moreover, if any level of customisation is required, it usually starts with optimistic estimates on effort, and eventually leads to changes and delayed deployment. In this context, off-the-shelf tools are preferred; but if customisation is required, effort in developing a clear customisation specification is imperative.

Utilising a specialised software platform today requires extensive technical training and possible certification. Training has been found to be one of the biggest barriers to efficient adoption of ICT in India (Ahuja et al., 2009) and Brazil (Scheer et al., 2007). As the training and certification process requires employees to be away from their regular work, many conventional organisations view this as a waste of employee time. However, with today's rate of technological change, periodic training is imperative. In addition to knowledgeable employees, ICT consultants should also be easily available to provide expert advice and undertake troubleshooting. Only in an ecosystem where such resources are available can technological implementation be successful. Developing a critical mass of people to support and sustain the use of software can be accomplished only if there is a genuine interest and need for the software from the construction industry.

Ultimately, culture plays a critical role in successful ICT adoption. Culture, in this context, refers largely to organisational culture, but is not independent of the work culture of the society in which the organisation functions. Organisations that are willing to learn and change will rapidly assimilate the knowledge of the processes required for successful ICT adoption. A society that is generally positive towards the use of technology will encourage and support technology-based organisations to flourish. Changing organisational culture is a slow process, and in many cases it has been found that culture change will only be significant over generations. In addition to the above factors, clear definition and understanding of end-user requirements and the security of the data and procedures are also of significant concern to the industry.

Many of the areas discussed in this chapter seemed futuristic when they were initially identified by ECTP and FIATECH. Over the last few years researchers, software developers and early adopters of the technology have demonstrated that the themes, technologies and applications identified in the roadmap can be implemented in industry. Although most implementations have been in developed countries, a few have been on projects in developing nations. Today, there is global awareness of the potential of these technologies, and the benefits of implementing the technologies outweigh the costs in many of the emerging economies. Progressive owners and contractors in these economies are already implementing appropriate ICT technologies to enhance the value of their projects. It is inevitable that the rest of the industry will follow.

Acknowledgement

The author would like to thank Dr. Mohan Kumaraswamy, Professor, The University of Hong Kong for his valuable inputs to this chapter and Dr. George Ofori, Professor, National University of Singapore for his exceptional editorial support.

References

Ahuja, V., Yang, J. and Shankar, R. (2009) Study of ICT adoption for building project management in the Indian construction industry. *Automation in Construction*, 18(4): 415–23.

Anderson, J.E. and Shivam, H. (undated) *Performing Structural Health Monitoring of the Naini Bridge in India Using the LabVIEW Real-Time Module*, http://sine.ni.com/cs/app/doc/p/id/cs-12653 (accessed on 27 December 2010).

Arellano, N.E. (2009) 3D video *reassures* San Franciscans about new Presidio Parkway. *itbusiness.ca*, 12 August, http://www.itbusiness.ca/it/client/en/home/News.asp?id=55635 (accessed on 27 December 2010).

Autodesk (2010) *Visualizing Future Transportation*, http://images.autodesk.com/adsk/files/parsons_brinckerhoff.pdf (accessed on 27 December 2010).

Chan, H.W. and Liu, C. (2007) Corporate portals as extranet support for the construction industry in Hong Kong and nearby regions of China. *ITcon*, Vol. 12: 180–92.

Construction 21 Steering Committee (1999) *Reinventing Construction*. Singapore: Ministry of Manpower and Ministry of National Development.

Construction Industry Review Committee (2001) *Construct for Excellence*. Hong Kong: Hong Kong Special Administrative Region Government.

CORE Projects and Technologies Ltd. (2010) *Asset Tracking for Petroleum industry – Case Study*, http://www.coreprojectstech.com/pdf/Reliance – Case Study.pdf (accessed on 27 December 2010).

Di Marco, M., Taylor, J.E. and Alin, P. (2010) Emergence and role of cultural boundary spanners in global. engineering project networks. *Journal of Management in Engineering*, 26(3): 123–32.

Egan, J. (1998) *Rethinking Construction: Report of the Construction Task Force*. London: Her Majesty's Stationary Office.

El-Mashaleh, M.S. (2007) Benchmarking information technology utilization in the construction industry in Jordan. *ITcon*, Vol. 12, Special Issue Construction Information Technology in Emerging Economies: 279–91.

Eppinger, S.D. (2001) Innovation at the speed of information. *Harvard Business Review*, 79(1): 149–60.

European Construction Technology Platform (2008) *Vision and Strategic Research Agenda Focus Area Processes and ICT*, http://www.ectp.org (accessed 18 May 2011).

FIATECH (2010) *Capital Projects Technology Roadmap Overview*, http://fiatech.org/tech-roadmap/roadmap-overview.html (accessed 27 December 2010).

Hannus, M. (2007) *Presentation*, http://fiatech.org/images/stories/techroadmap/fiatech_ectp_mapping (accessed 27 December 2010).

International Labour Organization (2001) *The Construction Industry in the Twenty-First Century: Its image, employment prospects and skill requirements.* Geneva: ILO.

International Telecommunications Union (ITU) (2010) *World Telecommunication/ICT Indicators Database*, http://www.itu.int/ITU-D/ict/statistics (accessed 7 March 2011).

Jacoski, C.A. and Lamberts, R. (2007) The lack of interoperability in 2D design – a study in design offices in Brazil. *ITcon*, Vol. 12, Special Issue on Construction Information Technology in Emerging Economies: 251–60. http://www.itcon.org/cgi-bin/works/Show?2007_17 (accessed 18 May 2011).

McDonald, C. (undated) *Performing Structural Health Monitoring of the 2008 Olympic Venues Using NI LabVIEW and CompactRIO*,
http://sine.ni.com/cs/app/doc/p/id/cs-11279 (accessed 27 December 2010).

Oladapo A.A. (2007) An investigation into the use of ICT in the Nigerian construction industry. *ITcon*, Vol. 12, Special Issue on Construction Information Technology in Emerging Economies: 261–77.

Roberts, J. (1987) The Business Roundtable Construction Industry Cost Effectiveness Task Force. *Construction Management and Economics*, 5(2): 95–100.

Rogers, E.M. (1962) *Diffusion of Innovations*. New York: Free Press.

Savage, R.L. (1985) Diffusion research traditions and the spread of policy Innovations in a federal system. Publius, 15 (Fall): 1–27.

Scheer, S., De Amorim, L.S.R., Santos, E.T., Ferreira, R.C. and Caron, A.M. (2007) The scenario and trends in the Brazilian IT construction applications' experience. *ITcon*, Vol. 12, Special Issue Construction Information Technology in Emerging Economies: 193–206.

Serpell, A. and Barai, S.V. (2007) Editorial on construction information technology in emerging economies. *ITcon*, Vol. 12, Special Issue Construction Information Technology in Emerging Economies: 165–6.

Sreepuram, P. and Rao, K. (2006) Build organization capabilities to utilize IT. *Proc. World Conference for Design and Construction*, INCITE/ITCSED 2006, Vol. 4. New Delhi, India, November 2006: 72–80.

Thomas, D. and Keithley, T. (2002) Knowledge management improves performance. *AACE International Transactions*, PM.17-1–17-4.

Ugwu, O. and Kumaraswamy, M.M. (2007) Critical success factors for construction ICT projects – some empirical evidence and lessons for emerging economies. *ITcon*, Vol. 12, Special Issue Construction Information Technology in Emerging Economies: 231–49.

Wong, A.K.D, Wong, F.K.W. and Nadeem, A. (2011) Government roles in implementing building information modelling systems: Comparison between Hong Kong and the United States. *Construction Innovation: Information, Process, Management*, 11(1): 61–76.

6 Safety and health in construction in developing countries

The humanitarian paradox

Richard Neale
University of Glamorgan, Cardiff, United Kingdom
and
Joanna Waters
Independent Consultant, Cardiff, United Kingdom

Introduction

> Every year about 250 million workers suffer accidents in the course of their work, and over 300,000 are killed. Taking account of those who succumb to occupational diseases, the death toll is over 1 million people a year. Yet international concern with awareness of health and safety at work remains surprisingly modest and action is limited.
>
> ("Decent Work", Report of the Director-General 87th Session International Labour Conference, Geneva, (ILO, 1999).

> "Accident": an event which is unforeseen or has no apparent cause.
>
> (Compact Oxford English Dictionary)

> Workers are killed, injured and made sick whilst carrying out routine jobs. The hazards are well known and so are the prevention measures. The overwhelming majority of "accidents" are absolutely predictable and preventable. They are caused by failure to manage risks, or by straightforward negligence on the part of the employer.
>
> (Building and Woodworkers International, http://www.bwint.org/default.asp?Issue=OSH&Language=EN)

These quotations "set the scene" for this chapter, which reviews the main contemporary issues in achieving effective Occupational Safety and Health (OSH) in the construction industry in developing countries. The literature shows that there can be many perspectives on OSH: legal, regulatory, procedural, technical, commercial (in the development of equipment), combative and adversarial (in the case of "accidents" and disputes), and humanitarian. In this chapter, many of these aspects of OSH are embraced, but the discussion and explanation have been set out within a broadly social context, with an emphasis on the fundamental expectations and cultures of a modern society. In outline, the sections of this chapter follow a course from establishing a social, ethical, economic and legal case for effective OSH, followed by a discussion on the barriers that need to be overcome in developing countries as to how to establish a "national culture of safety" that pervades society from a government and legislative level through to a "safety culture" that can be nourished at an organisational level to encompass workplaces and workers. The chapter concludes with some suggestions for further development and research.

Background to occupational safety and health in developing countries

In order to explore specific issues around *construction* safety and health in developing countries, it is essential to consider their *wider* cultural, social, political and economic contexts. The development of processes and attitudes to *general* occupational safety and health in developing countries is one that needs to be addressed first as a matter of priority; if governments do not embrace OSH generally, it will be even more of a challenge to improve OSH in the construction industry.

Historically, employers in developed countries became concerned about workplace safety and health during rapid industrialisation and urbanisation and the realisation that every "accident", injury and fatality had not only a human cost, but also high economic costs. An "accident" is described in this chapter as "an event which is unforeseen or has no apparent cause" (*Compact Oxford English Dictionary*). In this chapter, the authors use the term "accident" in inverted commas to show that many "accidents" are foreseeable and preventable. According to the International Labour Office (ILO) (2005) the construction industry accounts for around 7 percent of the world's employment but has on average over a third of fatalities across the world. India has one of the world's highest accident rates among construction workers; the on-line resource newKerala.com (2010) reports that "the average Fatal Accident Frequency Rate in the Indian construction industry is 15.8 percent of every 1,000 employees as against 0.23 percent in the US." This harshly illustrates the challenges legislators, employers and workers face. To avoid such fatalities a culture of safety, among others, is crucial, yet the "current role of training and certification is equivalent to only a drop in the ocean" (Sandhir 2009: 4). Meanwhile, Kartam and Bouz (1998) discuss how the construction industry is the most hazardous industry in Kuwait, accounting for 62 percent of all deaths in 1994, and suggest that many managers are simply unaware of how OSH programmes can save lives and project costs.

The ultimate objective of OSH is to maintain healthy and productive workers, free from risks of "accident", injury or illness. Occupational safety and health also underpins the social and economic wellbeing of working people and promotes motivated and productive employees. To achieve this requires "continuous improvement of the conditions of work and a comprehensive and multidisciplinary approach" (World Health Organization (WHO), 1995: 29). Some developing countries have responded to the need for effective OSH systems, especially with the intervention of trade unions, by designing safer machinery and introducing safety processes and procedures, including the use of personal protective equipment (PPE).

A complex combination of factors will influence occupational health and safety legislation, standards, policies and their enforcement in developing countries. Together, they shape the extent to which a country has a "national culture of safety" and how employers and workers embrace a "safety culture" at their places of work.

The humanitarian paradox

The ILO (1919) Constitution sets forth the principle that workers should be protected from sickness, disease and injury arising from their employment. Yet for millions of workers the reality is far from this ideal: "Of the approximately 3.2 billion people working in the world, about 2.4 billion workers, i.e. 80 percent, live in developing countries compared to about 600 million in the industrialized countries" (Rantanen et al. 2004: 65). The ILO (2010b) estimates that 160 million people suffer from work-related diseases a year, and there are an estimated 270 million fatal and non-fatal work-related "accidents" annually. Workers in high risk industries such as construction are often at an unreasonably high risk, with one-fifth to one-third of workers suffering occupational injuries or diseases annually, leading in extreme cases to a high prevalence of disability and in some cases to premature death (WHO 1995: 5). Even in developed countries in Europe, "fatal and non-fatal accidents in construction remain around twice as high as the EU sectoral average" (OSHA Europe 2004).

Focusing on the construction industry, at least 180,000 workers are killed globally each year in this industry, representing a huge humanitarian problem. The construction industry is one of the prime instruments of development, and infrastructure such as schools, hospitals, factories, homes, transport systems and many other physical elements required for a nation to develop economically and socially depend on it. "Humanity", in the context of "development programmes", means to be concerned with or seeking to enhance human welfare. Contrasting the huge numbers of deaths and injuries with this notion of "development" gives rise to a stark paradox: *consideration for human welfare is often irrelevant when constructing the infrastructure required to benefit and enhance human welfare.*

There has been a significant increase in both output and employment in the construction industry in developing countries in the past 30 years: construction is a major component of investment; hence expansion in construction activity is closely related to economic growth (ILO 2001b: 8). In many countries, demand for infrastructure and real estate is booming. However, many construction workers engaged in "development projects" are trapped in a cycle of dangerous work, poor pay and little job security, and the wellbeing of entire families is often critically dependent on the productivity of its working member/s (WHO 1995). The illness, injury or death of construction workers who provide for their family can have a devastating impact, at worst leading to impoverishment, and this example reveals a "trickle effect" that indirectly has an impact on the wellbeing, health and the subsequent economy of entire communities. Direct loss to the victim includes wages, medical costs, pain and suffering, family distress, lack of promotion, psychological damage and loss of future earnings (Ma and Chan 1999: 65). Although OSH is an essential component of any industrial economy, and indirectly contributes to social progress and economic growth, it is all-too-often under-formulated.

Ultimately, if countries do not have adequate health and safety provisions to protect a growing, and frequently inexperienced, workforce, the drive for more development will often equate to more deaths and injuries. The work of international organisations in this area has been extensive and they play a vital role in instigating change. It is crucial to remember that *every* (construction) worker has the right to safety and health at work in both developed and developing countries; it is "a part of basic human rights" (WHO 1995: 5).

A healthy, safe, motivated and productive workforce is one of a country's most valuable social and economic assets and, given that improving a developing country's infrastructure improves its wealth, as well as the costs incurred by individual enterprises if they neglect OSH, it is surprising that the construction industry remains insufficiently regulated and that the performance and achievements of the industry are compromised by a humanitarian catastrophe through injury, illness and death.

Conceptualising occupational safety and health

Before considering examples of ways to improve construction OSH in developing countries, it is useful to consider how OSH can be conceptualised. For example, Nuwayhid (2004: 1916) presents occupational safety and health as existing in two key domains:

- The internal environment, which focuses on the place of work, and, for example, its risks;
- The external-contextual domain, which focuses on the wider social context that the internal domain operates within.

These domains are depicted in Figure 6.1.

Nuwayhid (2004: 1916) suggests that "traditional workplace-oriented occupational health has proven to be insufficient in the developing world". This is because it fails to consider the wider social, cultural and political contexts that workplaces operate within. Therefore, Nuwayhid calls for a different approach to occupational health research in developing countries: he suggests that:

> ... instead of focusing on the workplace as an isolated entity and moving outward to the wider social and political arena as done in occupational health research in industrialized countries, occupational health research in the developing world should focus on the social and political issues and then move inward to address the particularities of the workplace (i.e., from the "external–contextual domain" to the "internal domain"). This approach builds a wider alliance up front with social scientists, economists, political scientists, unionists, non-governmental organizations, women's organizations, human rights groups, and others as an entry point into the occupational health field (2004: 1918).

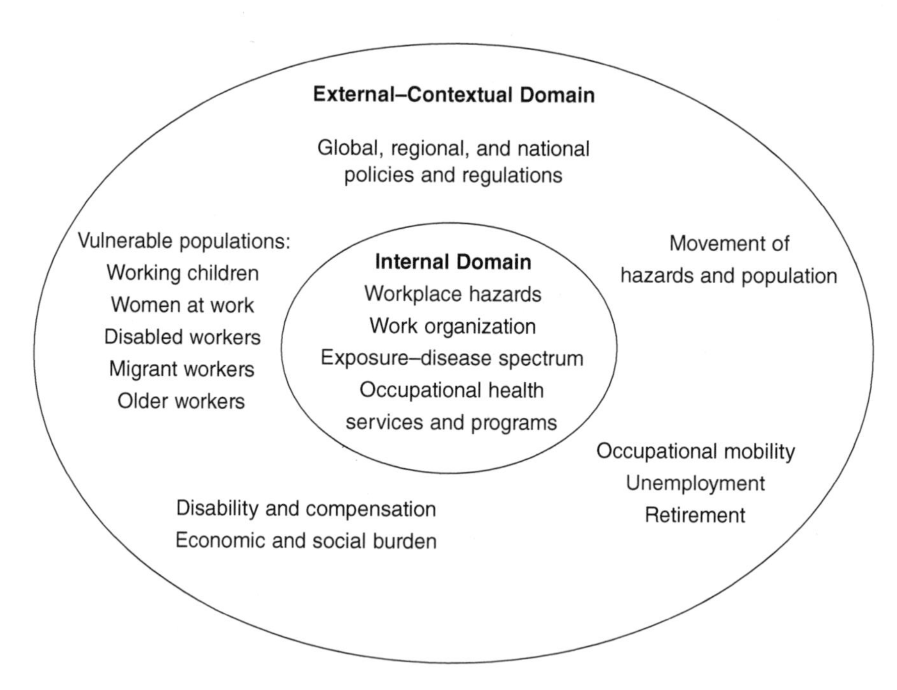

Figure 6.1 'Domains of occupational health research'
Source: Nuwayhid, I.A., 2004, Occupational health research in developing countries: A partner for social justice. *American Journal of Public Health*, 94, 11, 1916-1921. (Figure from page 1917). Copyright: American Public Health Association.

This approach offers a useful way to address wider social conventions, as well as cultural, political and economic ones, that shape attitudes to OSH in developing countries, and how these shape attitudes to OSH in the workplace. Crucially, instead of focusing on the *place of work* and *organisational characteristics*, including OSH procedures and workplace risks (the internal domain), the focus is on "the workers in their social contexts" (Nuwayhid 2004: 1918).

One may also think of the external domain as reflecting macro-level issues, with the internal domain reflecting micro-level issues. Governments, and their attitudes and actions, would fall within the macro level, and employers and employees within the micro level. This is a useful distinction because it lends itself to distinguishing between what is required to foster safer work practices at both levels or domains. On the micro level, there are a number of factors at play when considering how OSH manifests itself; for example, the debate over the extent to which organisational culture interacts with workers' values and beliefs, or how corporate culture vies with traditionally held values (see Suplido-Westergaard 2006).

Making health and safety a priority, i.e. by preventing Step 1 of the "cycle of neglect" (as presented in Figure 6.2) in developing countries will facilitate the development of effective OSH processes and lay the foundations for the creation of a "culture of safety" that is adopted by governments, employers and employees as a pervading "safety ethos".

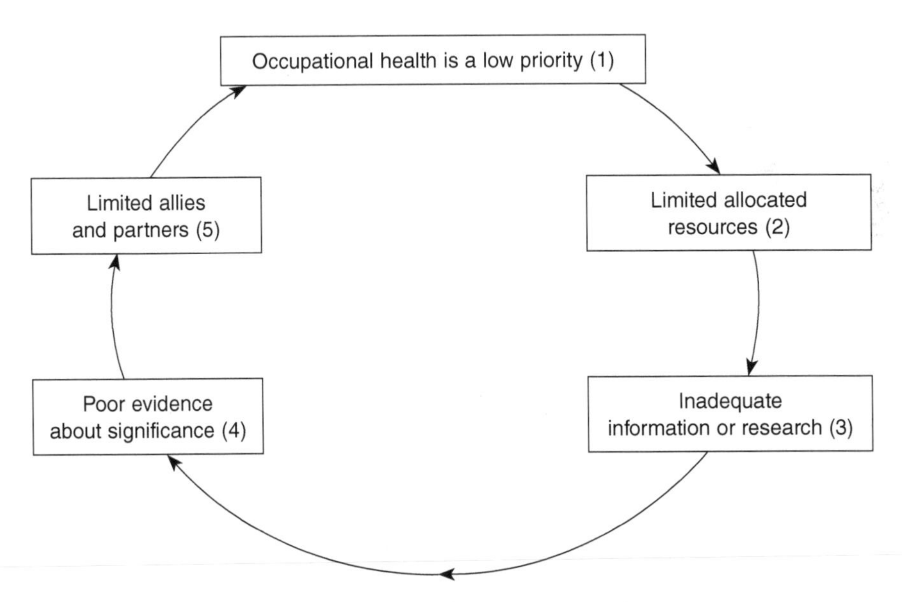

Figure 6.2 'The occupational health "cycle of neglect" in developing countries'
Source: Nuwayhid, I.A., 2004, Occupational health research in developing countries: A partner for social justice. *American Journal of Public Health*, 94, 11, 1916-1921. (Figure from page 1918). Copyright: American Public Health Association.

An example of OSH processes that focus on the "internal domain" is the risk assessment standard used in the UK. One reason for its moderate success is that the UK is a developed country that is no longer caught in the "cycle of neglect" depicted in Figure 6.2; OSH has long been made a priority issue in the UK and a "culture of safety" dominates British politics, and this is echoed in organisational "safety cultures" in workplaces across the country. A combination of factors will determine the number, types and enforceability of occupational health and safety legislation, standards and policies in developing countries.

Defining "safety culture" and "culture of safety"

This is a useful juncture to consider in more detail what is meant by the terms "safety culture" and "safety climate". It is important to distinguish between the "safety culture" of an organisational workplace and a "culture of safety" as an ethos promoted at government and national levels; but the concept of "safety consciousness" is useful to apply to individuals with a strong commitment to OSH, no matter what work they are engaged in.

"Safety culture" is usually used in an organisational context and has been described by the Chartered Institute of Building (CIOB 2009: 12) as follows:

> The safety culture of an organisation is the product of individual and group values, attitudes, perceptions, competencies and patterns of behaviour that determine the commitment to, and the style and proficiency of, an organisation's health and safety management. Organisations with a positive safety culture are characterized by communications founded on mutual trust, by shared perceptions of the importance of safety and by confidence in the efficacy of preventive measures.

This provides an insight into how commitment and shared responsibility are the lifeblood of a successful "safety culture" on an organisational level.

Gadd (2002: 3) further attests that there can exist a diversity of "safety cultures" within an organisation but the management of an organisation is most likely to have the decisive influence on the degree to which an organisation has a safety culture, its specific features and its relative success. Safety cultures in the developed world are often based on providing unified messages to the workforce on risk assessments and preventative activities. While organisational health and safety policies have the potential to inculcate in workers a strong safety ethos, such a managerial approach can be particularly dogmatic, and also be at the expense of being open to other ideas about safety, health and risk prevention. Given that a diversity of work is undertaken in many organisations, particularly in construction, it is not surprising that subcultures exist. One only has to consider the range of different roles on a single construction site to recognise this point. Ultimately, different workers will be exposed to different risks and hazards and subconsciously adapt to them in a certain way. As Gadd (2002: 23) attests, although they suggest a lack of one cohesive safety culture, "subcultures are not seen as undesirable and it can be argued that they provide useful contextual insight into the different risk and hazards experienced by workgroups".

In this chapter, it is useful to distinguish between a "safety culture" and what may be termed a "national culture of safety"; the former is typically used in relation to an organisation, while the latter refers to a safety ethos that can be shared and valued by the wider population. Also, while the term "safety climate" is associated with the psychological element of workers in an organisational safety culture, i.e., workers' safety values, attitudes, and perceptions (HSE 2005: iv), the terms "safety mind-set" or "safety consciousness" are particularly useful for describing either an individual or a collective positive safety ethos.

A "safety culture" is characterised by a number of human, social and cultural factors, such as attitudes, behaviours, values and norms. Misnan and Mohammed (2007: 14) define construction safety as a standard of quality, and current research indicates that the construction industry has been labelled as an industry with a low level of safety and health culture because it is perceived as being high-risk and having a high "accident" rate. Arguably, unless a "safety culture" develops, i.e. a natural, automatic desire to reduce risks, legislation alone will not make the industry much safer. Misnan and Mohammed (2007: 15) equate a safety culture with a set of belief systems, attitudes and social practices "that are concerned with minimising the exposure of individuals, within and beyond an organisation, to conditions considered dangerous or injurious". Such a "safety culture" exists on a continuum – all countries have some degree of a "safety culture" but the differences between those in developed and developing countries are quite evident. Fortunately, "safety culture" can be improved and developed. Creating a safe and healthy work culture requires the inculcation of safe and healthy practices as part of everyday life, at work and at home, among all workers (Misnan and Mohammed 2007: 19), and social change is a necessary priority before the situation in individual workplaces can be addressed.

Having developed the notion of a "safety culture", a number of key arguments for effective OSH in developing countries will now be presented.

The case for effective construction OSH in developing nations

In this section, a summary of the key arguments that collectively provide an argument for developing holistic and effective construction safety and health processes and systems in developing countries is presented.

The ethical argument

While sharing the basic premise of the humanitarian right to a safe working environment detailed above, the ethical case for health and safety on construction sites is based on an employer's duty of care to protect workers from the risks involved. However, in some cases, the ethical impetus is less of an incentive to some employers than the risk of liability costs they will incur if occupational health and safety is neglected and workers are injured, fall ill or die as a result. The welfare of workers is key and it is hoped that through gradual social reform employers will become genuinely morally motivated to protect their workers. London and Kisting (2002) present a discussion on attaining ethically acceptable global standards for OSH.

The legal argument

While legislation on safety is well established in developed countries, the situation in developing countries is often that the statutes are under-formulated, lacking in coherence and appropriateness and often unenforced, due to a lack of commitment from governments and/or to safety professionals being poorly equipped with time and knowledge. Indeed, Ahmed et al. (1999: 528) found that one of the most serious safety problems on construction sites was the inability of safety officers to enforce safety regulations. In contrast to this, while employers in developing countries should be proactive and anticipatory of safety and health, historically many have acted reactively in response to injuries or fatalities. In many developing countries, work regulation and labour legislation are in need of revising and are "either old, or not up-to-date, nor revised recently" (Ahasan 2001: 313). With rapid industrialisation sometimes comes the piecemeal appropriation of regulations due to pressures from international organisations.

Governments have the authority to direct public spending to support any of their nationally defined commitments, one of which should be safety and health in construction. In developing countries, it is evident that governments' commitment to safety and health in construction is often lacking (Coble and Haupt 1999: 903). Government involvement is essential but needs to be supported by international influence, assistance, regulation and enforcement. These play a key role in encouraging developing nations to adopt laws and policies to create healthy workplaces provided they are enforceable and enforced in the workplaces. However, although promising, multinational trade agreements have largely failed to ensure worker health and safety in developing countries thus far.

The economic argument

A lack of OSH standards can lead to less productivity with: an estimated reduction of 10–20 percent of gross national product (WHO 1995: 30);

and an estimated 4 percent reduction in the world's annual gross domestic product (GDP) as a consequence of occupational diseases and "accidents" (ilo.org 2010b). Thus, there is an irony in a company's lack of will to invest money in ensuring staff safety and training when construction work is a catalyst for socio-economic development and a major contributor to GDP. Box 6.1 presents an example of the economic cost of failing to allocate resources to OSH. Such investment can only contribute to a project's deliverability and quality when workers' health and safety is not compromised. The integration of safety and health measures into a total management system in the construction industry can contribute significantly to cost efficiency, quality assurance and environmental protection.

Box 6.1 The economic consequences of failing to invest in construction OSH

As an example of the economic costs of lack of investment in occupational health and safety, the construction industry in the EU is estimated to be worth €902 billion a year, and with accident and ill-health costs in the sector totaling 8.5% of project costs, that means that poor OSH standards in construction could be costing over €75 billion each year – nearly €200 for each member of the population.

Source: European Agency for Safety and Health at Work (2010), available online at: http://osha.europa.eu/en/topics/business-old/performance/benefits_construction_html

The business case

A growing body of data shows that the impact of occupational health is positive, not only at the level of the national economy, but also at the level of the economy of an individual enterprise (WHO 1995: 30). The benefits of good OSH in construction, to clients and employers, include quicker and more cost-effective completion of projects, higher profit margins, and greater likelihood of becoming the supplier of choice. This is the basis of the argument for a "business case" for effective OSH. Cost is often a factor that influences whether a workplace adopts effective interventions to address occupational hazards.

On a purely financial basis, it is difficult to understand why employers do not promote the safety of their workers due to a correlation between poor OSH and *their* economic losses; employers face indirect losses such as loss of productivity, legal costs, compensation claims, loss of reputation, absenteeism, and high insurance premiums. Indeed, in a survey of Hong Kong construction workers, Ma and Chan (1999: 64) found that as a result of a high "accident" rate, there were increasing numbers of "common law" claims by employees against their employers for negligence.

The negative effects of a major incident are demonstrated by a case study of an excavation collapse in Uganda that is presented in Box 6.2. Eight workers were killed, the project was delayed and the client, designers, project managers and contractor all suffered economically and their reputations were adversely affected. That case provides a good example of an "OSH incident" rather than an "accident". The case study shows that "accidents" are bad for business and demonstrates that good OSH performance can contribute to good business performance. That leads into the explanation of "the business case for OSH".

Box 6.2 Case study demonstrating the business case for good OSH performance*

The side of an excavation for the basement of a large building project in Uganda collapsed. Eight people were killed.

According to one worker interviewed on site, the "accident" occurred at around 11:30am on 14 October 2008. At the time of the "accident", workers were reinforcing the excavation with steel bars and timber, and a compactor was compacting soil close to the cave-in point. He also said that there had been a spate of soil cave-ins in the past, including one on 30 September 2008, in which part of the site offices collapsed into the excavation.

The government inspection team observed the following:

- The excavation extended almost to the site boundaries; it was about 15m deep and the sides were nearly vertical;
- Most parts of the excavation, from the base up to a height of about 7m, had been strutted with metal and timber, with the exception of the western part near to the site offices, and the eastern part, where there was the main gate and access ramp. In addition, the rest of the excavation above the strutting had a plaster covering of about 2 inches thick;
- At the western part there seemed to have been a previous cave-in, which had downed part of the site offices, as cleavage markings were observed from the site office structure;
- Near to the "accident" point was an excavator with fresh track marks on the ground, an indication that it was probably in use at the time of the "accident";
- Above the "accident" point was a house that also caved into the site with the soil cave-in;

- On site, seven bodies were recovered and two injured workers were taken to hospital. There were reports, that could not be independently confirmed by the investigation team, that one of the injured workers had also died in hospital;
- The identities of the dead and injured could not immediately be established;
- Furthermore, the terms and conditions of employment of the workers could not be established.

Contractor compliance with the OSH Act, 2006

- Pursuant to Section 40(2) of the OSH Act, the site was notified to the Commissioner for Occupational Safety and Health on 12 May 2008;
- In addition, the Contractor submitted to the Department of Occupational Safety and Health a Construction Phase Safety, Health and Environment Plan for the project in compliance with Section 14 of the OSH Act. The Construction Phase Safety, Health and Environment Plan for the project was reviewed for adequacy and a response requesting the Constructor to further submit safety method statements for particular operations among others was sent on 14 July 2008 as the Plan was inadequate. To date, the department has not received a response from the Contractor.

Issues that could have contributed to the occurrence of the "accident"

- The presence of a house is an indication of disturbed ground and therefore special attention was needed. Furthermore, the water run-off from the house enabled water percolation in the ground that could have been compounded during the rainy season;
- An excavation of such magnitude could have been undertaken progressively, i.e. section by section; there should also have been strutting and backfilling. The previous cave-in at the site was an indication of poor methods of work. The steps taken to prevent the reoccurrence are yet to be established;
- The methods for protecting the excavation were inadequate and did not provide protection to the ground level;
- The vibrations and rumbling of the excavator at the time of the "accident" could have triggered the chain of events.

Progress of investigations

Further investigations would be undertaken and there is a need to work with other stakeholders to establish the circumstances of the "accident" and propose actions to avoid reoccurrence of such events. In addition, an attempt would be made to establish how the safety, health and environment plan was operationalised on site.

Conclusions from this case study

This is a case of inadequate support to a major excavation and failure to make it safer by restricting the movement of plant and equipment at the surface near to the excavation.

The client's dream of an impressive building was tarnished by such loss of life and the project was delayed while the excavation was cleared, the OSH investigation took place and the excavation process was re-engineered to provide a safe method of working.

The designers allowed unsafe practices, which have damaged their reputation as competent supervisors of construction work on behalf of their clients. In addition, the question has to be asked about the need for such a deep basement in ground of this nature and whether a different design of the building would have provided similarly useful areas and facilities that were easier and safer to build. If the deep basement was necessary, the designers may have considered the use of such construction methods as contiguous piling.

The contractor suffered increased costs, delays, legal action and compensation costs, and might find it difficult to attract good workers to a site with such a reputation. In a regulated procurement system, the contractor would find it more difficult to get future work.

It was eight workers and their families who suffered the ultimate loss. This is a good example of the need for worker participation in the construction process.

Given that an "accident" is "an event which is unforeseen or has no apparent cause", it is clear that this was a safety "incident" not an "accident" because there are many indications that the cave-in had several causes and warning signs that were quite evident.

(*The authors are grateful to Evelyn Katusabe for providing this case study.)

Definition of "business case"

The term "business case" is capable of broad interpretation. In the context of OSH in construction, the following definition, from the Longman Business English Dictionary, is brief but appropriate:

> An explanation of how a new project, product etc. is going to be successful and why people should invest money in it.

This is an appropriate definition because it asks for a convincing justification of why an investment should be made and effective OSH does require investment. The case study above demonstrates that this can be justified as an investment rather than an optional overhead cost.

The "case for a business case" for effective OSH

The European Agency for Safety and Health at Work provides the following argument for a business case (see reference to OSHA web site (2010)):

> For enterprises, good OSH helps to:
> - Enhance "brand image" and "brand value" as a socially responsible business (which may affect investors' decisions);
> - Reduce absences and increase the productivity of workers;
> - Increase motivation and the commitment of employees to the business;
> - Reduce business costs, such as insurance premiums, and business disruption;
> - Enable enterprises to meet and exceed customer expectations;
> - Avoid the substantial costs of "accidents" and ill-health if businesses get OSH wrong.

For businesses, disruption, claims for damages, loss of goodwill and loss of confidence in management can sometimes lead to total collapse.

Structure and content of a business case for construction OSH

A business case is not solely an economic case. It should be a comprehensive document which explains the reasoning and principles on which the case is based and should include a statement of aims and objectives, the alternatives that were investigated, operational plans, all the resources required, how the investment will be managed, contingency plans and all other relevant information. A guide to preparing a business case has been published by the Australian Safety and Compensation Council (2007).

Corporate social responsibility (CSR)

In essence, corporate social responsibility (CSR) brings together the points made in this section. Because businesses work in a wider social environment they have a responsibility to the wider community. CSR refers to the responsibility that modern business organisations have towards creating a healthy and prosperous society. Businesses have a responsibility to operate ethically and consider the impact of their operations and activities on the people and environment internally and externally. Voyi (2006) discusses the relationship between globalisation, ethics and social responsibility in occupational safety and health, and London and Kisting (2002) explore the impact of globalisation on the international ethical practice of OSH, especially in relation to vulnerable groups. In simple terms, CSR can be expressed as the well-known "triple bottom line" of "people, planet, profit", which requires enterprises to report on their social and environmental performance.

The ILO definition of CSR would seem to be most relevant:

> [CSR is a] way in which enterprises give consideration to the impact of their operations on society and affirm their principles and values both in their own internal methods and processes and in their interaction with other actors. CSR is a voluntary, enterprise-driven initiative and refers to activities that are considered to exceed compliance with the law.
>
> (ILO Subcommittee on multinational enterprises, GB.295/MNE/2/1 Geneva, March 2006)

CSR can be seen as a major "New Perspective" in OSH for the developing world, especially within the context of large international companies operating in these countries.

It is now necessary to explore the main challenges developing countries face in formulating and instituting effective OSH systems.

Challenges to developing effective OSH systems in developing countries

Government complacency and inaction due to lack of resources

It is essential that governments in developing countries place construction safety and health high on the political agenda. Ahasan and Partanen (2001: 74) claim that:

> Officials, who are employed by the state, are not able to implement work regulations and labour legislation easily. Generally, they are not professionally trained and expert in the occupational health … or safety fields, and thus, successful application and implementation of control measures are lacking.

Some fiscally-challenged governments may be reluctant to invest in OSH because it would be impossible to fund even the basic but essential departments responsible for implementing OSH legislation and regulations. This situation can be compounded if OSH is culturally not perceived as a priority issue, and therefore considered unworthy of apportioning resources to.

Developing countries also exemplify extreme diversity in the way they are governed and how resources are allocated and distributed. Ahasan (2001: 312) argues that the implementation of proper safety and health action is hindered because "limited funds may be allocated in national budgets" to health and safety. This is echoed by Dainty et al. (2004) who, focusing on Ghana, assert that a lack of resources available for financial investment in formulating and enforcing OSH regulations is a major problem in developing countries.

Ideally, governments should promote the principle of *proactive* rather than *reactive* safety and health management. When legislation exists, a real contemporary problem is its lack of application and enforcement. Developed countries have extensive OSH legislation and enforcement processes yet still suffer numerous "accidents", many of them fatal. It is for this reason that this chapter emphasises the humanitarian and cultural aspects of OSH, which should manifest themselves through diligent application of the principles of CSR.

A lack of collaboration and cooperation among governments, agencies, international organisations and trade unions can hinder the development and implementation of OSH regulations. Coble and Haupt (1999: 903) argue that "governments can play a proactive role in promoting that safety and health management systems be integrated into the entire construction process". However, this commitment from government must be matched by commitment from all construction "actors" (clients, contractors, designers and construction workers) if safety and health policies are to be successfully introduced. OSH systems cannot exist without a supporting infrastructure, and the development of OSH legislation, regulation and standards requires as a necessity, for example, a labour department and an occupational health service.

In developing effective OSH processes and systems, it is vital for international organisations, such as the WHO and the ILO, non-governmental organisations (NGOs) and aid agencies, to work with government authorities "in order to improve the co-operation and minimise the social costs of health and safety" (Ahasan 2001: 315). In addition, some developing countries are characterised by political unrest and economic instability, so international agencies have to act with sensitivity and diplomacy in the processes of collaboration.

The proliferation of small and medium enterprises (SMEs) also makes the enforcement of OSH regulations problematic and in some cases these types of endeavours are the most in need of effective OSH processes because their development may be hindered by a lack of resources. Indeed, according to the European Agency for Safety and Health at Work (2004), workers for SMEs are at a higher risk of suffering an "accident" than other enterprises.

Employer complacency or inaction

OSH problems exist in developing countries at different stages of development, but the prevalence and effectiveness of preventive measures vary substantially within and among countries. Much needs to be done to improve OSH in developing countries and, as Gibb and Bust (2006: 1) acknowledge, it is a process that is particularly complex, being influenced by, among other things, "physical characteristics such as infrastructure, security, politics, and weather, and human elements such as language, literacy and skills." This is reiterated by the WHO, which identifies a number of obstacles to achieving effective OSH in developing countries, including "a lack of political will, legislation, inspection, education, training, information and even awareness of the importance of occupational health and of its positive impact on socioeconomic development" (WHO 1995: 32). According to Murie (2007: 5): "International health and safety standards to protect construction workers are already in effect but are often ignored by management."

Haupt and Smallwood (1999: 49), in a survey of construction workers in rural South Africa, found that 96 percent of workers had not received any OSH training while working on their current project; and 83 percent had not received any OSH training. Koehn and Reddy (1999: 44), in a discussion on the construction industry in India, claim that, "in general, few training programs exist and, therefore, no orientation for new staff or workers is conducted, hazards are not pointed out, and no safety meetings are held." For workers new to a project, a safety induction is essential; research shows that workers new to construction are at a greater risk of being injured than more experienced workers (Occupational Safety and Health Administration (US) 1993).

Ahasan (2001, 313) also argues that "It is also common knowledge that new entrepreneurs, and private-sector industries do not have the competence for, nor willingness to learn national law or regulations." Vilanilam (1980) argues that labour is cheap and there is an abundance of workers looking for employment, so employers are not overly concerned with OSH (see also Sandhir 2009 for a discussion on the shortage of construction workers in India, despite the high rates of economic growth and development occurring there).

Vilanilam (1980) asserts that, since the level of unemployment in developing nations is so high, workers are prepared to accept any job, irrespective of the dangers involved. Using India as an example, he draws attention to the general lack of concern for public health and poverty, and indifference on the part of employers and politicians to OSH. This situation is echoed by Koehn and Reddy (1999: 42): "high unemployment rates allow contractors to easily replace workers who are not performing at a satisfactory level. Therefore, workers often take risks which tend to increase the level of accidents and fatalities on the job site."

Lack of recorded incident statistics and evaluation/auditing processes

How is this issue a barrier to developing effective OSH processes and systems? Well, collecting, maintaining and analysing data are crucial in monitoring and evaluating the efficacy of any risk prevention and safety control systems that are in use, as well as for the development of best practice. However, incident reports are largely lacking in developing countries and there is a need for mechanisms to collate, evaluate and assess construction OSH data. Many cases of occupational disease are under-diagnosed and under-reported, yet preventive actions are not taken (WHO 1995: 29). In many countries, basic manual systems may exist but data often does not facilitate ease of analysis or act as a tool for positive change. Another inhibitor in failing to provide data is that they can be compared with those of other countries. Indeed, it may be considered undesirable to place OSH data in the public domain because of international tensions and the "politics" that exists between some countries (see Warternberg and Greenberg 1992; Burgess and Harrison 1993). OSH is basically a performance indicator that can be checked at any stage of the construction process, and it is relatively straightforward to introduce procedures to monitor incident data as a corollary of implemented OSH systems; the risk workers face in their work is typically measured by the number of injuries and fatalities that a group of workers have. Eiff (1999) makes a pertinent point when stating that a culture of reporting and collating incident data is the very foundation of a "safety culture".

Globalisation: a double-edged sword?

While growing trade relations between developed and developing countries present many prospects for rapidly industrialising nations and their labour force, the new "corporate" climate can also bring with it difficulties. An example is that developing countries need to reconcile this economic, market-driven growth with appropriate and effective health and safety solutions for the mass of workers who facilitate it. The consequences of globalisation have led to a new economic landscape for many developing countries and, as demand for domestic and commercial buildings, for example, grows, the construction industry is thriving. A different view is given by Madhok (2005: 28): "the situation of construction workers is worse today than before, as globalisation has brought with it the perils of mechanisation and loss of jobs. Although there has been substantial growth in the construction sector in recent years, this growth has actually led to a decline in the availability of jobs." This ties in with the notion of expendable jobs (and workers).

As a nation develops, its infrastructure needs to be improved, and usually this means that it requires modern and complex technology to construct, so introducing new challenges to the workforce. Safe use of this technology requires the workforce to be thoroughly and competently trained and supervised. The technology introduces new and potentially severe hazards, which have to be properly assessed. A good example is the project in Dar es Salaam, Tanzania shown in Figure 6.3, which is for a multi-storey, in-situ reinforced concrete framed building, in an area of similar buildings. The main means of transporting materials and components are a static concrete pump supplied by a batching plant and a tower crane.

Modern technology can also bring safer working. Machinery and equipment are now usually quality assured and "fit for purpose". Furthermore, the widespread use of digital electronics has led to the development of devices explicitly designed to aid safe working. A good example is a crane safe-load indicator, shown in Figures 6.4 and 6.5. This device is based on a number of load cells, which weigh the load as it is about to be lifted. All the information from the load cells comes to the driver on one device, giving clear and useful information.

Figure 6.3 Modern construction in Dar es Salaam Tanzania (photo by Richard Neale)

777 Multi-line load cell installation and placement guide for load cell links

Figure 'A' - Pin the **BLUE** load cell between the wire rope socket (becket) and the dead end on the main boom of the crane. **>MAIN<**
Figure 'B'- Pin the **YELLOW** load cell between the wire rope socket (becket) and the headache ball on the auxiliary sheave (rooster) on the cranes boom tip. **>AUX 1<** (Note: for two line load systems only)
Figure 'C'- Pin the **ORANGE** load cell between the wire rope socket (becket) and the headache ball on the jib extension. **>AUX 2<** (Note: for three line load systems only)
Figure 'D' - Location of external magnetic mount antenna on operator cab of crane.

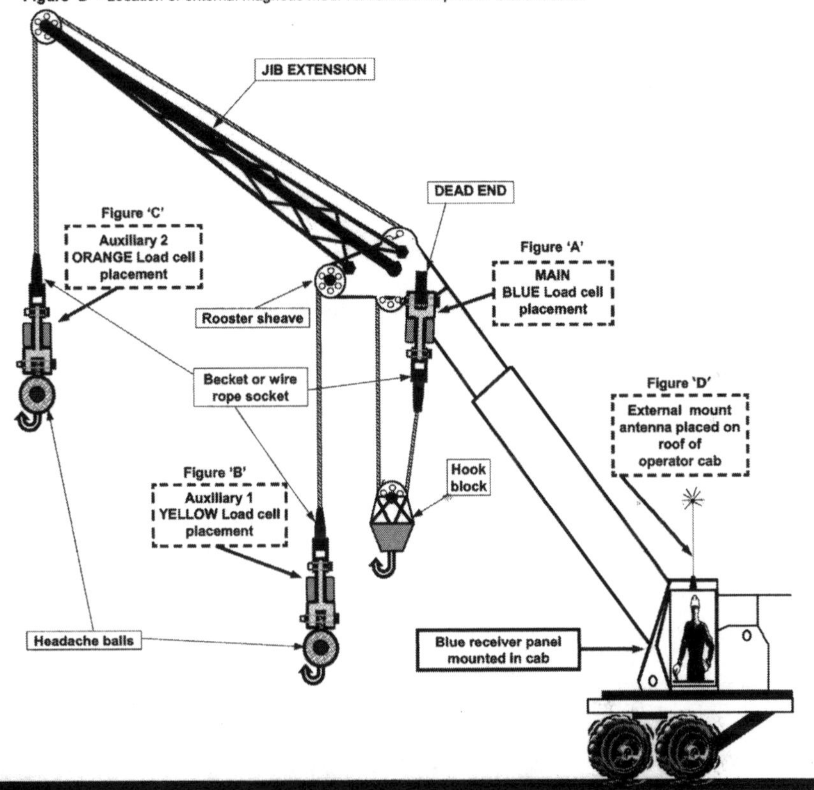

Figure 6.4 Crane load monitoring system

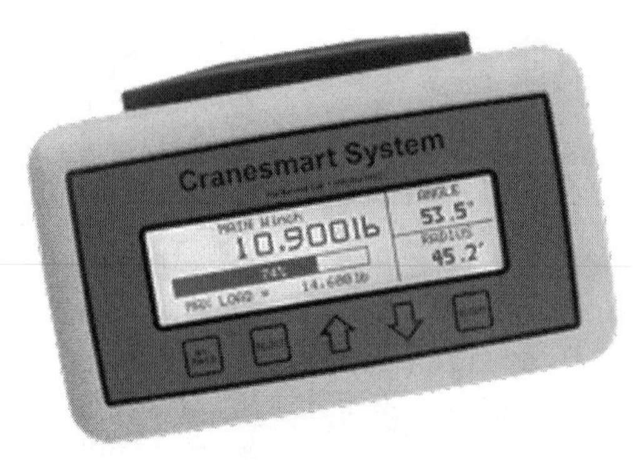

Figure 6.5 Driver's information from crane monitoring system

A condition for the safe use of modern technology is a good inspection, maintenance and repair capability. This emphasises, once more, that OSH has to be holistic and embedded within a competent project management framework and systems, rather than being treated as a number of separate and distinct issues.

A dispersed industry of dispersed responsibility?

Cheah (2007: 81) highlights aspects of the nature of construction work that can hamper effective OSH: "Firstly, the industry is highly fragmented, which marginalises efforts to safeguard safety and health (S&H) standards. Secondly, unlike manufacturing, construction site activities are physically dispersed across various locations; thus supervising and monitoring S&H issues in the workplace is much more challenging." Indeed, subcontracting means that responsibility for health and safety is often unclear and is operationally difficult, potentially hindering compliance with OSH processes and regulations. Koehn and Reddy (1999: 39) assert that "Construction projects ... are complex undertakings" and "The building of a structure involves many people, skills, materials, equipment, and literally hundreds of other operations." The number of workers on construction sites also presents difficulties in forging a sense of belonging as a result of the transient nature of their employment.

Construction activity necessarily involves rapid change and, as progress of work increases, the working environment changes hour by hour and day by day. Indeed, change can be a cause of unsafe behaviour (Coleman 2004). Figure 6.6 is an illustration of the wide range of those who may be involved

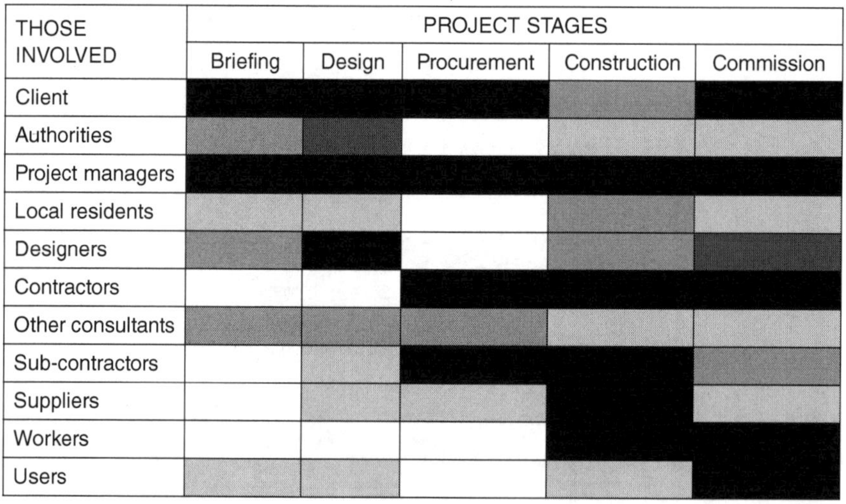

THOSE INVOLVED	PROJECT STAGES				
	Briefing	Design	Procurement	Construction	Commission
Client					
Authorities					
Project managers					
Local residents					
Designers					
Contractors					
Other consultants					
Sub-contractors					
Suppliers					
Workers					
Users					

Figure 6.6 Range of persons who may be involved in a construction project, whose safety and health must be safeguarded. The intensity of the shading is an indication of the intensity of the involvement (source: ILO 2010a)

in a construction project, whose safety and health must be safeguarded. Construction sites are complex environments; there is a diverse workforce and this can lead to a lack of ownership of responsibility for safety issues.

The construction industry has a diverse workforce

Diversity among workers and PPE issues

Diversity on construction sites is common, with, for example, workers of different genders, ages and physiques. There are also cultural issues to address: consider personal protection equipment (PPE), designed for male bodies and being unsuitable for women and men of smaller builds; and women feeling uncomfortable wearing masculine clothing. There are also issues around country-specific traditions and practices and attitudes to safety to consider. Gibbons and Hecker (1999: 374) found that foremen and superintendents generally order one-size-fits-all harnesses. Not only is this a gender issue but also one for different *types* of workers: for example, carpenters, who wear harnesses all day, reported that a poor fit caused them major discomfort. Adequate supply and use of PPE is a rarity in developing countries.

Many workers are involved in unsafe acts, due perhaps to flexible attitudes of management and workers, PPE often being inaccessible; and the challenge of working in hot climates exacerbates this. This is illustrated by a study of construction workers in South Africa by Haupt and Smallwood (1999: 51) which found that: "Only 4.3% of construction workers required and received any … PPE" and "Not a single worker was wearing a hard hat". The two findings are particularly significant given that research has shown that having access to free PPE, along with instruction in how to wear it properly and safely, are among the most important determinants of safety performance improvements (Sawacha et al. 1999). The importance of properly fitting PPE was also illustrated in a recent British survey by the Women's Engineering Society (2010), which found that only 8 percent of women wear PPE designed for women and over half of the respondents claimed that ill-fitting PPE hampered their work in some way.

Workers and cultural issues

Coble and Haupt (1999) discuss how cultural issues have an impact on construction activities. They argue that many safety and health practices that would be frowned upon in developed countries may actually be acceptable and considered the norm in developing ones. A good example is the use of bamboo scaffolding that is preferred in developing countries in Asia because of its strength and flexibility but may be considered inappropriate or unsafe in developed countries. A more contentious example is workers who prefer to work barefoot, which is not only normal in several developing countries, but is permitted and often considered to be good practice. However, in developed countries, this would not be permissible because it would be classed as an easily preventable risk; and the risk is considerable when one considers what damage could be done to an exposed foot on a construction site. It is a good example of how cultural issues need to be taken into account when formulating safety and health programmes in developing countries. Coble and Haupt (1999) conclude that workers' customs, practices and values need to be considered when developing a safety and health programme; it should be adapted, or at the very least be flexible to accommodate these characteristics. It is also important to consider that workforces are often diverse in terms of their ethnicity. For example, Malaysia, Singapore and many of the countries in the Middle East have construction workers from several Asian countries. Suplido-Westergaard (2006: 20) attests that 77 percent of Filipino migrants worked in Asia with over a third of them working in high-risk jobs like construction. Similarly, Koehn and Reddy (1999: 42) assert that "Communication problems related to differences in language, religion and culture tend to inhibit safety on the work site."

The informal workforce

One of the biggest challenges in ensuring that construction workers in developing countries receive adequate health and safety training is the large "informal workforce". The informal sector consists of often large numbers of men, women and children who are classed as casual workers with no employment contract. They often work in groups or families or informal enterprises but crucially they do not obtain regular salaried employment or other benefits. This poses significant problems for educating them in construction safety and health.

Women workers in the construction industry

According to SEWA (2000), of the female labour force in India more than 92 percent were in the informal sector and there were more than 500,000 women workers engaged in the construction industry in the state of Gujarat. Here, women construction workers would gather at different "nakas" (street corners) early in the morning in search of work, waiting to be picked up by the contractors. The majority of the women workers (69 percent) worked as loaders on construction sites, and 90 percent of the women were unskilled labourers. Some 70 percent of the women workers complained about chronic body aches after they joined the construction industry, and 51 percent of women workers reported that they had sustained physical injuries during on-site work, while the incidence of injuries was much lower in the case of male workers (13 percent).

Baruah (2010) identified the opportunities and constraints faced by female construction workers in urban India and argues for a "gender sensitive" approach to health and safety at work. Girija and Geetha (1989) undertook research on female construction workers in Tamil Nadu, India (see Madhok, 2005). They explored the type of work undertaken by women in construction and its intensity, and their findings included the observations that, in concreting, "In an 8-hour shift … an incredible 32,000kg would have passed through a woman worker's hands", while women carrying out earth work "carried on their head 15kg of mud and walked 30 feet to deposit the mud and return. In an hour this was repeated 180 times" (Madhock 2005: 11). It is important to consider, too, that in many developing countries there is a culture whereby children work long hours in often physically exhausting roles. Estimates suggest that only one in five working children is in paid employment, with the majority being unpaid family workers.

Worker's attitudes to construction OSH

Vilanilam (1980) asserts that, owing to a lack of education, many workers in the developing world are unaware of the hazards of their occupations. Haupt and Smallwood (1999) carried out a study of a rural community low-cost housing project in South Africa and found that the educational level of workers would affect their amenability to formal health and safety training and education and this barrier could be overcome by varying the training methods to include visual aids, role-playing and demonstrations. A good example is the Construction OSH training package (ILO 2010a). Another issue is that construction is known to be a risky job, and some workers believe "accidents" to be part of the job. This was borne out by Enshassi et al. (2008: 51) who focused on the Palestinian construction industry. Their study showed that hazards were considered a normal part and consequence of working in the construction industry, and contractors seemed to be unaware of their legal obligations to promote the safety of workers. Koehn and Reddy (1999: 242) claim that workers consider "accidents" a product of their negligence and place no onus on managers. Much needs to be done to change such attitudes.

The appropriateness of transferring models of OSH from developed nations to developing countries

It is a worthwhile exercise to reflect on whether it is appropriate or feasible for developing countries to "import" the OSH processes, procedures, legislation and regulation used in developed nations. Developed countries have made great strides in the prevention and control of occupational hazards, and various models of risk analysis and management exist. Such countries also show how comprehensive legislation, managerial commitment and good systems can reduce risks, but are they appropriate to be transferred to developing countries?

Careful consideration must be given to recognising that developing countries are diverse and vary considerably in terms of their traditions, customs and cultures and represent a wide spectrum of economic, political and social development. Essentially, they go through a transition period as they become accustomed to the products of globalisation, rapid industrialisation, urbanisation and economic development. Moreover, while developing countries can *learn* from OSH legislation and systems that are successful in developed nations, governments and legislators must ensure that they develop appropriate country- and culturally-specific occupational health processes for their own needs. In line with the previous discussion on OSH paradigms, it would seem that "importing" safety and health models from the developed world is unsuitable because, as the UK example above showed, they are traditionally focused on the *internal* domain, i.e., the workplace, the workers, and risks found in the work environment.

Policy development in developing countries also needs to consider that informal workers form a large portion of the workforce. These "invisible" workers will have no contracts of employment, additional benefits and, crucially, will have only limited access to OSH services or trade unions (see Levy 1996; De Alwis 2003).

Kheni et al. (2008) explored construction health and safety management in SMEs in Ghana and found that: "In most developing countries such as Ghana, the construction industry relies on procedures, regulations and practices that are legacies of colonial administration. Procedures have not been updated to reflect the social, cultural and economic milieus of developing countries" (Kheni et al. 2008: 1166). Therefore, health and safety systems which reconcile what are new industrial, economic, political, social and urban landscapes with the individual country's cultural character should be developed and implemented.

Improving construction OSH in developing countries – recommendations and good practice

Nourishing a "culture of safety" and a "safety culture" in developing countries

It is necessary for a "cultural (and attitudinal) shift" to occur that recognises the importance of protecting workers' safety and health at work at government, employer and employee level. Culturally, workers in developing countries may not perceive risks in the same way as their counterparts do in developed nations – a natural consequence of a lack of a "safety consciousness" among workers and the general population. What is needed as a matter of urgency is an integration of OSH principles and requirements as key elements in national and international priorities. This means sustaining mechanisms for the continuous and committed improvement of safety and health systems that foster a "safety culture" at an organisational level and become an integral part of wider societal culture and economic development as a "safety consciousness". To achieve this, commitment, time and investment are required.

Understanding how construction safety and health in developing countries is shaped by the countries' own cultural, political, economic and social landscapes is an essential first step, and this has to underpin any legislation and standards along with a "fundamental change in the *attitude* (emphasis added) toward the day-to-day exposure to risk" (Bayer 2000: 532). An OSH system is explained, below.

Organisational "safety cultures" present problems in the construction industry because of the transient and physical nature of construction, with a host of different workers constantly moving from job to job, changing location and often doing dangerous work. However, if a government advocates a robust "national culture of safety" then this can trickle down to individuals and the formation of "safety consciousnesses".

Coleman (2004) discusses the importance of a "safety culture" at an organisational level. This should be promoted on construction sites as a matter of routine. Ensuring safety on a construction site is a complex challenge: as Coleman (2004) argues: "The development of a safety culture must start at an early age – Pre-school". The basic premise is reflective of the need for occupational health and safety to be taught as a natural mindset that can be applied when people enter the world of work, reinforced by legislation, policies and industry standards, as well as in workplaces by employers. This "national safety culture" would ensure that safety would be considered from the outset of a construction project and not be factored into it during the construction stage, especially important given that 60 percent of fatal "accidents" are attributable to decisions and choices made before work begins (Coleman 2004). Different developing countries have their own traditions and cultural and social norms. Thus, in developing a "safety culture", it is a delicate process to reconcile people's cultural characteristics, values and traditions with safety and health systems and processes.

For a "safety culture" to exist, an holistic approach is needed which encompasses a proactive government attitude to general occupational safety and health across all types of worker and construction job type, so that it filters from the top down. It requires concerted mobilisation of all available means to raise awareness, through education and training, to increase knowledge and understanding of the concepts of occupational hazards and risks and how they may be prevented and controlled.

Tam and Chan (1999: 118) provide an example of how a "safety culture" can be nourished in the construction industry of Hong Kong, which historically has operated an "enforcement" approach to occupational safety and health legislation. The Hong Kong government initiated safety improvement programmes with the main aim of improving construction safety performance by changing the "safety culture" of the industry. These included making it mandatory for contractors to hire safety officers and supervisors on site. This was the first step in encouraging the formation of a self-regulatory framework in safety management, and these efforts reduced the number of construction "accidents" in Hong Kong (Tam and Chan 1999: 121).

International agreements, national policies and national laws

Rantanen et al. (2004) suggest that collaboration with international organisations is a key factor in determining the success and longevity of OSH processes in developing countries. This could help developing nations to develop effective construction OSH at both practical and cultural levels.

Although the main theme of this chapter is that effective OSH should be founded on principles of humanity and the need to establish strong OSH attitudes, laws, codes of practice and project-based agreements provide an important foundation and framework for application. The starting point for many developing countries is the legislative framework that was left behind by previous colonial powers. The governments of these countries have subsequently customised this as they have developed their individual nationhood.

In parallel with national development, the international system offers some good advice and recommendations, an example being the ILO's Convention C167 Safety and Health in Construction Convention (ILO 1988). Although it requires updating to take into account quite significant changes during the past 22 years, it remains a comprehensive and useful compendium of advice that provides a good basis for national legislation and codes of practice.

Another basis of legal OSH requirements lies in the contractual requirements of international agencies such as the World Bank that are under increasing pressure to insist on good-quality OSH provisions in their contracts and procurement procedures. This can raise the standard of OSH on individual projects and has the potential to enhance standards in the industry more generally.

Wells and Hawkins (2009: 2) suggest that "many measures are needed to improve OSH, including an appropriate legal framework, an effective inspectorate, training of workers and supervisors, restrictions on working hours and wide availability of occupational health services." They argue that the provision for effective OSH can be significantly enhanced through construction project contracts: "procurement procedures and contract documents have the potential to act as important mechanisms to remind the parties to the contract of their obligations under the law" (Wells and Hawkins, 2009: 2). Therefore the client must ensure that tendering bids make adequate provision for the cost of safety and health measures during the construction process in order to minimise risks.

As an example, the structure of international legal requirements for OSH for member states of the ILO is shown in outline in Figure 6.7.

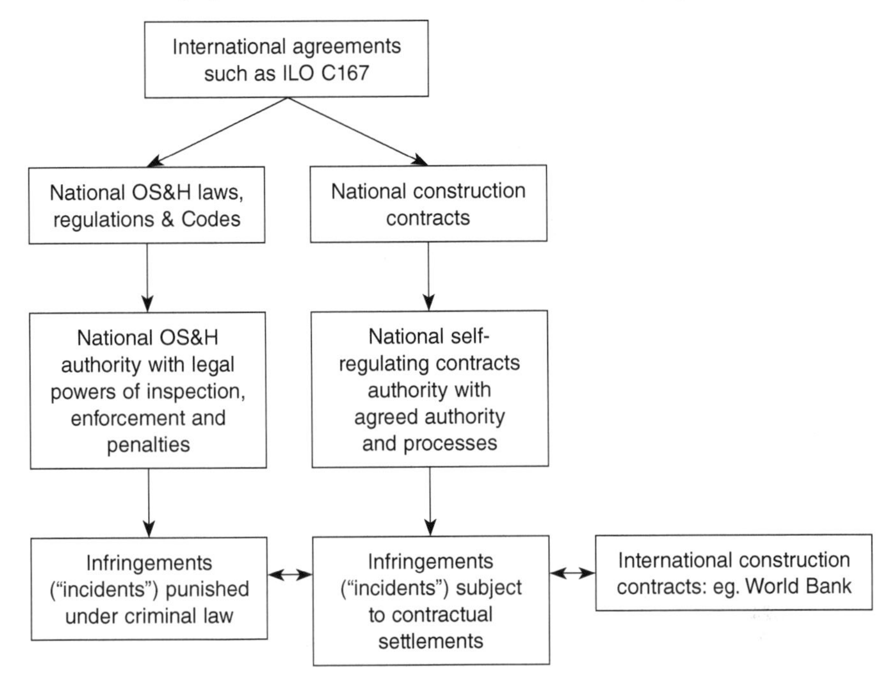

Figure 6.7 An international and national OS&H legal system

Systematic analysis of hazards and risk

The basis of hazard and risk analysis

International and national safety and health organisations adopt a consistent approach to safety management, based on the principles of assessing and managing hazards and risks. A good example from the ILO (2001a) is given below.

> A *hazard* is the inherent potential to cause injury or damage to people's health.
> *Hazard assessment* is a systematic evaluation of hazards.
> *Risk* is a combination of the likelihood of an occurrence of a hazardous event and the severity of injury or damage to the health of people caused by this event.
> *Risk assessment* is the process of evaluating the risks to safety and health arising from hazards at work.

These definitions provide the basis for a systematic approach to OSH in organisations.

Table 6.1 gives an overview of the likely hazards that may be commonly encountered on construction projects. This provides examples of risks, but

Table 6.1 Examples of hazards commonly encountered in construction work

Category of hazard	*Examples*
Hazards that may affect the project due to its location	Weather, flooding, active utilities, difficult access, aggressive neighbours
Hazards that may affect the location due to the project	Pollution from site activities, danger to public from site activities and traffic
Hazards that may be caused by project briefing and design (by actions or negligence)	OS&H not considered at the outset, client and designers only consider end result not process of construction
Hazards that may be caused by project management and organisation (by actions or negligence)	Lack of senior management awareness and commitment, failure to implement diligent OS&H practices
Hazards inherent in construction methods ("active hazards")	Safety of workers not considered in the method statement, unsafe equipment
Hazards inherent in construction components and materials ("embedded hazards")	Materials contain injurious chemicals, components heavy or require excessive force or special techniques
Hazards that may be caused by human behaviour	OS&H not taken seriously by managers, financial pressures on workers and supervisors
Wholly unpredictable or "latent" hazards. (Only such hazards cause "accidents", all others cause preventable incidents)	Chemical or structural defects which were quite unknown at the briefing or design stage

each construction site will have its own unique set of hazards and so must be assessed specifically. A European Directive provides good guidance on risk assessment (see European Council 1989).

The systematic OSH process: elements and linkages

A detailed and systematic process is required to put the principles of risk assessment into practice. Figure 6.8 brings together the main requirements.

The Method Statement is of crucial importance to this process. This should comprise, as a minimum requirement, a clear, fully documented and agreed statement of the way in which a specific construction element shall be built.

The difficulty of managing OSH holistically and comprehensively is illustrated by an example from a construction project in Dar es Salaam, Tanzania. All parties to this project were positive about good OSH and had established policies, systems and facilities that were, generally, exemplary. A good example is the safety supervisor, a vigorous and enthusiastic young man, pictured in Figure 6.9, and the facilities provided for him. During a tour of the site by one of the present authors, the general standard was shown to be very high, nevertheless it was still possible to find the poor conditions in the workplace shown in Figures 6.10 and 6.11. This is a good

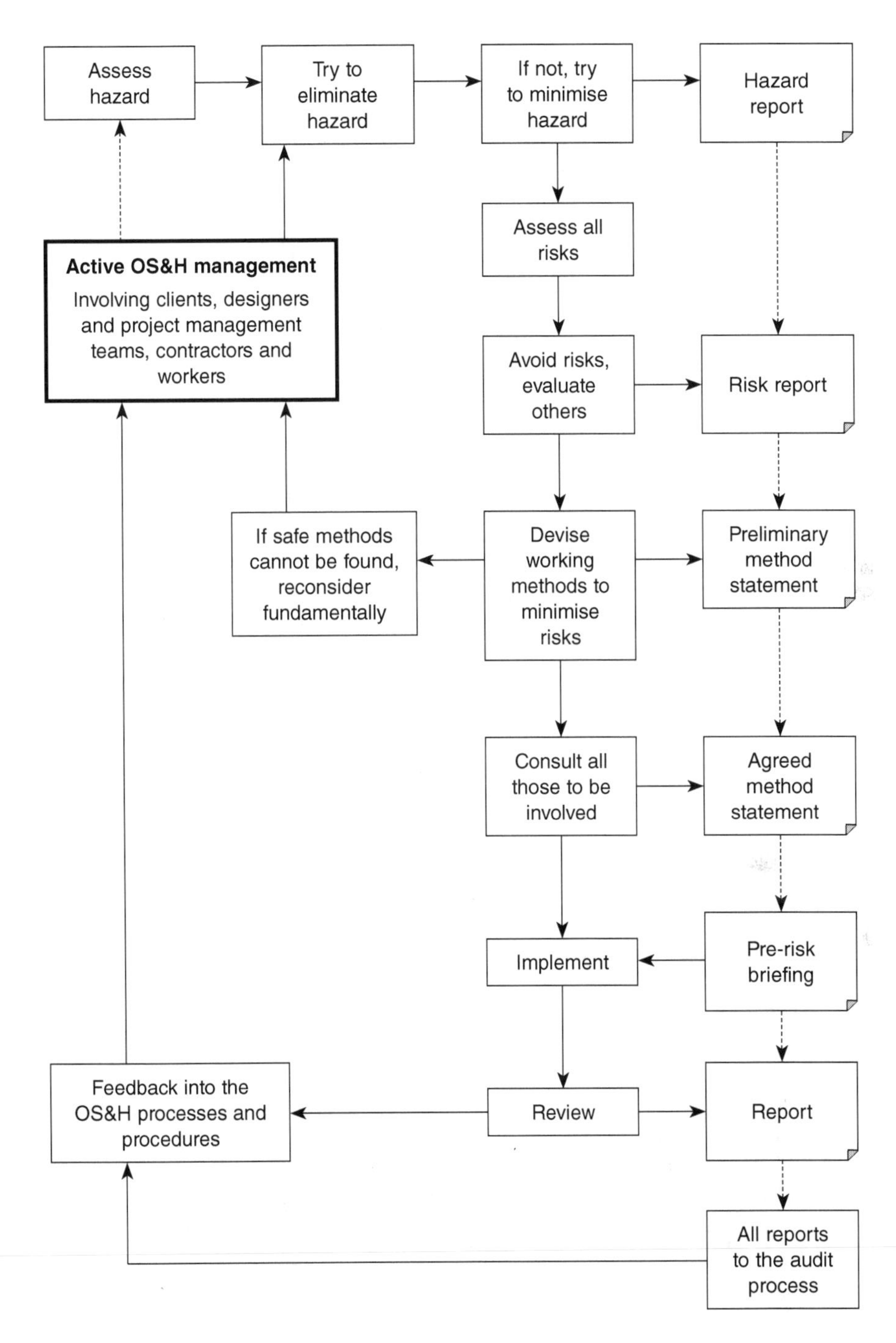

Figure 6.8 Systematic OS&H process: elements and linkages

Figure 6.9 Safety representative and facility on site in Dar es Salaam, Tanzania (photo by Richard Neale)

Figure 6.10 Unsafe practice on the site shown in Figure 6.9 (photo by Richard Neale)

Figure 6.11 Detail of Figure 6.10 (photo by Richard Neale)

example of the need for an overall organisational "safety culture", to which everyone, including the workforce, is committed.

Effective communication

As has been discussed above, communication is essential at an organisational level to develop a "safety culture"; there needs to be communication with, and among, clients, contractors, designers and workers, and information, training and awareness-raising for safety to be successful (MacKenzie et al. 1999: 423). Indeed, Ahmed et al. (1999: 528) found that one of the most serious safety problems on construction sites was inefficient communications between different "actors" due to multi-level subcontracting. It is crucial, too, that information is communicated to workers about their rights to gain access to, and join, a trade union (see http://www.hazards.org for a discussion on "the union safety effect"). Communication among government, legislators, safety inspectors and representatives, and construction companies is also essential.

Worker participation

Workers must be empowered to contribute their values and practical knowledge and experiences of construction site safety to a "safety culture". Workers should also be consulted regarding the development of OSH processes, since they are more familiar with the working environment than some managers.

Peckitt et al. (2004) compared construction in the Caribbean with that in the UK. The cultures are found to be quite different; the Caribbean culture was found to be more relaxed, with more emphasis on general quality of life. The following quotations capture the essence of the paper and provide an example of the contribution that workers can make:

> Caribbean contractors did not formulate or implement formal safety management systems and paid little regard to health and safety regulations …

but

> Workers have a strong sense of locus of control over safety, generally trusted management and looked out for each other.

There is a useful body of literature on full worker participation. An example is a study conducted by Page (2002) in Australia that provides a good review of current practices, makes suggestions for active worker participation, and also provides detailed guidance. The model advocated is based on the principle of "co-regulation", which refers to a shared involvement in risk control decision-making between employers and employees with the aim of ensuring workers' safety and health.

Full worker participation is also advocated in a paper by Soehod (2008: 15), in the abstract of which it is noted:

> … safety practitioners and observers have widely agreed that the traditional belief that employers are solely responsible for the workers' safety at work should have a new paradigm. To create a safe working condition, workers should be allowed to participate actively in safety and health matters and cooperate with the employers. … Evidence showed that various benefits could be yielded if workers work together with employers including the reduction of death and injury rates at work. However, to make workers' participation in this field effective, several criteria are crucial. They are the legal support, management support, trade union support, training and the positive quality of the workers involved.

A "safety culture" in which everyone has a part to play

Who should be responsible for worker safety and health on construction projects? Historically, this was solely the contractor's ultimate responsibility; now, other key "actors" have been required to play their role in construction OSH. For example, it has been recognised that designers can play a key role in improving construction site safety through reflecting on how their design decisions determine which construction methods and materials will be used. Each of these methods and materials will involve specific risks. MacKenzie et al. (1999: 419) argue that "Poor investment of management practices, particularly within the design phase, have been identified as a prime cause of the unacceptable 'accident' rate within the construction industry". Wells and Hawkins (2009) assert that risks could be "designed out" at the planning and design stage of a project by carrying out a thorough risk assessment, which allows potential dangers to be identified at a stage when alternative choices can be made with minimal impact on costs, both human and financial. Thus, designers need to be taught to recognise how their designs have an impact on worker safety. This can be followed by the development of an OSH plan that serves "to mitigate the risks with clear allocation of responsibilities" (Wells and Hawkins 2009: 4).

Many designers will find taking into account all the safety and health implications of their designs cumbersome, but education is crucial (Hinze et al. 1999: 395), as is undertaking suitable research, and providing guidance on "designing for safety" (Gambatese et al., 2005). Indeed, in time, it should become an inherent, routine part of a designer's role.

The need for enhanced and effective education and training in OSH

In this chapter, OSH has been set within a broad framework of construction project management, because it should be considered as an integral part of the processes and procedures required to manage construction works successfully.

Rantanen (1997) and Ahasan (2000a, 2000b) emphasised that training and skill development programmes should be launched to provide basic knowledge to both the workers and managers who may contribute much to the improvement of working conditions. To meet broader needs, such as persons working in rural areas, distance learning courses are to be commended, provided the prospective trainees have access to appropriate facilities. Cost is often a factor that influences whether or not a workplace adopts effective interventions to address occupational hazards. Training of supervisors and workers may be difficult because of impediments such as educational proficiency and language barriers, and the applicability of training materials to local contexts. Training is central, and it is needed at all levels: by clients, designers, contractors, sub-contractors, technical experts, as well as employers, workers, and representatives from the Labour and Health Departments. This can be assisted by guidelines and technical assistance from international organisations (an example is given below).

However, OSH can be a dense, legalistic and technically detailed subject to teach. The challenge for teachers and trainers is to bring this subject alive, to make it interesting, to give the learners something that they consider to be relevant, and to motivate them to take care of themselves and to look after others. This requires the adoption of good teaching and learning methodologies, relevant teaching materials and active participation in real projects. One such approach to teaching and learning was given by Neale and Murie (2009), which is founded on the well-known aphorism of Confucius:

> I hear and I forget
> I see and I remember
> I do and I understand

Modern digital technologies provide excellent aids to achieving these basic educational requirements. Educators in OSH can acquire, relatively cheaply, a digital camera which is easy to use and can produce good-quality still images and short video films. Construction work is often all around us and can provide useful images that can be used to bring reality and interest to a topic. Construction companies are usually willing to support educators, providing that their own OSH can be assured. Furthermore, the internet is a huge source of information and educational aids. The best way to engage students and trainees in practical project work is by using actual, live projects. This is usually quite straightforward in the context of "safety by design", because designers are usually helpful and their offices are usually safe and accessible.

A free, comprehensive, international, digital training package in occupational safety and health for the construction industry

With the aim of providing support for teachers and trainers, a major educational resource has recently been offered free of charge by the ILO (2010a). The material has been structured into 15 subjects, or "themes", as shown in Table 6.2. Each theme provides the tutor with an illustrated body of information, from which training course materials can be selected and edited. "Theme PowerPoint Presentations" are provided, and there are more than 800 PowerPoint slides to assist in teaching them. There is also guidance on teaching the subject and a substantial body of background information.

Towards zero incidents

A pertinent development in construction theory and practice is that "the avoidance of accidents and failures in the construction industries worldwide has emphasized managerial forces rather than the action of an individual" (Atkinson 1999: 325). However, some construction companies are moving to adopt a strategy of "zero incidents" as a key objective in their work environment. Hinze (2002) offers ten key strategies and techniques that can assist in achieving zero incidents.

Seeking to have "zero incidents" is a contentious issue; humans are not infallible and, while aiming to achieve "zero incidents" is entirely commendable, it does fail to consider that genuinely unforeseeable accidents do occur and that latent or active human error may play a role. However, Watanabe and Hanayasu (1999: 55–6) showed that, in Japan, "zero incidents" could be attained through construction safety management. It was also found that, in order to have "zero incidents" as a site aim, all "actors", including sub-contractors, workers, clients, designers, planners, project managers and prime contractors, would need to develop their own safety management capabilities. Therefore, flexibility and adaptability are crucial.

Table 6.2 Subjects or "themes" in the ILO's construction OS&H training package

Fundamental themes
Fundamental principles General duties Safe and healthy working environment Workers' perspectives

Project management themes
Principles of safe project management Project planning and control for OS&H Processes and systems Welfare and project site

Technical themes
Personal protective clothing & equipment (PPE) General plant and equipment Vertical movement Horizontal movement Working at or below ground level Working at height

Integration and conclusion
Project, concluding case study, evaluation

Within an organisation, achieving "zero incidents" is reflective of a total safety culture. However, its achievability is difficult even in developed nations. For example, in the UK, there were 41 construction-related fatal injuries in 2009–10, a rate of two deaths per 100,000 workers; and there were 3286 major injuries in 2008–9, a rate of 254.1 per 100,000 employees (http://www.hse.gov.uk). To reiterate a statement made above, "importing" OSH from the developed world is unsatisfactory.

Case studies of construction OSH systems in developing countries

Case study one: Ghana

Kheni et al. (2008) discuss construction health and safety management in developing countries by focusing on SMEs in Ghana. The sector is one of the most hazardous, with frequent accidents and health-related problems (Kheni et al. 2008: 1159). According to Kheni et al. (2008: 1164), "Annual turnover is positively associated with five health and safety practices adopted by the SMEs, namely: accident investigation procedures, accident reporting procedures, use of health and safety posters, documentation of method statements, and health and safety inductions." OSH regulation in the construction industry in Ghana is still based heavily on practices from its colonial past, and procedures and processes have not been updated to reflect the social, political, cultural and economic changes that the country has encountered. Wells (1986) has questioned the appropriateness of continuing to use safety and health procedures and regulations that are rooted in the past. It is clear that they do not reflect the social and political changes within the country over recent decades, including new levels of industrialisation and urbanisation, as well as corresponding social and cultural change.

Case study two: Bhutan

Dorj and Hadikusumo (2006) explored safety management practices in the Bhutanese construction industry and carried out a survey at 40 construction sites and government departments responsible for the construction industry to identify management practices and perceptions. They conclude that many problems exist, including safety not being perceived as a priority and few safety regulations and standards in place at government level. From a construction site perspective, safety is clearly not considered a high priority; incident data are rarely recorded and there is minimal safety training or application of safety management systems. OSH in Bhutan's construction industry is at a very basic level and it highlights how passing legislation for safety cannot reduce "accidents" or incidents alone; all construction "actors" need to be proactive, and embrace safety as something that is of benefit to them, their colleagues, the construction company, and also to wider society. Dorj and Hadikusumo (2006: 54) sum it up well as follows: Bhutan "lacks the three 'E's – engineering, education and enforcement – of safety".

Case study three: Palestine

Enshassi et al. (2008) studied the safety performance of subcontractors in the Palestinian construction industry. They identified, evaluated and ranked factors that influence the safety performance of subcontractors working in the Gaza Strip and found a correlation between the use of dated and unsafe equipment and reported "accident" rates. In Palestine, safe working practices were not being adopted for a number of reasons: "employers and employees are unwilling to spend or invest in safety measures, equipment or practices; hazards are considered a necessary part and consequence of construction; employees cannot afford to purchase their own safety equipment (such as safety shoes) and fear they might be penalised if they request PPE from their employers. Additionally, contractors are not aware of their legal responsibilities in relation to health and safety issues" (Enshassi et al. 2008: 52).

Case study four: Malaysia

Hamid et al. (2008) explore the causes of accidents in the Malaysian construction industry and argue that a rapid and dramatic overhaul of safety practices and processes is needed. The fatality rate in the Malaysian construction industry increased by 58.3 percent, from 60 cases in 1995 to 95 cases in 2003 (Hamid et al. 2008: 243). The findings of this study also revealed that the critical factors that led to "accidents" included: "unsafe methods, human element, unsafe equipment, job site conditions, management, and the unique nature of the industry" (Hamid et al. 2008: 242). The construction industry is described as high-risk and contingent on four key factors: time, cost, quality and safety. In Malaysia, it appears there has been greater emphasis on the first three aspects of a construction project at the expense of the fourth, safety, with grave human and economic consequences. The authors suggest that in Malaysia the emphasis appears to be on profit over safety (Hamid et al. 2008: 257). The lack of adherence to safety requirements has led to increased risks to workers, resulting in a high chance of accidents and, worse, fatalities.

Case study five: South Africa

Construction plays a vital role in South Africa's economic and social development. However, Haupt and Smallwood (1999) found a construction industry beset with problems. They studied health and safety practices on community projects in South Africa and found, for example, that safety inductions and/or staff training are rarely given to workers; there is little in the way of safety policies to provide a supporting framework for workers; PPE is seldom provided; and inspections are not conducted. The rural nature of the location of the study does bias the findings in that such areas are likely to be far removed from more government-centric activities in urban areas. However, South Africa offers inspiration in the case of designers who have to consider any safety risks their designs could cause – a role and responsibility that designers have undertaken there for some years.

Case study six: India

Koehn and Reddy (1999) discuss safety issues and labour requirements of building construction in India, "a country with 94 million people involved in the construction industry" (Koehn and Reddy 1999: 40). The government has recognised the importance of market-oriented approaches to economic and social development and has seized on the role construction can play in boosting its economy. Construction workers in India are generally uneducated, unskilled and migrate in groups, and are prepared to work in risky environments for poor wages, so desperate are they for money to support themselves and their families. Safety training is a rarity, new workers are unlikely to have a safety induction, and no hazards are pointed out to them. It seems a case of "just get on with it". Many workers consider "accidents" to be down to negligence on *their* part; they accept that they work in a dangerous environment and the norm is that contractors pay the family of a person who died at work around 3 months wages. As Koehn and Reddy (1999: 44) point out, this is not a strong monetary incentive for contractors to provide funds to develop programmes in order to increase safety on construction sites.

Conclusion

In this chapter, some of the barriers that developing countries face when trying to establish effective and sustainable OSH systems and management practices have been discussed. One of the most significant inhibitors is the absence of organisational "safety cultures", along with "a culture of safety" that develops when the state and policy-makers recognise OSH as a priority issue which warrants investments and commitment, and so filters down to employers and workers. As in the UK, this would lead to more and more people, including workers, having an occupational "safety consciousness" that eventually becomes a natural mindset.

Focusing on the macro and external domain means there should be consideration of how a country's cultural, social, economic and political landscape have historically shaped attitudes to, and perceptions of, OSH, and more specifically construction OSH, and how these might be changed to reduce risks on construction sites. In many developing countries, OSH is not a priority issue. However, the risks that result from this are not acceptable. This state of affairs leads to a humanitarian paradox: economic development and rapid industrialisation have brought many opportunities to developing countries, particularly in construction, but not only are the interests of companies and workers compromised when one considers the high cost they face in terms of risk and the level of injuries and fatalities that occur, but also there is an adverse impact on the wider society that benefits from the opportunities development brings.

It is not possible to offer a prescriptive guide to OSH in developing countries, but an holistic approach which encompasses the suggestions above is recommended. The governments of developing countries can play a crucial role in overcoming barriers to an OSH culture and to reducing construction-related injuries, illnesses and fatalities, and they must play a positive and active role. Leaders should send strong messages that OSH is a crucial element in national development. This can be achieved by legislating, publicising and reiterating the importance of workplace OSH.

Organisational "safety mind-sets" will flourish if iterated by influential people and reinforced in the workplace. The change will also be seen among wider populations.

There should be a broad social and cultural approach to safety management that evolves around country-specific values, attitudes, norms and behaviours. This must be approached carefully and diplomatically and supported by education, training and awareness-raising and the commitment and efforts of the international community, such as the ILO and the WHO. In this way, OSH will eventually become a behavioural and attitudinal norm, echoing a general "national culture of safety" that values human life.

Recognising and embracing the context is important. Legal structures, practices and procedures of OSH used in the "developed countries" may not easily be transferable to the "developing countries". Moreover, there are examples of good practice on construction sites in developing countries. Therefore, developing effective OSH processes and a "safety consciousness" in this group of countries is not unachievable. Indeed, as the example in Dar es Salaam, Tanzania showed, an effective OSH culture is not only a realistic goal, but also one that can be attained provided it is applied and embraced comprehensively.

Finally, by way of recapitulation, the chapter closes with a list of ten important contemporary issues that have been discussed and some recommendations for research and development in future.

1. Occupational safety and health should be anchored within the context of national economic and social development;
2. Comparative studies of the OSH environment and practices in a range of developing countries would be useful in identifying the importance and relevance of such possible key factors as culture, climate and differences between urban and rural environments;
3. A comprehensive, generalised model on the business case for OSH should be developed;
4. Workers should be more directly involved in planning and implementing safe and decent work. There should be some social research on how this may be achieved;
5. There should be further development of OSH clauses in international agreements and national laws;
6. More research and development is required so that "safe by design" can become underpinned by a comprehensive design methodology;
7. Better (iterative) analytical models of hazard and risk methods should be developed, based on proper design and/or problem-solving theory, to replace the current linear models;
8. Further research is required into the "diversity of the international workforce", including women, men of various physiques, attitudes, cultures, and impact of HIV/AIDS so that OSH principles and practices which respond to such diversity can be developed;

9 The technology to improve OSH, including better control and warning systems, communication devices and better (safer) machines, should be further developed;
10 Effective education and training in OSH is required globally; it should be designed in such a way as to measurably enhance attitudes, skills and knowledge.

References

Ahasan, M.R. (2000a) Occupational health and hygiene in Bangladesh. In proceedings of the 3rd Nordic Health Promotion Research Conference: Outcomes in health promotion: Key questions for research and policy. 6–9 September 2000, Tampere, Finland.

Ahasan, M.R. (2000b) Global corporate policy and financing health services in the developing countries – crisis for structural adjustment. In Sohn, (ed.) *The 5th International Congress on Physiological Anthropology: Human space time-environment.* 1–5 October 2000, Seoul, Korea: 79–88.

Ahasan, M.R. (2001) Legacy of implementing industrial health and safety in developing countries. *Journal of Physiological Anthropology and Applied Human Sciences*, 20(6): 311–19.

Ahasan, M.R. and Partanen, T. (2001) Occupational health and safety in the least developed countries – a simple case of neglect. *Journal of Epidemiology*, 11: 74–80.

Ahmed, S.M., Tang, S.L. and Poon, T.K. (1999) Problems of implementing safety programmes on construction sites and some possible solutions (Hong Kong experience). In Singh, A., Hinze, J. and Coble, R.J. (eds) *Implementation of Safety and Health on Construction Sites.* Proceedings of the Second International Conference of CIB Working Commission W99, 24–27 March 1999, Honolulu, Hawaii: 525–9.

Atkinson, A.R. (1999) Implications for construction safety of studies of the management of human error. In Singh, A., Hinze, J. and Coble, R.J. (eds) *Implementation of Safety and Health on Construction Sites.* Proceedings of the Second International Conference of CIB Working Commission W99, 24–27 March 1999, Honolulu, Hawaii: 325–32.

Australian Government (2007) *Guidance on Preparing a Simple OHS Business Case.* Canberra: Australian Safety and Compensation Council.

Baruah, B. (2010) Gender and globalization: Opportunities and constraints faced by women in the construction industry in India. *Labor Studies Journal*, 35(2): 198–221.

Bayer, R. (2000) Editor's note: Whither occupational health and safety? *American Journal of Public Health*, 90: 532.

Burgess, J. and Harrison, C.M. (1993) The circulation of claims in the cultural politics of environmental change. In Hansen, A. (ed.) *The Mass Media and Environmental Issues.* Leicester: Leicester University Press; 198–221.

Chartered Institute of Building (2009) *Health and Safety in the Construction.* Ascot, UK: Chartered Institute of Building.

Cheah, C.Y.J. (2007) Construction Safety and Health Factors at the Industry Level: The Case of Singapore. *Journal of Construction in Developing Countries*, 12(2): 81–99.

Coble, R.J. and Haupt, T.C. (1999) Construction safety in developing countries. In Singh A., Hinze J. and Coble R.J. (eds) *Implementation of Safety and Health on*

Construction Sites. Proceedings of the Second International Conference of CIB Working Commission W99, 24–27 March 1999, Honolulu, Hawaii: 903–90.

Coleman, M. (2004) Health and Safety at Work in the Construction Industry. Presentation given at the 40th Meeting of the European Council of Civil Engineers. 1–2 October, Zagreb, Croatia.

Dainty, A.R.J., Ison, S.G. and Root, D.S. (2004) Bridging the skills gap: a regionally driven strategy for resolving the construction labour market crisis. *Engineering, Construction and Architectural Management*, 11(4): 275–83.

De Alwis, R.D. (2003) An approach to providing occupational health services to the informal sector. *Asian-Pacific Newsletter on Occupational Health Safety*, 10: 14–16.

Dorj, K. and Hadikusumo, B.H.W. (2006) Safety management practices in the Bhutanese construction industry. *Journal of Construction in Developing Countries*, 11(2): 53–75.

Eiff G. (1999) Organizational safety culture. Proceedings of the Tenth International Symposium on Aviation Psychology, Columbus, Ohio, Department of Aviation: 1–14.

Enshassi, A., Choudhry, R.M., Mayer, P.E. and Shoman, Y. (2008) Safety performance of subcontractors in the Palestinian construction industry. *Journal of Construction in Developing Countries*, 13(1) 51–62.

European Agency for Safety and Health at Work (2010) *Benefits of good OSH in construction and infrastructure procurement*. Available online at: http://osha.europa.eu/en/topics/business-old/performance/benefits_construction_html

European Agency for Safety and Health at Work (2004) *Construction accident rates continue to fall but remain unacceptably high, according to Eurostat data published by the Agency*. Press release. Available online at: http://osha.europa.eu/en/press/press-releases/041021_Construction_magazine/

European Council (1989) European Council Directive 89/391/EEC of 12 June 1989 on the introduction of measures to encourage improvements in the safety and health of workers at work. http://osha.europa.eu/en/data/legislation/1/

Gadd, S. (2002) Safety Culture: A review of the literature. HSL/2002/25Health and Safety Laboratory. Crown Copyright 2002. Available online at: http://www.hse.gov.uk/research/hsl_pdf/2002/hsl02-25.pdf/

Gambatese, J.A., Behm, M. and Hinze, J.W. (2005) Viability of designing for construction worker safety. *Journal of Construction Engineering and Management*, 131(9): 1029–36.

Gibb, A. and Bust, P. (2006) *Construction Health and Safety in Developing Countries*. The European Construction Institute (ECI) in conjunction with Loughborough University. Loughborough: ECI.

Gibbons, B. and Hecker, S. (1999) Participatory approach to ergonomic risk reduction: Case study of body harnesses for concrete work. In Singh, A., Hinze, J. and Coble, R.J. (eds) *Implementation of Safety and Health on Construction Sites*. Proceedings of the Second International Conference of CIB Working Commission W99, 24–27 March 1999, Honolulu, Hawaii: 373–80.

Girija R. and Geetha R. (1989) Socio-Economic Conditions of Construction Workers in Tamil Nadu. Report submitted to the Indian Council of Social Science Research (ICSSR) (Mimeo)

Hamid, A.R.A., Majid, M.ZA. and Singh, B. (2008) Causes of accidents at construction sites. *Malaysian Journal of Civil Engineering*, 20(2): 242–59.

Haupt, T. and Smallwood, J. (1999) Health and safety practices on community projects. In Singh, A., Hinze, J. and Coble, R.J. (eds) *Implementation of Safety and Health on Construction Sites*. Proceedings of the Second International Conference of CIB Working Commission W99, 24–27 March 1999, Honolulu, Hawaii: 47–54.

Health and Safety Executive (HSE) (2005) A review of safety culture and safety climate literature for the development of the safety culture inspection toolkit. Research Report 367. Bristol: HSE Books.

Health and Safety Executive (HSE) (2010) *Fatal Injury Statistics*. http://www.hse.gov.uk/statistics/fatals.htm (accessed 25 August 2010).

Hinze, J.W. (2002) *Safety Plus: Making Zero Accidents a Reality*, Construction Industry Institute Research Report. The University of Texas at Austin, Texas, March 2002. Available online at: http://web.dcp.ufl.edu/hinze/Final-Safety%20Plus-5737.htm (accessed 14 June 2010).

Hinze, J., Coble, R.J. and Elliott, B.R. (1999) Integrating construction worker protection into project design. In Singh, A., Hinze, J. and Coble, R.J. (eds) *Implementation of Safety and Health on Construction Sites*. Proceedings of the Second International Conference of CIB Working Commission W99, 24–27 March 1999, Honolulu, Hawaii: 395–401.

ILO (International Labour Organisation) (1919) *Constitution*. Geneva: ILO. Available online at: http://www.ilo.org/ilolex/english/iloconst.htm (accessed 28 May 2010).

ILO (International Labour Organisation) (1988) Convention C167: Safety and Health in Construction. Geneva: ILO.

ILO (International Labour Organisation) (1999) *Decent Work*. Report of the Director General, ILO 87th Session, June. Geneva: ILO.

ILO (International Labour Office) (2001a) *Guidelines on occupational safety and health management systems ILO-OSH 2001*. Geneva: ILO.

ILO (International Labour Office) (2001b) *The construction industry in the twenty-first century: Its image, employment prospects and skill requirements*. Tripartite Meeting on the Construction Industry in the Twenty-first Century: Its Image, Employment Prospects and Skill Requirements. Geneva: ILO. Available online at: www.ilo.org/public/english/dialogue/sector/techmeet/tmcit01/tmcitr.pdf/

ILO (International Labour Office) (2005) *Global Estimates of Fatal Work Related Diseases and Occupational Accidents, World Bank Regions 2005*. Geneva: ILO.

ILO (International Labour Organisation) (2006) *Subcommittee on multinational enterprises*. Geneva; ILO.

ILO (International Labour Office) (2010a) *Construction OS&H*. Geneva: ILO. Available online at: http://www.ilo.org/

ILO (International Labour Office) (2010b) *Occupational safety and health*. Available online at: http://www.ilo.org/global/What_we_do/InternationalLabourStandards/Subjects/Occupationalsafetyandhealth/lang--en/index.htm (accessed 28 May 2010).

Kartam, N.A. and Bouz, R.G. (1998) Fatalities and injuries in the Kuwaiti construction industry. *Accident Analysis and Prevention*, 30(6): 805–14.

Kheni, N.A., Dainty, A.R.J. and Gibb, A. (2008) Health and safety management in developing countries: a study of construction SMEs in Ghana. *Construction Management and Economics*, 26: 1159–69.

Koehn, E. and Reddy, S. (1999) Safety and construction in India. In Singh, A., Hinze, J. and Coble, R.J. (eds) *Implementation of Safety and Health on Construction*

Sites. Proceedings of the Second International Conference of CIB Working Commission W99, 24–27 March 1999, Honolulu, Hawaii: 39–45.

Levy, B.S. (1996) Global occupational health issues: working in partnership to prevent illness and injury. *Journal of the American Association of Occupational Health Nurses*, 44: 244–7.

London, L. and Kisting, S. (2002) Ethical concerns in international occupational health and safety. *Journal of Occupational Medicine*, 17(4): 587–600.

Ma, T.Y.F. and Chan, A.P.C. (1999) The attitudes of workers towards construction site safety – A study of site safety in Hong Kong. In Singh, A., Hinze, J. and Coble, R.J. (eds) *Implementation of Safety and Health on Construction Sites*. Proceedings of the Second International Conference of CIB Working Commission W99, 24–27 March 1999, Honolulu, Hawaii: 63–8.

Mackenzie, J., Gibb, A.G.F. and Bouchlaghelm, N.M. (1999) Communication of health and safety in the design phase. In Singh, A., Hinze, J. and Coble, R.J. (eds) *Implementation of Safety and Health on Construction Sites*. Proceedings of the Second International Conference of CIB Working Commission W99, 24–27 March 1999, Honolulu, Hawaii: 419–26.

Madhok, A.S. (2005) Report on the status of women workers in the construction industry. New Delhi: National Commission for Women.

Marshall, G. (1994) *Safety Engineering*, 2nd edn. Des Plaines, IL: American Society of Safety Engineers.

Misnan, M.S. and Mohammed, A.H. (2007). Development of safety culture in the construction industry: a conceptual framework. In Boyd, D. (ed.) Proceedings from the 23rd Annual Association of Researchers in Construction Management (ARCOM) Conference, 3–5 September, Belfast: 13–22.

Murie, F. (2007) Building Safety – An international perspective. *International Journal of Occupational and Environmental Health*, 13: 5–11.

Neale, R.H. and Murie, F. (2009) *Construction OS&H: A Comprehensive Training Package by the International Labour Office*. Presented at the CIB W099 Conference, Working together: planning, designing and building a healthy and safe construction industry, Melbourne, Australia, October.

newKerala.com (2010) Construction sector accounts for highest accident rate. Available online at: http://www.newkerala.com/news/fullnews-114752.html (accessed 19 September 2010).

Nuwayhid, I.A. (2004) Occupational health research in developing countries: a partner for social justice. *American Journal of Public Health*, 94(11): 1916–21.

OSHA (1993) *Improving workplace protection for new workers: New worker, high risk!* Fact Sheet No. OSHA 93-07. Available online at: http://www.osha.gov/pls/oshaweb/owadisp.show_document?p_table=FACT_SHEETS&p_id=145 (accessed 17 June 2010).

OSHA (2010) *Why is health and safety important?* European Agency for Safety and Health at Work. Available online at: http://osha.europa.eu/en/topics/business/performance/index_html/why_html (accessed 5 June 2010).

Page, S. (2002) *Worker Participation in Health & Safety: A Review of Australian Provisions for Worker Health and Safety Representation*, a paper for the Health and Safety Executive. London: HSE.

Peckitt, S.J., Glendon, A.J. and Booth, R.T. (2004) Societal influences on safety cultures in the construction industry. In Rowlinson, R. (ed.) *Construction Safety Management Systems*. New York: Taylor and Francis; 17–54.

Rantanen, J. (1997) Occupational health and safety training as part of life long education. *The African Newsletter on Occupational Health and Safety*, 7: 52–5.

Rantanen, J., Lehtinen, S. and Savolainen, K. (2004) The opportunities and obstacles to collaboration between the developing and developed countries in the field of occupational health. *Toxicology*, 198: 63–74.

Sandhir, S. (2009) Education and professional skills: The labor shortage is for real. *Times Journal of Construction and Design*, 27 March. Available online at: http://www.rics.org/site/scripts/documents_info.aspx?categoryID=637&documentID=858&pageNumber=6 (accessed 10 September 2010).

Sawacha, E., Naoum, S. and Fong, D. (1999) Factors affecting safety performance on construction sites. *International Journal of Project Management*, 17(5): 309–15.

Self Employed Women's Association (SEWA) (2000) *Labouring Brick by Brick: A Study of Construction Workers*. Ahmedabad, India: SEWA.

Soehod, K. (2008) Workers' participation in safety and health at work, *Jurnal Kemanusiaan bil*. 11 June 2008.

Suplido-Westergaard, M.L. (2006) Recognizing national culture as a determinant of safety subculture. *Asian-Pacific Newsletter on Occupational Health and Safety*. 13(1): 19–23.

Tam, C.M. and Chan, A.P.C. (1999) Nourishing safety culture in the construction industry in Hong Kong. In Singh, A., Hinze, J. and Coble, R.J. (eds) *Implementation of Safety and Health on Construction Sites*. Proceedings of the Second International Conference of CIB Working Commission W99, 24–27 March 1999, Honolulu, Hawaii: 117–22.

Vilanilam, J.V. (1980) A historical and socioeconomic analysis of occupational safety and health in India. *International Journal of Health Services*, 10(2): 233–49.

Voyi, K. (2006) Is globalization outpacing ethics and social responsibility in occupational health? *Medicina del Lavoro*, 97(2): 376–82.

Warternberg, D. and Greenberg, M. (1992) Epidemiology, the press and the EMF controversy. *Public Understanding of Science*, 1: 383–94.

Watanabe, T. and Hanayasu, S. (1999) Philosophy of construction safety management in Japan. In Singh, A., Hinze, J. and Coble, R.J. (eds) *Implementation of Safety and Health on Construction Sites*. Proceedings of the Second International Conference of CIB Working Commission W99, 24–27 March 1999, Honolulu, Hawaii: 55–62.

Wells, J. (1986) *The Construction Industry in Developing Countries: Alternative strategies for development*. Beckenham, Kent: Croom-Helm.

Wells, J. and Hawkins, J. (2009) *Promoting Construction Health and Safety through Procurement: A briefing note for developing countries*. London: Institute of Civil Engineers Engineers Against Poverty.

WHO (World Health Organisation) (1995) *Global Strategy on Occupational Health for All: the way to health at work*. Recommendation of the second meeting of the WHO Collaborating Centres in Occupational Health, 11–14 October 1994, Beijing. Geneva: WHO.

Women's Engineering Society (2010) *PPE clothing survey results 2010*. Available online at: http://www.wes.org.uk/?q=content/ppe-clothing-survey-0 (accessed 17 September 2010).

Part III

Strategies

7 Sub-contracting and joint venturing in construction in developing countries

Abdul-Rashid Abdul-Aziz
and
Izyan Yahaya
University of Science Malaysia

Introduction

The definitions of strategic alliances vary, depending on what the scholars give attention to. Some definitions highlight the resource-seeking nature of strategic alliances (Teece, 1992), others focus on their use to achieve competitive advantages (Das and Teng, 2000), and yet others consider the blurring of organisational boundaries (Mintzberg et al., 2009). There are various types of equity and non-equity based alliances which can be located along a continuum with short-term pure market transactions at one extreme and long-term complete ownership solutions at the other extreme (Peng, 2006) (see Figure 7.1).

Of all the forms of strategic alliances and networks in the construction industry, sub-contracting and joint venturing are the most frequently

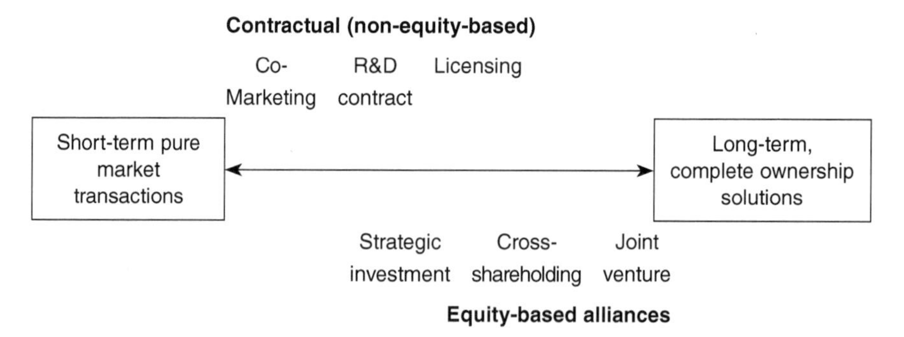

Figure 7.1 A continuum of the different strategic alliance forms (source: Peng, 2006)

exercised. Sub-contracting is common in countries of varying cultural and historical backgrounds such as Japan (Reeves, 2002), Germany (Syben, 2000) and China (Fang et al., 2004). The level of sub-contracting has risen over the years owing to the increase in the volume and extent of complex and specialist construction works such as air-conditioning, lift and escalator installation, and building automation and control systems (Yik et al., 2006). Labour-only sub-contracting is also both long established and widespread in the construction industry (Wells, 2006). There is evidence from around the world that the recruitment of labour through intermediaries has been increasing (ILO, 2001). As for joint ventures, those between foreign and local contractors have been growing worldwide at an increasing pace (Lim and Liu, 2000; Kumuraswamy et al., 2000).

Sub-contracting can account for as much as 90 percent of the total value of the project in some cases (Hinze and Tracy, 1994; Kumaraswamy and Matthews, 2000). Yet, as Karim et al. (2006) note in their study on construction quality, the significance of sub-contractors has not received as much attention from researchers as it should. Studies on international construction joint ventures have so far been conducted by only a few scholars (Ozorhon et al., 2008), despite recommendations by some authors such as Ofori (2000) that there should be more intensified studies on this subject. Given the important contribution of sub-contracting and joint venturing to the construction industries of developing countries, both real and potential, it is pertinent to discuss them, and consider the policy implications.

This chapter begins by elaborating on the nature of sub-contracting and joint venturing. Next, sub-contracting and general forms of strategic alliance are discussed using transaction cost theory as a framework. The benefits and problems of sub-contracting and joint venturing are then considered. Finally, the policy implications of sub-contracting and joint venturing are discussed.

Mechanics of sub-contracting and joint venturing

Nature of sub-contracting

Whenever a main or general contractor secures a construction project, the company searches from its network of contacts to find sub-contractors which are able and willing to undertake specific tasks on the project (Abdul-Aziz, 1995; ILO, 2001). The main contractor might already have communicated with the sub-contractors prior to that when seeking their quotations in preparing its own tender to secure the job. On some occasions, the client may nominate certain sub-contractors to perform particular duties on the project. The sub-contractors, in turn, may sub-contract portions of their work to other firms as "sub-sub-contractors" and so on, giving rise to a hierarchical sub-contracting arrangement.

Excluding nominated sub-contractors, the relationship between a main contractor and its sub-contractors may not be sealed by formal contracts. Many of the small contractors exist in the informal sector of the industry. Therefore, a significant portion of the productive work in the construction industry is carried out in the informal economy. As Ganesan (1982) observes, there is dynamism in the construction industry with two-way movement of productive units across the formal–informal divide; those who fail to secure jobs in the formal sector may be driven to find work normally associated with informal contractors while those in the informal sector may seek recognition as contractors to secure jobs from public clients, as has happened in Tanzania (see Jason, 2008). Dynamism in the community of sub-contractors also manifests itself by way of the emergence of new production units, as skilled and experienced workers leave their employers to become sub-contractors themselves, while those owned by ageing employers cease to operate. Fang et al. (2004) observe that may of the sub-contractors in China come from the rural areas in search of employment in the urban areas.

A spirit of both competition and collaboration prevails in the sub-contracting arrangement.

Formal and informal contractors may compete intensely for projects. The former may obtain small works considered the preserve of the informal contractors, while "established informal contractors" sometimes undertake projects that are sufficiently large to be carried out by formal contractors (Mlinga and Wells, 2002). At the same time, because of the mutual dependence of the main contractors and certain reliable sub-contractors, long-standing alliances may be forged as the success of the former depends on the quality of the output of the latter, while the latter depends on the former for their livelihood and their technical and managerial development (Eccles, 1981; Ganesan, 1982; Mlinga and Wells, 2002). The cost of work packages is directly negotiated between the members of the alliance, thereby saving on tendering costs on both sides. For the sake of maintaining the alliance, the sub-contractors may even be willing to follow the main contractor to undertake projects outside their locality. Moreover, the main contractor may treat their sub-contractors paternalistically (Bennett et al., 1987). Bennett et al. (1987) attribute the success of Japan's building industry to such long-term paternalistic alliances. The sub-contractor trusts the main contractor to protect the sub-contractor's interests, and to offer it continuous employment and fair compensation. By highlighting how much Chinese culture values relationships, Ang and Ofori (2001) draw attention to the amenability of certain national cultures to subcontracting alliances.

As a consequence of the high proportion of sub-contracting, large construction companies manage the construction process rather than undertake actual productive functions (ILO, 2001). The divorce from actual physical work leads to their workforce consisting mainly of white-collar workers. Some firms have become service companies, finding clients and marketing products that are produced by sub-contractors. Some have expanded their operations by taking responsibility for other service activities up and down the supply chain, such as funding privatised infrastructure projects and managing facilities. Others have diversified their activities away from their core area into other sectors.

In most countries, the sub-contractors generally pass down their trades under an informal apprenticeship arrangement that has prevailed for as long as the construction industry has existed (ILO, 2001). Young apprentices are recruited through the social network of the employers. This gives rise to the labour composition of sub-contractors being skewed towards certain ethnic, clan or village groups. With the aversion of local youths to careers in the construction industry owing to the "dirty, difficult and dangerous" image of construction, but more importantly to poor terms of employment (ILO, 2001; Rogerson, 1999), sub-contractors have been driven to engage and train foreign workers, even if they are not regularised in terms of their immigration status. For example, in South Africa, construction workers come from Mozambique, Zimbabwe, Swaziland and Botswana (Rogerson, 1999). In Malaysia, the bulk of foreign workers are Indonesians and Bangladeshis (Abdul-Aziz, 1995). Construction labour migration has become a global phenomenon (Wells, 1996). The detection and apprehension of unregulated foreign construction workers becomes difficult when they are engaged by the myriad small and medium-sized contractors scattered geographically. Experience in many countries has shown that, once foreign labour becomes a feature of the construction industry in a particular country, it becomes difficult to reverse the trend (Abdul-Aziz, 1995).

There is wide variation in the quality of sub-contractors in China (Fang et al., 2004). Many of them do not have any technical training. Their managerial personnel also generally have low education. Labour disputes often occur between the sub-contractors and their workers due to delayed or non-payment of wages. Therefore, the mobility of the labour force is high. All these impair the quality and progress of projects. It is pertinent to note that, for trades which are regulated and licensed by governments in some countries, such as scaffolding, electrical wiring, plumbing, and lift and escalator installation, the sub-contractors can be expected to be less varied in terms of quality.

Wells (2006) asserts that the extent of labour-only sub-contracting has been increasing over the years. Other researchers have found that, whether in China (Lu and Fox, 2001), Namibia (Jauch and Sakaria, 2009), Malaysia (Abdul-Aziz, 1995) or the UK (Winch, 1998), there is a trend towards increasing utilisation of labour-only sub-contractors. Wells (2006) points out that labour outsourcing provides a means to circumvent regulations prohibiting the hiring and firing of workers, although she notes that in countries where labour regulations are non-existent or ineffectual (such as those in Sub-Saharan Africa) labour sub-contracting is still common, thus suggesting that there are other, equally compelling forces at play.

Nature of joint ventures

Joint ventures take many forms, and are established for many reasons. Essentially, joint ventures are set up when legally separate entities form a jointly owned entity in which they invest and engage in various decision-making activities (Geringer, 1991). As for international joint ventures, developing countries are aware of the contribution foreign contractors can make towards the realisation of their national development objectives (Drewer, 1980). Yet they are reluctant to let the foreign firms dominate the local construction activities (Sornarajah, 1992). Some of the goals of foreign contractors are incongruent with those of the governments of host countries (Ofori, 1996). By requiring foreign contractors to form joint ventures with, or sub-contract part of the work to, their domestic counterparts, the governments of developing countries hope that advanced technology and managerial skills would be transferred to domestic contractors (Ofori, 1996; Luo et al., 1998). Multilateral agencies also encourage foreign contractors to form joint ventures with local companies. For example, on World Bank and Asian Development Bank projects, domestic contractors may be given a 7.5 percent margin of preference when the bids of prequalified domestic and foreign contractors are being compared. Joint ventures between domestic and foreign contractors are also eligible for the preference. From a study of Sino-foreign joint ventures, Luo (2001) found that their structures more or less replicate the organisational forms of the foreign partners' parent companies. Even if the foreign partners hold a majority equity share, less participation in daily operational management could reduce their level of control of the joint ventures. Evidence from the big names in international construction shows that the relationship between joint venture partners may last for many years, and span several disparate projects, because of the compatibility of the partners and their expectation of mutual gains (Abdul-Aziz and Cha, 2008).

Transaction cost approach and sub-contracting

The transaction cost approach devised by Williamson (1975, 1979) has been used by some authors such as Eccles (1981), Gunnarson and Levitt (1982), Reve and Levitt (1984) and (Winch (1989) when analysing the prevalence and growth of sub-contracting in the construction industry.

The transaction cost approach regards the firm and the market as alternative governance structures within which transactions can take place (Williamson, 1975, 1979). The model predicts that, under certain situations, transaction cost is low, thereby making market exchange more attractive, whereas under different situations, transaction cost is high, hence promoting the internalisation of the exchange. The cost of transaction is determined by two human and three environmental factors. The two human factors are:

1 Bounded rationality: Humans have limited memories (they cannot absorb all the information available) and limited cognitive competence (they cannot possibly anticipate all possible outcomes that may arise from a transaction);
2 Opportunism: There is a tendency for humans to act in a self-interested manner.

The three environmental factors are:

1 Uncertainty or complexity: Uncertainty or complexity of the environment compounds the problems associated with rationality and opportunism;
2 Small number: The absence of competitive situations results in behavioural inter-dependencies between actors;
3 Asset specificity: The value of an asset associated with a particular transaction. Asset specificity can take the form of site specificity, physical asset specificity, human asset specificity, dedicated assets and brand name capital.

The transaction cost approach predicts that environmental uncertainty or complexity, coupled with bounded rationality, leads to contracting problems in that it becomes virtually impossible or excessively costly to prepare a contract which takes into account all contingencies (i.e. a contingent claims contract) (see Figure 7.2). It also predicts that small numbers combined with opportunism can lead to "hold up" problems as the party in a more advantageous position tries to obtain more favourable terms of exchange. In both scenarios, the cost of commercial transaction becomes high, thereby rendering internal transaction more attractive.

When applied to sub-contracting in the construction industry, transaction cost theory suggests that the internal rather than the external market should be used by main contractors; this is at odds with reality. Thus, many scholars have tried to reconcile the theory–reality mismatch. They have done this in many different ways.

The contribution of Eccles (1981) lies in using the transaction cost approach to explain that the market and internal hierarchy can exist in

Environmental factors		Human factors		Outcome
Uncertainty/ complexity	+	Bounded rationality	=	Contracting problems
Small numbers	+	Opportunism	=	'Hold up' problems

Figure 7.2 Situations in which internal transaction becomes more attractive (source: adapted from Gunnarson and Levitt, 1982: 523)

the construction industry under different conditions. He notes that hold up problems arising from small numbers and opportunism are remote. Except for highly unusual projects, there is usually a large number of trade contractors competing for work, normally on a competitive bidding basis. The time available for the general contractor to suffer from the opportunistic behaviour of the sub-contractors is limited by the temporary nature of projects. In a situation where the general contractor chooses to negotiate with a sub-contractor rather than obtain competitive bids, those keen to enjoy such a relationship are motivated not to be opportunistic. Eccles (1981) rationalises this by suggesting that uncertainty/complexity and bounded rationality exist simultaneously for construction. The uniqueness of projects prevents contractors and sub-contractors from exploiting their past experiences to win the bid for the next project. Therefore they face a high degree of uncertainty. Bounded rationality prevents small firms especially from directly supervising all of the trades. The simultaneous existence of uncertainty/complexity and bounded rationality suggests that sub-contracting is more common than vertical integration.

However, Eccles (1981) argues that Williamson's (1975, 1979) model emphasises how hierarchical transactions can reduce risk, ignoring how market transactions can reduce risks as well. Consequently, his model implies that integration is the response to bounded rationality and uncertainty/complexity. Eccles (1981) posits that whether the response to this pair of factors is integration or disintegration depends at least partly on whether it is certainty of output (price) (because of uncertainty of demand) or uncertainty of input (cost) (because of unique projects in variable conditions) that prevails. Sub-contracting is preferred when demand and prices are uncertain but the firm is relatively sure of obtaining the necessary resource inputs. However, if demand and prices are more certain and matters pertaining to resources less certain, backward integration is preferred when bounded rationality renders contingency claims contracts prohibitive. Eccles (1981) is particularly acknowledged for identifying a relationship that exists between the main contractor and sub-contractor that is a hybrid of the market and hierarchy – the "quasifirm" – which is characterised by stability and continuity over fairly long periods, and not fashioned by competitive bidding.

Gunnarson and Levitt (1982) use transaction cost theory to analyse the situation across several projects as well as within a single project. For the across-projects scenario, they, like Eccles (1981), note that construction work is fraught with uncertainties. Site-related issues and the construction team cannot be known in advance. For complex projects, design work may be far from complete when actual construction work begins. Bounded rationality that renders it impossible or at least extremely costly to specify responses to all contingencies then leads to contracting problems, which the industry overcomes by resorting to well-prepared standard sub-contract documents. Gunnarson and Levitt (1982) concur with Eccles (1981) that hold up problems are unlikely to occur as there are usually many sub-contractors in the market from which the main contractor can select. Combining these two reasonings, they posit that across several projects the market option is the most preferred, as it reduces production costs without creating excessive transaction costs. In addition, they note that the market option offers the advantage of allowing for economies of scale to be fully exhausted by both the general contractor and the sub-contractors – as the former can work on many projects by engaging specialist sub-contractors when necessary, while the latter can provide full-time employment to their workers by working for several general contractors. Secondly, the general contractor can benefit from the sharing of risks with its sub-contractors.

As for within a single project, Gunnarson and Levitt (1982) concede that the threat of hold ups is real. The non-appearance of one sub-contractor can have serious repercussions on timely delivery. The industry addresses hold up problems by introducing strong incentives to ensure that the sub-contract is honoured, the recourse of arbitration being one of them. Despite arguing for the market mode, Gunnarson and Levitt (1982) note that, in reality, a typical construction project is neither a pure market nor a pure hierarchy. Sub-contracting conveys a relationship based on the market. However, by assuming the responsibility for the coordination of the sub-contractor's personnel on site, the general contractor establishes a hierarchical relationship. They suggest that the market and hierarchy modes exist simultaneously.

Gunnarson and Levitt (1982) point to the inadequacy of the transaction cost theory for describing the administration on the construction job site, a point repeated in a subsequent paper by Reve and Levitt (1984: 18): "… hierarchy per se does not necessarily produce all the transactional advantages suggested by the transaction cost analysis in its crudest form." Reve and Levitt (1984) agree with Eccles (1981) on the possible existence of a "quasifirm" or clan-type relationship in the contractor–subcontractor relationship, based on the economies of familiarity and which may be viewed by some as being anti-competitive.

Winch (1989) criticises Eccles (1981) for using the speculative house-building industry to examine the transaction cost theory. While it is of general relevance, Winch feels it cannot be directly applied to contracting. Winch (1989) also criticises Gunnarson and Levitt (1982) and Reve and Levitt (1984) for using the project rather than the firm as the object of their analysis. Winch identifies four types of uncertainties normally faced by contractors: task uncertainty, natural uncertainty, organisational uncertainty, and contracting uncertainty. He then points out that construction projects can be complex. Like previous authors, Winch (1989) notes that the ingredients of uncertainty/complexity and bounded rationality are present in the construction industry. Winch (1989) agrees with Eccles (1981) and Gunnarson and Levitt (1982) that small numbers situations are limited. However, once the contract is let, there is room for opportunism.

Hence, according to transaction cost analysis, the rational response to the confluence of all these is the hierarchy. Winch (1989) reconciles the incongruity between theory and reality by pointing to the propensity of contractors to strive for flexibility rather than efficiency in the face of high uncertainty by opting for sub-contracting. Sub-contracting results in low commitment of the general contractor to fixed capital and in the casualisation of labour. It also enables some of the costs of uncertainty to be passed on to others.

Hughes et al. (2001) suggest that the efforts made in using the transaction cost approach to examine the decision of main contractors on whether or not to sub-contract have not been fruitful. Lee et al. (2009) adopted the approach to analyse the conditions under which a contractor adopts competitive (i.e., project-by-project) or strategic (i.e., long-term) relationships. They, too, conclude that: "However, because this paper assumes hypothetical transaction-cost-based conditions, the research results are limited in their applicability to real projects" (2009: 1239).

Other forms of construction strategic alliances and transaction cost analysis

No other forms of strategic alliances in construction have been analysed using transaction costs theory to the same extent as sub-contracting. Here, other forms of strategic alliances are discussed in the context of the theory, where possible, in relation to the construction industry.

Just as with sub-contracting, Grant and Baden-Fuller (2004) and Douma and Schreuder (2008) note that all other forms of strategic alliance are hybrids of the market and hierarchy. Even equity joint ventures, which more closely replicate some of the features associated with organisational hierarchies than do other alliances (Gulati, 1995), are no exception (Kamminga et al., 2007). One of the limitations of the transaction cost approach when applied to strategic alliances is that it ignores social control. Goshal and Moran (1996) state: "The strength and seductiveness of the markets and hierarchy argument lies in the parsimony of its narrow assumptions of human nature and its equally narrow interpretation of economic objectives" (1996: 41). By observing an alliance for an infrastructure construction project in Australia during the formation stage, Langfield-Smith (2008) concludes that the transaction cost economics prescriptions need to be interpreted in the light of the control mechanisms which the parties to the alliance might put in place. In situations where the theory recommends internal hierarchy but the market is resorted to instead, a control package combining rational and social measures can manage contractual and hold up hazards effectively. The control package may include monitoring and control mechanisms; incentive alignment measures, trust-building programmes; agreement that emphasises mutual collaboration, shared responsibility and risks; and embedded risk mitigating processes. Referring to project partnering, Kadefors (2004) adds project-wide communication in the early phase, and systems to monitor relations and manage conflicts as other social mechanisms.

Dacin et al. (2007) note that reputation (just like trust) is partially a socially constructed phenomenon that has a profound effect on the integrity and effectiveness of strategic alliances. To avoid ex-post transaction cost problems, partners are selected based on their reputation (Kirmani and Rao, 2000). Reputation may represent a credible signal that a firm intends to deal honestly. Ex-post transaction cost can also be reduced through the use of a relational contract which promises gains from future exchange relations, thereby motivating the partners to behave honestly and curb the threat of opportunism (Eccles, 1981; Gibbons, 2005).

Proponents of the transaction cost theory have also been criticised for ignoring the role of trust in transactions. For example, Van der Meer-Kooistra and Vosselman (2000: 56) state that: "By treating social embeddedness as an environmental factor Williamson denies the possibility of using trust as a means to reduce opportunism. Trust is a variable which can be brought into action in a transactional relation and which can be strengthened by the parties by taking the right actions." Trust is the mutual confidence that no party to an exchange will exploit another's vulnerabilities (Sabel, 1993). Kadefors (2004) points out that a higher level of trust can enhance project performance. From their study of architect–contractor partnerships in Hong Kong, Lui and Ngo (2005) reaffirm the central role of trust in inter-firm cooperation.

According to Sako (1991, 1992) there are three types of trust: contractual trust, competence trust and goodwill trust. Contractual trust revolves around the question of whether the other party will fulfill its promises. It rests on a shared moral norm of honesty and promise keeping. Competence trust pertains to the capability of the other party to carry out what it says it will do. It requires a shared understanding of professional conduct and technical and managerial standards. Goodwill trust addresses the issue of whether the other party will make an open-ended commitment to take initiatives for mutual benefit while refraining from taking unfair advantage. It can only exist when there is consensus on the principle of fairness. In their study of equity joint ventures, Kammiga et al. (2007) posit that different types of trust are required for different levels of control complexity. For joint ventures with little control complexity, contractual trust suffices: for those with medium control complexity, the contractual, competence and goodwill forms of trust are required, while for joint ventures with high control complexity, all three types of trust are required, and to a higher degree. Joint ventures with little complexity face low uncertainty because the operating environment is stable and the risks for opportunistic behaviour are low. The opposite applies with respect to joint ventures which require high control complexity. From his study of construction alliances in Botswana, Ngowi (2007) found that the partners start with contractual trust which then develops to goodwill trust as the number of projects executed by the alliance grows. Hence, the types of trust and the degree to which each of them is present varies according to the characteristics of the activities and/or the environment in which the joint venture operates, as well as the length of time over which the partners have been together.

Moreover, the form of strategic alliance may also change over time as the relationship is prolonged. By examining American, European and Japanese firms in the biopharmaceutical, new materials, and automotive sectors, Gulati (1995) found that the alliance takes on a looser form as the relationship continues, as the partners grow increasingly more confident of one another. To a degree, trust can supplant contractual safeguards as a control mechanism.

Benefits of sub-contracting and joint venturing

The two main benefits of sub-contracting highlighted in the literature are mentioned above: the main contractor achieves economies of scale by engaging expertise on the basis of need, thereby lowering overhead costs (Eccles, 1981; Gunnarson and Levitt, 1982) while the sub-contractor achieves economies of scale by offering its services to a number of main contractors (Eccles, 1981).

With reference to the first point, uncertain and uneven construction demand encourages main contractors to adopt flexible production and labour strategies (Rogerson, 1999). Competitive tendering compounds the volatility for the main contractors. Construction products are not homogenous (ILO, 2001). Therefore, the skills required from one project to another may not be entirely similar. By sub-letting specialised tasks, the main contractors can free themselves from high overheads involved in maintaining skilled workers and special machinery (Yik et al., 2006). The specialist sub-contractors are better able to organise resource inputs for the works such as purchasing materials at the right time and at the right price.

Under sub-contracting, labour becomes a variable cost. The cost of labour includes social security coverage, paid vacation, medical coverage and such like (Wells, 2006). Also, labour sub-contracting relieves the main contractor from supervising the labour force (Leonard, 2000). Both trades sub-contracting and labour sub-contracting simplifies the estimating process, thereby lowering further the overhead cost. Basically, sub-contracting gives a contractor the capacity to undertake work beyond its scope in terms of size and specialisation. Overall, the main contractor enjoys greater financial benefits, higher efficiency and greater flexibility (Winch 1989, 1998; Ng, 2009).

With regards to the second point, sub-contracting has expanded due to the increasing specialisation of trades (Yik et al, 2006). Because sub-contractors are specialists in their own technical fields, they can be expected to produce high-quality workmanship (Mbachu, 2008). Their specialism extends to controlling labour, possibly across several dispersed locations (ILO, 2001). Gray and Flanagan (1989) point out that, by working alongside the same specialist sub-contractors on many projects, main contractors can accumulate skills and knowledge from them. By forging long-term relationships with capable and compatible sub-contractors, the main contractors benefit from effective communication, less conflict and lower risks (Haksever et al., 2001).

Joint ventures between construction companies enable mutual benefits to be derived: these are mainly the pooling of financial and technological strengths, specialists and experienced staff, and networks (Kumuraswamy et al., 2000). Local contractors can provide knowledge of local practices, host country information and strong client relations to their foreign counterparts (Kumaraswamy et al., 2000; Ozorhon et al., 2008). Moreover, technology and knowledge transfer takes place, not only in one direction, but in both directions between the joint venture partners (Kumuraswamy, 2006). Sub-contracting has also benefited local contractors by providing them with opportunities to learn from international contractors, as has happened in many countries, including Turkey (Katsarakis et al., 2007).

By being joint venture partners of, or working as sub-contractors with, foreign contractors, the domestic entities in developing countries can learn from them, and later adopt some of their internationalisation strategies (Abdul-Aziz, 1993). The two parties may even forge an informal long-term relationship which enables the latter to penetrate markets overseas by working alongside their international partners, something which they are unlikely to be able to do on their own (Abdul-Aziz and Wong, 2010).

Disbenefits of sub-contracting and joint venturing

Sub-contracting increases the main contractor's management cost, as sub-contractors need to be supervised, coordinated and monitored (Yik et al., 2006). Close monitoring of some sub-contractors is required as they normally work on multiple projects concurrently (Kim and Paulson, 2003). Indeed, the work of the sub-contractors involved in a particular project can influence its successful completion (Parfitt and Sanvido, 1993). Findings from the study by Long et al. (2004) to identify common problems on large projects in Vietnam support the observation that excessive use of sub-contractors results in complex relationships, and can increase coordination costs. The costs incurred in managing sub-contractors escalates when conflict and disputes arise, for example, owing to a mismatch of goals (Gray and Flanagan, 1989).

Regardless of the extent of monitoring, the reality is that the main contractors have little control over the sub-contractors' workers. Thus, sub-contracting arrangements can lead to problems such as delay of on-site operations, untidy site conditions and lower overall efficiency; this might affect the sub-contractors' own profit (Skoyles and Skoyles, 1987). Excessive sub-contracting also increases the risk of work-related accidents for the manual workers, as found by Kartam et al. (2000) in Kuwait. In many cases, small contractors have neither the time nor the inclination to keep abreast with the legal requirements concerning safety, as they often struggle to survive (Wong and So, 2002). Sub-contractors have a high bankruptcy rate because they are vulnerable to shocks such as market fluctuations (Schaufelberger, 2003). By being highly reliant on sub-contractors, the main contractor might place itself at greater risk in the event that any one of them fails (Gray and Flanagan, 1989).

Labour-only sub-contractors have also been associated with the high incidence of construction accidents (Tham and Fung, 1998), higher levels of wastage of materials (Tam et al., 2007), and poor work quality and productivity (Debrah and Ofori, 2001). Labour-only sub-contractors have been criticised by some authors and observers for being exploitative (Wells, 2006), as their workers receive low wages and have poor working conditions. Jauch and Sakaria (2009) note that some foreign contractors operating in Namibia evade their legal and social obligations towards their employees by engaging them through labour sub-contractors or non-registered companies. By outsourcing labour, the main contractors free themselves from the cost of abiding by labour and welfare legislation, but this is at the expense of the workers (ILO, 2001). Increased fragmentation of the construction industry has also undermined collective bargaining in many countries. Collective agreements, where they exist, are applied to a small and decreasing proportion of the workforce.

Dubois and Gadde (2000) found that, unlike in other industries where partnering with suppliers leads to mutual adaptation in terms of technical solutions (which connect the production operations of supplier and customer), logistics (which improves the flow of materials) or administrative routines (which streamline the information exchange), such positive outcomes are unusual in the construction industry primarily because of the emphasis on competitive tendering and efficiency on individual projects. As relations are often confined to the duration of individual projects, Dubois and Gadde (2000) contend that little benefit is gained from shared learning. This also hampers industry-wide efficiency and innovation gains.

Low investment in research and development (R&D) as a consequence of the fragmentation of the construction industry has also been noted by authors such as Ofori (2006). Such investments become even less urgent when cheap migrant workers are readily available. Winch (1998) notes that, faced with the more expeditious and less costly option of engaging labour sub-contractors, main contractors are reluctant to invest in fixed and human assets which would stimulate innovation. There is also less attention to training. Indeed, in a highly competitive environment, the self-employed or very small firms cannot afford to lose the productive services of experienced operatives as they attend training courses.

As for the good intentions of the host governments in some developing countries which make it mandatory for foreign contractors to joint venture with domestic contractors in order to realise technology transfer (Ofori 1996; Luo et al., 1998), the relationship forged may be one of convenience, which enables the former to increase the chances of securing jobs while the latter does not acquire any new technology (Kumuraswamy, 2006). Policies which are aimed at promoting the development of the domestic construction industry run the risk of being subverted in this manner. For example, in Malaysia, which has an affirmative action policy to encourage the emergence of Malay entrepreneurs, it has long been observed that some Malays willingly become "sleeping partners" in cross-ethnic joint ventures, contributing only their political influence and connections (Wah, 2008). They gain nothing, other than easy money, for literally doing nothing.

Another potentially negative aspect of alliancing is collusion among the ostensibly competing construction companies during tendering, as highlighted by Zou (2006) in a study on China, and by Reeves (2002) in work on Japan. In 2001, a former director of a large Dutch construction company exposed the common practice of tender price collusion in the industry (Graafland, 2004). Following a parliamentary investigation, it was estimated that the costs of public works in the Netherlands were about 8.8 percent higher than they would normally have been. Corruption in the construction sector leads to cost increases, poor-quality construction, and insufficient maintenance, factors which impair economic return to investments and carry high human costs in terms of injury and death (Kenny, 2007). Kenny (2006) found that the impact of stealing one dollar's worth of supplies from a road construction project is as much as four times higher than the impact of a dollar increase in contract costs due to collusion. Collusion is a "network offence" committed by parties that come together for deceitful purposes emboldened by their arrogance that they can act above the law unpunished (Van den Heuvel, 2005). Such network abuse damages public welfare and harms the image of the construction industry.

Success factors

The success factors of sub-contracting and joint venturing can be considered at two levels: the parties involved, and the process.

Some authors have identified certain attributes that sub-contractors should possess: manpower with the appropriate qualifications or competence, with a low turnover rate (Parfitt and Savido, 1993; Chua et al., 1999). In a study in South Africa, Mbachu (2008) proposed key criteria for assessing sub-contractors' eligibility to be invited to tender, and to be selected, and their subsequent performance during the construction stage. He found that the quality record is the most influential criterion for selecting high-performing sub-contractors at the pre-qualification stage, and for assessing their construction performance. During the sub-contract award phase, the tender price exerts the most significant influence. As for the process, Kumaraswamy and Matthews (2000) in their study of main contractors and sub-contractors found that project success can be achieved through positive attitudes, compliance with commitments, correct and quick responses, increased trust and fair dealing with sub-contractors, and earlier involvement of sub-contractors.

As for international joint ventures, Shen et al. (2001) warn that the choice of the wrong partner can lead to disaster. Partner selection is therefore one of the highest risks in Sino–international joint ventures. Adnan and Morledge (2003) also studied international joint ventures involving Malaysian parties and put stress on the importance of proper selection of partners. The partners should have international joint venture experience and should be financially stable. Bing and Tiong (1999) indicate that the local partners' strong financial and managerial ability, industrial relationships and relationships with local government agencies can reduce the risks of operating in Asian countries. From an analysis of information on Turkish contractors involved in international joint ventures, Ozorhon et al. (2008) found that the partners should have compatible technical and managerial skills, and complementary experience in projects similar to the one currently being undertaken, as these attributes generate commitment, cooperation and trust between the partners. The partners should also be compatible in terms of financial capability, size, management systems and workload. They note that inter-partner relations strongly influence the joint venture's performance.

Bing and Tiong (1999) recommend that the international joint venture should adopt a one-partner-dominant style of management whenever one partner is strong enough to handle major construction works on its own. The partners must all be committed to the joint venture, which can only stem from the perception that they are all deriving value and benefits from being together (Kwok et al., 2000). There must be mutual trust, sharing of information and confidentiality (Sridharan, 1997). The differences in management styles have to be reconciled. Adnan and Morledge (2003) add the following: mutual understanding, mutual trust, well prepared joint venture agreement, commitment, cooperation, coordination, good communication, management control and profitable venture. Bing and Tiong (1999) emphasise that each partner's responsibility and liability must be clearly defined in the agreement, using simple language that can be understood by the partners' employees. In addition, Bing and Tiong (1999) note that frontline personnel must be unbiased and experienced in joint management, and should master the local language.

Loosemore and Muslamani (1999) indicate that the cultural difference between the partners must be reconciled to avoid communication breakdown. Studies have shown that it is possible to manage cultural differences. In two case studies of international joint venture projects in India and another two in Taiwan, Mahalingam et al. (2005) found that there were few instances where differences in national cultural values or beliefs led to a significant impact on those projects. They attributed the absence of cultural clash to the long, prior international experience of the staff of the foreign contractors and the standardisation of the engineering "language" that acts as a bridge. Mahalingam et al. (2005) came across professionals who work on a freelance basis on construction projects in developing countries, acting, amongst other things, as intermediaries whenever disagreements arose. Ozorhon et al. (2008) also found that cultural dissimilarities did not adversely affect performance or inter-partner relations on international joint ventures involving Turkish contractors.

Policy implications and recommendations

Sub-contracting and joint venturing practices are entrenched in countries around the world. Therefore, policy interventions to improve the construction industry should build on these practices (Wells, 2006; Ofori, 1994a). In this section, those policies impinging on sub-contracting and joint venturing are discussed. Specifically, they relate to employment generation and poverty alleviation, labour welfare including trade unionism and work safety, innovation including worker and sub-contractor training, the avoidance of collusion, and the prescription of mandatory joint ventures.

Firstly, the labour intensity of construction should be capitalised upon by developing countries that tend to have high levels of unemployment (ILO, 2001). The construction industry is one of the few sectors which provide job opportunities to unskilled migrants from rural to urban areas. Thus, construction activity can be used to generate employment and alleviate poverty if properly implemented (McCutcheon 2001) and oriented towards maximising these economic and social gains.

As for protection of workers, those engaged by labour-only subcontractors and foreign workers are particularly vulnerable. Wells (2006) notes attempts by some countries to provide social security to labourers engaged by labour-only sub-contractors, although she adds that not all of them are successful. Such attempts must continue to be made. As for foreign construction workers, the International Labour Standards should be adopted and enforced by the governments of developing countries. In essence, the standards spell out the core rights and freedoms to which all migrant workers (and their families) should be entitled, irrespective of their legal situation (Rogerson, 1999). The basic right to be treated no less favourably than national workers in respect of conditions of work and terms of employment should be the main guiding principle. Many developing countries acting as hosts to these foreign construction workers only pay lip service to equity issues, and continue to marginalise them (Wells, 1996). Trade unions can play a role in championing the rights of construction operatives. There is a need for trade unions to admit as members casual, temporary and even migrant workers, so that their welfare can be protected through collective actions (ILO, 2001). Major construction employers should be encouraged to allow trade union membership for this segment of the construction workforce, as improvements in their welfare will help to enhance the image of the construction industry (Debrah and Ofori, 2001).

Sub-contracting has been associated with poor construction safety performance. Suggestions that multi-layer sub-contracting be reduced are likely to meet stiff resistance from industry players, as was found in Hong Kong (Wong and So, 2002). Also in Hong Kong, safety training is not provided to site operatives not directly engaged by the contractor (Tam and Fung, 1998), although it would appear that such workers need safety training. Enshassi et al. (2008) studied sub-contractors' safety performance in Palestine, and recommended that sub-contractors and their workers be made to attend safety programmes on a regular basis before being allowed to work on site. In Malaysia, the Construction Industry Development Board's (CIDB) Green Card programme, launched in 1997, requires everyone working on construction sites to undergo a one-day construction safety programme. Once an individual obtains a green card, that person is automatically insured against any injury.

Labour and sub-contractor training are closely related to innovation in the construction industry (Winch, 1998). As pointed out earlier, those already engaged in the construction industry are not keen to take up these training programmes, less so when their employers discourage them from doing so because of loss of earnings and productive time. Also, training costs money, which the employers are unwilling to invest for fear of losing skilled workers after they complete their training (ILO, 2001). The small and very small contractors may not even be able to afford to make such a sacrifice. As the flight of trained workers to another employer is a vexing issue with training, the ILO (2001) suggests that collective agreements between the social partners be struck for joint absorption of the cost to ensure that all contractors put training costs in their bids so that the industry can avoid "free rider" problems.

As for the training of small contractors, Enshassi and Shaath (2007) have shown that formal training is effective in Gaza, Palestine. Trainers supplemented by guest speakers from the industry conduct lectures, discussions, site visits and exercises. The trainees, comprising owners and managers of small construction enterprises, were generally satisfied with the programme. However, Enshassi and Shaath (2007) suggest that a post-training advisory service should be provided to help ex-trainees to solve problems they encounter.

Sub-contractors in the informal sector play an important role in the construction process – for example, by providing labour for registered sub-contractors. Mlinga and Wells (2002) argue that the linkage between the formal and informal sectors should be strengthened for the latter to contribute to the national economy up to its full potential. They propose that a parallel registration system for informal contractors should be set up in tandem with the registration system for formal contractors so that programmes to assist and develop the former can be implemented. Their recommendation reinforces Ofori's (1994b) observation that programmes for the development of the construction industry should include the development of its informal sector.

Maximising competition in bidding should reduce the scope for collusion (Kenny, 2007). The World Bank employs bid tracking and collusion detection software in the e-procurement system. This software may not be suitable for developing countries. Publishing detailed information about contract awards including winning bids for government projects is a simple, less costly and yet effective alternative to identify padded prices. The governments of developing countries must therefore make clear their intolerance of all forms of corruption. The contractor–sub-contractor alliance should be allowed to flourish, and must not be the basis for collusion. Anti-trust laws should be passed and enforced so that network abusers can be prosecuted (Van Den Heuvel, 2006). The authorities should blacklist construction companies found to be involved in collusion. Intolerance of corruption must manifest itself in the serious implementation of anti-corruption laws and regulatory regimes. Various international organisations such as Transparency International have offered many recommendations to prevent corruption on construction projects around the world. An often-suggested measure is the completion of an anti-corruption agreement signed by the client and the tenderers.

Various authors recommend that international construction companies be made to form joint ventures with their local counterparts in developing countries, or to provide opportunities for the latter to be involved in projects through sub-contracting, leading to capacity-building of the local contractors (Zawdie and Langford, 2002). It is also suggested that, on large projects involving foreign contractors, work packages should be reduced in size to maximise the participation of local specialist contractors to promote technology transfer (Kumuraswamy, 1997). Ofori (1996) observes that the implementation of the requirement for foreign contractors to form joint ventures with local firms or to sub-let minimum proportions of projects to local sub-contractors has not been entirely successful. Two issues need to be addressed. First, effective technology transfer depends on the level of knowledge of the local joint venture partners or sub-contractors, as they need to have a basic level of knowledge in order to absorb and utilise the transferred technology (Wang et al., 2004). Also, the local contractors must have adequate numbers of qualified technical and managerial personnel, a good business structure, high financial capacity, and adequate equipment. Devapriya and Ganesan (2002) found that, in Sri Lanka, this may not necessarily be the case. Second, foreign contractors are naturally disinclined to train and develop potential competitors (Ofori, 1994b). Devapriya and Ganesan (2001) suggest that there should be contractual provisions for the transfer of technology to take place, and financial and time allocations must be made accordingly. The technology that is to be transferred must also be appropriate in the local context to ensure sustainable application and further development. They suggest that these provisions cannot be left to the contracting parties to resolve; they require government intervention. In Singapore, Sridaharan (1995) observes that, under a government scheme which was introduced in 1980 and abrogated in 1988, the transferor was offered a monetary incentive for transferring technology, and given a monetary penalty if it failed to do so. A formal statement of the technology to be transferred and the programme for the transfer were produced, and processes for monitoring attainment were also put in place.

Conclusion

The importance of the construction industry to developing economies has been noted by several authors over a long period of time. For example, investment in physical infrastructure has been found to raise the rate of economic growth of developing economies (United Nations, 1990; Canning and Fay, 1993). Moreover, during that development phase, construction contributes more to the national economy than does the manufacturing sector (World Bank, 1984). The complex set of inter-relationships (Ofori, 1990) stimulates other economic sectors with which the local construction industry has either backward or forward linkages (Bon and Minami, 1986; Kirmani, 1988). Unfortunately, in many developing countries, the construction industry is not fulfilling its full potential in development (Nordberg, 2004). Several sources point to capacity building, if necessary through institutional interventions as the central issue which needs to be addressed (Bentall et al., 1999; Ofori, 1990). As this chapter highlights, how the construction industry participants organise themselves in the delivery of their services can contribute towards capacity building or adversely affect it. In short, the impact can reach far beyond the life of a single project. This chapter also underscores the active role which the authorities must play in ensuring that the benefits of strategic alliancing are reaped and the insidious effects are suppressed. The outcome of strategic alliances should not be left to chance, as the same opportunity, such as a one-off mega-project, may not be available again.

References

Abdul-Aziz, A.-R. (1993) The trend towards globalization in business: lessons for contractors in developing countries. *Habitat International*, 17(3): 115–24.

Abdul-Aziz, A.-R. (1995) Site operatives in Malaysia: Examining the Foreign-Local Asymmetry, unpublished report for the International Labour Organisation, Geneva.

Abdul-Aziz, A.-R. and Cha, S.-Y. (2008) Patterns in strategic joint ventures of selected prominent cross-border contractors for 1999–2003. *Canadian Journal of Civil Engineering*, 35: 1009–17.

Abdul-Aziz, A.-R. and Wong, S.S. (2011) Business networks and internationalisation of contractors from developing countries: an exploratory study. *Engineering, Construction and Architectural Management*, Vol. 18, Issue 3, 282–96.

Adnan, H. and Morledge, R. (2003) Joint venture in Malaysian construction industry: factors critical to success. *19th Annual ARCOM Conference*, University of Brighton, 3–5 September, Vol 2, 765–74.

Ang, A.Y. and Ofori, G. (2001) Chinese culture and successful implementation of partnering in Singapore's construction industry. *Construction Management and Economics*, 19(6): 619–32.

Bennett, J., Flanagan, R. and Norman, G. (1987) *Capital & Counties Report: Japanese Construction Industry*. Reading, UK: Centre for Strategic Studies in Construction, University of Reading.

Bentall, P. Beusch, A. and De Veen, J. (1999) *Employment Intensive Infrastructure Programmes: Capacity Building for Contracting in the Construction Sector*. Geneva: International Labour Organisation

Bing, L. and Tiong, R.L.K. (1999) Risk management model for international construction joint ventures. *Journal of Construction Engineering and Management*, 125(5): 377–84.

Bon, R. and Minami, K. (1986) The role of construction in the national economy: a comparison of the fundamental structure of the US and Japanese input-output tables since World War II. *Habitat International*, 10(4): 93–9.

Buckley, P.J. and Enderwick, P. (1988) Manpower Management in the Domestic and International Construction Industry, Discussion Paper in International Investment and Business Studies No 111, University of Reading.

Canning, D. and Fay, M. (1993) *The Effect of Infrastructure Networks on Economic Growth*, New York: Columbia University, Department of Commerce.

Chua, D.K.H., Kog, Y.C. and Loh, P.K. (1999) Critical success factors for different project objectives. *Journal of Construction Engineering and Management*, 126(3): 191–200.

Dacin, M.T., Oliver, C. and Roy, J.P. (2007) The legitimacy of strategic alliances: an institutional perspective. *Strategic Management Journal*, 28: 169–87.

Das, T.K., and Teng, B.-S. (2000) A resource-based theory of strategic alliances. *Journal of Management*, 26(1): 31–61.

Debrah, Y.A. and Ofori, G. (2001) Subcontracting, foreign workers and job safety in the Singapore construction industry. *Asia Pacific Business Review*, 8(1): 145–66.

Devapriya, K.A.K. and Ganesan, S. (2002) Technology transfer through subcontracting in developing countries. *Building Research & Information*, 30(3): 171–82.

Douma, S. and Schreuder, H. (2008) *Economic Approaches to Organizations*, Harlow: Pearson Education.

Drewer, S. (1980) Construction and development: a new perspective. *Habitat International*, 5(3/4): 395–428.

Dubois, A. and Gadde, L.-E. (2000) Supply strategy and network effects – purchasing behaviour in the construction industry. *European Journal of Purchasing & Supply Management*, 6: 207–15.

Eccles, R.G. (1981) The quasifirm in the construction industry. *Journal of Economic Behavior and Organisation*, 2: 335–57.

Enshassi, A., Choudury, R.M., Mayer, P.E. and Shoman, Y. (2008) Safety performance of subcontractors in the Palestinian construction industry. *Journal of Construction in Developing Countries*, 13(1): 51–62.

Enshassi, A. and Shaath, K. (2007) Training assessment of Palestinian contractors. *Journal of Construction Engineering and Management,* 12(1): 39–57.

Fang, D., Li, M., Fong, P.S.-W. and Shen, L. (2004) Risks in Chinese construction market-contractors' perspective. *Journal of Construction Engineering and Management*, 130(6): 853–61.

Ganesan, S. (1982) *Management of Small Construction Firms*, Tokyo: Asian Productivity Organisation.

Geringer, J.M. (1991) Strategic determinants of partner selection criteria in international joint ventures. *Journal of International Business Studies*, 22(1): 41–62.

Gibbons, R. (2005) Four forma(izable) theories of the firm? *Journal of Economic Behaviour and Organisation*, 58(2): 200–45.

Goshal, S. and Moran, P. (1996) Bad for practice: a critique of the transaction cost theory. *Academy of Management Review*, 21(1): 13–47.

Graafland, J. (2004) Collusion, reputation damage and interest in code of conduct: the case of a Dutch construction company. *Business Ethics: A European Review*, 13: 127–42.

Grant, R. and Baden-Fuller, C. (2004) A knowledge accessing theory of strategic alliances. *Journal of Management Studies*, 41(1): 61–84.

Gray, C. and Flanagan, R., *The Changing Role of Specialist and Trade Contractors*, Ascot: CIOB.

Gulati, R. (1995) Does familiarity breed trust? The implications of repeated ties for contractual choices in alliances, *Academy of Management J*ournal, 38(1): 85–112.

Gunnarson, S. and Levitt, R.E. (1982) Is a building construction project a hierarchy or a market? *Seventh World Congress on Project Management*: 521–9.

Haksever, A.M., Demir, I.H. and Giran, O. (2001) Assessing the benefits of long-term relationship between contractors and subcontractors in the UK. *International Journal of Construction Marketing*, 3(1): 1–10.

Hinze, J. and Tracy, A. (1994) The contractor–subcontractor relationship: the subcontractor's view. *Journal of Construction Engineering and Management*, 120: 274–87.

Hughes, W., Hillebrandt, P., Lingard, H. and Greenwood, D. (2001) The impact of market and supply configurations on the cost of tendering in the construction industry, Paper presented at the CIB World Building Congress, April, Wellington, New Zealand.

ILO (International Labour Organisation) (2001) *The Construction Industry in the 21st Century: Its Image, Employment Prospects and Skill Requirements*, Geneva: Sectoral Activities Department, ILO.

Jason, A. (2008) Organising informal workers in the urban economy: the case of the construction industry in Dar es Salaam, Tanzania. *Habitat International*, 32(2): 192–202.

Jauch, H. and Sakaria, L. (2009) *Chinese Investments in Namibia: A Labour Perspective*. Windhoek: Labour Resource and Research Institute (LaRRI). Available online at http://www.larri.com.na/files/Inequality%20in%20Namibia%20 2009%20final.pdf/

Kadefors, A. (2004) Trust in project relationships – inside the black box. *International Journal of Project Management*, 22: 175–82.

Kamminga, P.E. and Van Der Meer-Kooistra, J. (2007) Management control patterns in joint venture relationships: a model and an exploratory study. *Accounting, Organizations and Society*, 21: 131–54.

Karim, K., Marosszeky, M. and Davis, S. (2006) Managing subcontractor supply chain for quality in construction. *Engineering, Construction and Architectural Management*, 13(1): 27–42.

Kartam, N.A., Flood, I. and Koushki, P. (2000) Construction safety in Kuwait: issues, procedures, problem, and recommendations. *Safety Science*, 36: 163–84.

Katsarakis, Y., Rezk, A., Sazak, E., Shaydullin, H. and Yadikar, B. (2007) Turkey & the Construction Services Cluster, *Microeconomics of Competitiveness*, Spring. Available online at http://www.isc.hbs.edu/pdf/Student_Projects/Turkey_ConstructionCluster_2007.pdf/

Kenny, C. (2006) Measuring and reducing the impact of corruption in infrastructure. World Bank Policy Research Working Paper 4099, Washington, DC.: World Bank.

Kenny, C. (2007) Construction, corruption and developing countries. World Bank Policy Research Working paper 4271, Washington, DC.: World Bank.

Kim, K. and Paulson, B.C. (2003) Agent-based compensatory negotiation methodology to facilitate distributed coordination of project schedule changes. *Journal of Computing in Civil Engineering*, 17(1): 10–18.

Kirmani, A. and Rao, A. (2000) No pain, no gain: a critical review of the literature on signaling unobservable product quality. *Journal of Marketing*, 64: 66–79.

Kirmani, S. (1988) *The Construction Industry in Development: Issues and Options*, World Bank Discussion Paper, Washington D.C.: World Bank.

Kumaraswamy, M. (1997) Repackaging construction mega projects and redefining technology transfer. CIB W78 Conference: 215–24.

Kumuraswamy, M. (2006) Accelerating construction industry development. *Journal of Construction Industry in Developing Countries*, 11(1): 73–96.

Kumaraswamy, M. and Dulaimi, M. (2001) Empowering innovative improvements through creative construction procurement. *Engineering, Construction and Architectural Management*, 8(5/6): 325–34.

Kumaraswamy, M. and Matthews, J. (2000) Improved subcontractor selection employing partnering principles. *Journal of Management in Engineering*, 16(3): 47–57.

Kumaraswamy, M., Palaneeswaran, E. and Humphreys, P. (2000) Selection matters – in construction supply chain optimization. *International Journal of Physical Distribution & Logistics Management*, 30(7/8): 661–80.

Kwok, H-C.A., Then, D. and Skitmore, M. (2000) Risk management in Singapore construction joint ventures. *Journal of Construction Research*, 1(2): 139–49.

Langfield-Smith, K. (2008) The relations between transactional characteristics, trust and risk in the start-up phase of a collaborative alliance. *Management Accounting Research*, 19: 344–64.

Lee, H.-S., Seo, J.-O. Park, M., Ryu, H.-G. and Kwon, S.-S. (2009) Transaction-cost based selection of appropriate general contractor-subcontractor relationship type. *Journal of Construction Engineering and Management*, 135(11): 1232–40.

Leonard, M. (2000) Coping strategies in developed and developing societies: the workings of the informal economy. *Journal of International Development*, 12: 1069–85.

Lim, E.C. and Liu, Y. (2001) International construction joint venture (ICJV) as a market penetration strategy – some case studies in developing countries.

3rd International Conference on Construction Project Management, Singapore: 377–89.

Long, N.D., Ogunlana, S., Quang, T. and Lam, K.C. (2004) Large construction projects in developing countries: a case study from Vietnam. *International Journal of Project Management*, 22: 553–61.

Loosemore, M. and Muslmani, H.S. (1999) Construction project management in the Persian Gulf: intercultural communication. *International Journal of Project Management*, 17(2): 95–100.

Lu, Y.-J. and Fox, P.W. (2001) *The Construction Industry in China: Its Image, Employment Prospects and Skills Requirements* (Working Paper 180). Geneva: Sectoral Activities Programme, ILO.

Lui, S.S., and Ngo, H.-Y. (2005) An action pattern model of inter-firm cooperation. *Journal of Management Studies*, 42(6): 1123–53.

Luo, J. (2001) Assessing management and performance of Sino–foreign construction joint ventures. *Construction Management and Economics*, 19: 109–17.

Luo, J., Whitehead, M. and Gale, A. (1998) JVs in construction with economies in transition: the development of Sino–foreign construction JVs. In C. Preece (ed.) *Opportunities and Strategies in a Global Marketplace*. Leeds: University of Leeds; 183–94.

Mahalingam, A., Levitt, R.E. and Scott, W.R. (2005) Cultural clashes in international infrastructure development projects: which culture matter? CIBW92/T23/W107 International Symposium on Procurement Systems: The Impact of Cultural Differences and Systems on Construction Performance, 7–10 February, Las Vegas.

Mbachu, J. (2008) Conceptual framework for the assessment of subcontractors, eligibility and performance in the construction industry. *Construction Management and Economics*, 26: 471–84.

McCutcheon, R. (2001) An introduction to employment creation in development and lessons learnt from employment creation in construction. *Urban Forum*, 12(3–4): 263–78.

Mintzberg, H., Ahlstrand, B. and Lampel, J. (2009) *Strategy Safari*. London: Prentice Hall/Financial Times.

Mlinga, R.S. and Wells, J. (2002) Collaboration between formal and informal enterprises in the construction sector in Tanzania. *Habitat International*, 26: 269–80.

Ng, S.T., Tang, Z. and Palaneeswaran, E. (2009) Factors contributing to the success of equipment-intensive subcontractors in construction. *International Journal of Project Management*, 27: 736–44.

Ngowi, A. (2007) The role of trustworthiness in the formation and governance of construction alliances. *Building and Environment*, 42(4): 1828–35.

Nordberg, R. (2004) Building sustainable cities, *Habitat Debate*, 5(2), United Nations Human Settlement Programme, New York. Available online at http://www.housingfinance.org/uploads/Publicationsmanager/Misc_BuildingSustainableCities.pdf/

Ofori, G. (1990) *The Construction Industry: Aspects of its Economics and Management*. Singapore: Singapore University Press.

Ofori, G. (1994a) Practice of construction industry development at the crossroads. *Habitat International*, 18(2): 41–56.

Ofori, G. (1994b) Construction technology development: role of technology transfer. *Construction Management and Economics*, 12, 379–92.

Ofori, G. (1996) International contractors and structural changes in host-country construction industries: case of Singapore. *Engineering, Construction and Architectural Management*, 394: 379–92.

Ofori, G. (2000) Globalisation and construction industry development: research opportunities. *Construction Management and Economics*, 18: 257–62.

Ofori, G. (2006) Revaluing construction in developing countries: a research agenda. *Journal of Construction in Developing Countries*, 11(1): 1–16.

Ozorhon, B., Arditi, D., Dikmen, I. and Birgonul, M.T. (2008) Effect of partner fit in international construction joint ventures. *Journal of Management in Engineering*, 24(1): 12–20.

Parfitt, M.K., and Sanvido, V.E. (1993) Checklist of critical success factors for building projects. *Journal of Management Engineering*, 9(3): 315–27.

Peng, M.W. (2006) *Global Strategy*. Cincinnati, OH: South-Western Thomson.

Reeves, K. (2002) Construction business systems in Japan: general contractors and subcontractors. *Building Research & Information*, 30(6): 413–24.

Reve, T. and Levitt, R.E. (1984) Organisation and governance in construction. *Project Management*, 2: 17–25.

Rogerson, C.M. (1999) *Building Skills: Cross-Border Migrants and the South African Construction Industry*. Migration Policy Series No. 11. Cape Town: Idasa.

Sabel, C.F. (1993) Studied trust: building new forms of cooperation in a volatile economy. *Human Relations*, 46: 1133–70.

Sako, M. (1991) The role of trust in Japanese buyer–supplier relationship. *Ricerche Econokmiche*, 41: 449–74.

Sako, M. (1992) *Prices, Quality and Trust: Inter-Firm Relations in Britain and Japan*. Cambridge: Cambridge University Press.

Schaufelberger, J.E. (2003) Causes of subcontractor business failure and strategies to prevent failure. *Construction research congress, winds of change: integration and innovation in construction*. 19–21 March, Honolulu: ASCE Press; 593–9.

Shen, L.Y., Wu, G.W.C. and Ng, C.S.K. (2001) Risk assessment for construction joint ventures in China. *Journal of Construction Engineering and Management*, 127(1): 76–81.

Skoyles, E.R. and Skoyles, J.R. (1987) *Waste Prevention on Site*. London: Mitchell.

Sornarajah, M. (1992) *Law of International Joint Ventures*. Singapore: Longman.

Sridharan, G. (1995) Cultural influence on joint venture performance: a study of the Singapore MRT project, unpublished Ph.D. thesis, University of London.

Sridharan, G. (1997) Factors affecting the performance of international joint ventures – a research model. 1st International Conference on Construction Industry Development, Singapore, Vol 2: 84–91.

Syben, G. (2000) Contractors take command: from a demand-based towards a producer-oriented model in German construction. *Building Research & Information*, 28(2): 119–30.

Tam, V.W.Y., Shen, L.Y. and Tam, C.M. (2007) Assessing the levels of material wastage affected by subcontracting relationships and project types with their correlations. *Building and Environment*, 42: 1471–7.

Teece, D.J. (1992) Competition, cooperation, and innovation: organizational arrangements for regimes of rapid technological progress. *Journal of Economic Behaviour and Organization*, 18: 1–25.

Tham, C.M. and Fung, I.W.H. (1998) Effectiveness of safety management strategies on safety performance in Hong Kong. *Construction Management and Economics*, 16: 49–55.

United Nations (1999) *Economics and Social Survey of Asia and the Pacific*. Bangkok: United Nations.

Van Den Heuvel, G. (2005) The parliamentary enquiry on fraud in the Dutch construction industry: collusion as concept between corruption and state–corporate crime. *Crime, Law & Social Change*, 44: 133–51.

Van Der Meer-Kooistra, J. and Vosselman, E. (2000) Management control of inter-firm transactional relationships: the case of industrial renovation and maintenance. *Accounting, Organizations and Society*, 25(1): 51–77.

Wah, C.Y. (2008) The evolution of Chinese Malaysian entrepreneurship: from British Colonial rule to post-New Economic Policy, *Journal of Chinese Overseas*, 4(2): 203–37.

Wang, S.Q., Dulaimi, M.F. and Aguria, M.Y. (2004) Risk management framework for construction projects in developing countries. *Construction Management and Economics*, 22(3): 237–52.

Wells, J. (1996) Labour migration and international construction, *Habitat International*, 20(2): 295–306.

Wells, J. (2006) Labour subcontracting in the construction industries of developing countries: an assessment from two perspectives. *Journal of Construction in Developing Countries*, 11(1): 17–36.

Wharton, A. and Payne, D. (2003) Promoting innovation in construction SMEs: and EU case study. *UNEP Industry and Environment*, April–September, 76–9.

Williamson, O.E. (1975) *Markets and Hierarchies: Analysis and AntiTrust Implications*. New York: Free Press.

Williamson, O.E. (1979) *Transaction Cost Economics: Analysis and Antitrust Implications*. New York: Free Press.

Winch, G.M. (1989) The construction firm and the construction project: a transaction cost approach. *Construction Management and Economics*, 7(4): 331–45.

Winch, G.M. (1998) The growth of self-employment in British construction. *Construction Management and Economics*, 16(5): 531–42.

Wong, R. and So, L. (2002) Restrictions of the multi-layers subcontracting practice in Hong Kong – is it an effective tool to improve safety performance of the construction industry? Presented at the CIBW99 Triennial Conference on Implementation of Safety and Health on Construction Sites: One Country–Two Systems, 7–10 May, Hong Kong.

World Bank (1984) *World Development Report 1994: Infrastructure for Development*. Washington, DC: World Bank.

Yik, F.W.H., Lai, J.H.K., Chan, K.T. and Yiu, E.C.Y. (2006) Problems with specialist subcontracting in the construction industry. *Building Services Engineering Research and Technology*, 27(3): 183–93.

Zawdie, G. and Langford, D.A. (2000) The state of construction and infrastructure in Sub-Saharan Africa and strategies for a sustainable way forward. Paper presented at the 2nd International Conference in Developing Countries: Challenges Facing the Construction Industry in Developing Countries, 15–17 November, Gabarone, Botswana.

Zou, P.X.W. (2006) Strategies for minimising corruption in the construction industry in China. *Journal of Construction Industry in Developing Countries*, 11(2): 15–29.

8 Risks, opportunities and strategies in international construction

A case study of China

Patrick X.W. Zou
University of New South Wales, Australia
and
Dongping Fang
Tsinghua University, China

Introduction

International construction

The world's population is increasing rapidly. It has grown from 2.5 billion in 1950 to 6.5 billion in 2007, and it is expected to keep growing for several decades (Bongaarts and Sinding, 2009). This growth, coupled with expansion in global economic activity and falling trade barriers, has led to massive increases in construction demand around the world, and encouraged the construction firms to expand internationally in order to meet the demand for infrastructure projects worldwide (Javernick-Will and Scott, 2010). Moreover, it is estimated that developing countries will spend US$22 trillion on infrastructure in the next ten years. Some firms have taken up these opportunities, and have had significantly increasing revenues. For instance, the revenues of the top 225 international contractors from projects outside their home countries increased by 18.5 percent between 2005 and 2006 (Javernick-Will and Scott, 2010).

An international construction project is defined as: "one that is located outside the country where the architectural, engineering and construction firm's headquarters is based" (Ling et al., 2008). More recent definitions are wider in their conceptualisation of an international project. One of these (Grisham, 2010: 5) defines such a project as:

> ... one that utilizes resources from or provides services in more than one country, physically or virtually. At one extreme, a project performed in a single country utilizing local people who have international backgrounds is an international project. At the other extreme, a project that utilizes resources from, and provides services in, multiple countries is also an international project.

From these definitions, especially the second one, it is clear that international construction markets and projects are complex due to differences that exist between countries in relation to government policies, legal, financial and tax systems, techniques and culture. To succeed in such markets and on such projects, firms must address certain factors and possess certain resources and corporate strengths. These include (Ofori 2006): technical expertise; managerial expertise; financial resources and expertise; risk management skills; ability to manage cultural issues; and ability to deal with political issues.

When entering a new international construction market, firms need to consider the strengths, threats (risks) and opportunities of the company (Gunhan and Arditi, 2005b). The opportunities (and benefits) of the market include: increased long-term profitability; ability to maintain shareholders' return; opening of more markets; and availability of new service areas. The company's strengths in this context include: track record, specialist expertise, capability to deliver project management services, and a network of contacts that can provide critical information (Gunhan and Arditi, 2005a). The threats (risks) that the company might face include: loss of key personnel, shortage of financial resources, inflation and currency exchange rate fluctuations, and interest rate increases.

Many researchers have attempted to analyse different aspects of international construction. The aspects which have been studied include: future business drivers (Flanagan 2005); decision models (Gunhan and Arditi, 2005b); business entry modes (Mohamed, 2003; Xu and Greenwood, 2006); competitiveness (Ofori, 2003; Firat and Huovinen, 2005); success factors (Gunhan and Arditi 2005a); risks (Wang et al., 2007; Fang et al., 2004); and management of culture (Chevrier, 2003).

This chapter focuses on the risks, opportunities and appropriate strategies in international construction, using China as a case study. Therefore, the rest of the chapter will focus on China as a market for international design, construction and project management.

The Chinese international construction market

China's economy has been developing rapidly over the last decade, and there has been increasing demand for new buildings and rising aspirations for excellence in design. China, as one of the largest international construction markets, offers a multitude of growth opportunities for foreign design and construction management businesses. However, foreign investors and practitioners may be hesitant in entering the Chinese market, as the legal system, business and social culture, language and traditions are profoundly different from those of the West. The country's particularly complex social, governmental and economic systems are also hindering the entry of foreign firms into the Chinese market. Before entering the Chinese market, it is important to understand the scale of the opportunities and the associated risks and develop appropriate business strategies in order to maximise the chances of business success in the country.

Researchers have studied different aspects of China's international construction markets and projects from a variety of perspectives. The aspects include: entry strategies and business strategies and success factors (Ling et al., 2005; Ling et al., 2006; Xu and Greenwood, 2006); forms of collaboration, including strategic partnering (Xu et al., 2004; Lu and Yan, 2007), joint ventures (Gale and Luo, 2004); foreign firms' strategic and project management practices (Ling et al., 2007a, 2007b); strengths, weaknesses, opportunities and threats (SWOT) analysis of the market (Ling and Gui, 2009); benefits from operating in China (Ling, 2005); risks which might be encountered by firms (Fang et al., 2004; Zou et al., 2007; Liu et al., 2010; Zou and Zhang, 2009); and culture, relationships (*Guanxi*), trust and coordination (Low and Leong, 2000; Wang, 2007; Chen and Partington, 2004; Jin and Ling, 2005a, 2005b).

This chapter begins with an overview of the major opportunities in the Chinese property, design and construction market, such as high-end design and green building technology. It then considers the major risks in the market including those in the categories of policy, social, cultural, safety, cost and quality as well as relationship-based businesses environment. The strategies for responding to the risks are next discussed. Finally, the key success factors and strategies for establishing and running construction businesses in China are proposed.

Opportunities

With the increase in demand for new construction, the aspiration for better and more sustainable design, as well as the accession of the country to the World Trade Organization (WTO), China offers attractive opportunities for foreign businesses with respect to investments, design and construction management. In this section, such opportunities are considered.

Demand for housing and infrastructure

China is considered to be the largest building market in the world, and it is estimated that nearly half of the world's new construction will take place in that country during the next five to ten years (Austrade, 2008). The construction spending in China increased by 165 percent in the last four years (up to 2007), and has been growing at 23 percent annually (NBSC, 2007). In China's 11th Five-Year Plan, 2006–10, it was estimated that about 2 billion square metres of new buildings would be constructed every year during the plan period. China's process of urbanisation is expected to lead to increased demand for housing; the continuing fast pace of internationalisation will give rise to high-quality industrial and commercial properties; and more infrastructure and new village development are also expected (MOC, 2007a). In addition, the demand for sophisticated construction equipment and technology is expected to grow as the development of China's infrastructure continues, especially in the western region of the country.

Demand for better design and quality

There is a significant shift in design awareness and greater appreciation of design in China (China–Britain Business Council, 2005; Navaratnam, 2002). An increasing number of domestic developers are looking for high-quality architectural services and construction firms and practitioners with experience in the delivery of large-scale projects. Foreign design and construction businesses should be able to find a good market niche in high-end projects.

Demand for sustainable design and green building technologies

There are also emerging opportunities for exporting sustainable design and green technologies and products to China (Austrade, 2008; Austrade 2009a, 2009b). The rapid pace of development in China is resulting in increasing pressure from the consumption of resources and energy as well as environmental and ecological deterioration. As a result, sustainable development has become the top priority of the Chinese government and several policies have been launched to encourage enterprises and industries to strive to attain sustainable development. For example, China had the following plans: an investment of approximately US$400 billion in energy efficiency projects before 2010; refurbishment of existing buildings in medium and small cities; and tax rebates and related policies, and other financial incentives, for the construction and purchase of energy efficient buildings (Austrade, 2008).

China's accession to the WTO

The impact of China's accession to the WTO has been, and will continue to be, a progressive reduction in its high import and investment barriers. The latest government policy, Decree 114, issued by the Ministry of Construction (MOC, 2007b) in January 2007, allows foreign-invested design enterprises (FIDEs) to apply for architecture, engineering and design licences in China, and the Science and Technology Improvement Law of China encourages enterprises to import advanced technology and equipment through technology consultation.

Specific areas providing opportunities for overseas enterprises

The export opportunities indentified by Austrade (2009c) for Australian (and also other overseas) design and construction firms include: pre-construction services such as spatial information, architecture, design and engineering; sustainable building design; construction services and project management; construction materials such as timber and steel products; building products and tools; architectural products; safety products; heating, ventilation and air-conditioning (HVAC) and lighting systems; building security and maintenance systems; and energy and water efficiency products.

Table 8.1 presents a summary of the business opportunities in China which are highlighted in the above discussion.

Risks

While China's economic development has been increasing rapidly, there are certain associated risks. For example, the foreign direct investment (FDI) that flows into the country is almost exclusively focused on the coastal belt, which has led to a widening gap between that part of the country and the western areas in social and economic development. The political and economic risks from the possible results of this disparity such as social

Table 8.1 Business opportunities in China

- China's strong infrastructure demand (China's 11th Five Year Plan 2006–11)
  - Foreign investment is needed to sustain China's infrastructure development
- China's strong housing demand (Austrade, 2008; NBSC, 2007)
  - To improve the living standards of its people
  - To meet the demands of the country's growing population
  - To address the implications of demographic changes (i.e. ageing population and fewer persons per household)
- Rising aspirations for better designs (China–Britain Business Council, 2005)
  - Growing design awareness
  - Greater appreciation of good design
- Reforms (MOC, 2007)
  - China's WTO accession is creating a more conducive investment environment for foreign enterprises
- Desire for sophisticated construction equipment and technologies in China (Austrade, 2008, 2009c)
  - The country is looking for alternative methods to improve construction quality and productivity
- China's desire for sustainable development (Austrade, 2008, 2009b)
  - Increasing demand for sustainable design and technologies

instability and unrest can disrupt the operations of organisations undertaking business in China. To gain a better understanding of the challenges within China's design and construction industry, it is useful to examine some characteristics and risks of the industry.

China's design and construction market remains highly competitive due to the large number of players (there are over 47,500 large construction enterprises and some 10,800 design institutes), resulting in low profit margins in the industry (Navaratnam, 2002). Local companies have three major advantages over foreign players in China: the ability to offer competitive rates, contacts (*Guanxi*), and good knowledge of the local environment. China's implementation of strict regulations and complex arrangements hinders the repatriation of funds. Since China's entry into the WTO, the State Administration of Foreign Exchange has been undertaking gradual relaxation of some of the country's foreign exchange control measures. However, under current rules, all capital account transactions are still subject to approval by that agency (Aslam, 2002).

Many foreign firms express concern about the complexity of the Chinese national and local laws and the uniqueness of the social, economic and political systems in the country. Although efforts are being made to improve the legislation, foreign firms still have doubts about the comprehensiveness and enforceability of the laws (PricewaterhouseCoopers, 2004). The differences in leadership and management styles between the Chinese and their counterparts from Western countries (Chen and Lee, 2008) is also a potential risk.

Social risks such as cultural misunderstanding can lead to missed business opportunities. *Guanxi* is a form of social networking in China. Relationship building is a fundamental strategy in Chinese business (So and Walker, 2006). In Chinese business, relationships are one of the most important company assets. Therefore, understanding *Guanxi*, and how to cultivate and manage it, can be a valuable tool for foreign companies doing business in China.

Other risks that foreign enterprises might face are the inadequate capabilities of the local personnel, the lack of management experience, and the cultural differences between local and foreign personnel. These factors may lead to human resources challenges.

Zou and Wong (2008) identified the main risks for foreign firms operating in China as: difficulties with the repatriation of capital and earnings from China; the delays in debt collection which may lead to problems with cash flow; problems in securing personnel from the home country to work long-term in China; selecting the right business structure; and negotiating through China's complex tax system.

Table 8.2 shows a summary of the typical risks that foreign firms offering design and construction management consultant services may face when operating in China. An example of the importance of risk management is presented in the discussion of the development of the Beijing National Olympic Stadium (see Case Study 1).

Table 8.2 Risks that might face foreign firms operating in China

- **Political risks**
  - Possible governmental instability
  - Socio-economic conditions
  - Investment profile
  - Corruption
  - Law and order, and related issues. (Hill, 2005).
- **Economic risks**
  - Weak international standards
  - Government involvement (Hill, 2005)
- **Competition risks**
  - Large number of players resulting in low profit margins for the industry as a whole
  - Local players have contacts and knowledge of local environment (Navaratnam, 2002)
  - Chinese government imposes stringent restrictions on foreign investors (Navaratnam, 2002).
- **Complex tax regime, exchange and regulation risks**
  - Discrepancies among central and provincial and local government tax regulations could hamper foreign investment (US–China Business Council, 2006).
  - Exchange controls prevent foreign firms operating in China from repatriating income and capital gains (Zou et al., 2007; Ling and Lim, 2007; Wang, 2007).
- **Human resource risks**
  - Differences in social values, language and culture between Western and Chinese people (Wilfred, 1997).

Data collection and analysis

The data for the study were collected via a postal questionnaire-based survey, which comprised two sections: (i) key risks; and (ii) business strategies. On the risks, the respondents were asked to rate the likelihood of occurrence and then the level of impact of a risk that was indicated, using a 4-point scale – "minimal", "low", "moderate" and "high". These were assigned the following values: 0.1, 0.3, 0.6 and 0.9 respectively. The likelihood of occurrence and level of impact values were then multiplied with one another to obtain the "Risk Significance Index" (see Table 8.3).

Table 8.3 Risk significance index matrix

Likelihood of occurrence	*Level of impact*			
	Minimum (0.1)	*Low (0.3)*	*Moderate (0.6)*	*High (0.9)*
Minimum (0.1)	0.01 (low)	0.03 (low)	0.06 (low)	0.09 (low)
Low (0.3)	0.03 (low)	0.09 (low)	0.18 (low)	0.27(medium)
Moderate (0.6)	0.06 (low)	0.18 (low)	0.36 (medium)	0.54 (high)
High (0.9)	0.09 (low)	0.27(medium)	0.54 (high)	0.81 (high)

As shown in Table 8.3, a Risk Significance value equal to, or more than, 0.54 is deemed to be high (H); a value in the range of 0.27 to 0.54 is medium (M); and a value of less than 0.27 is considered as low (L). This is in line with ISO31000/AS4360.

For the strategies section, a 7-point scale was used with 1 as "least effective"; and 7 representing "most effective".

Case study 1: Risks in Beijing Olympic Stadium "Bird's Nest" development

The Beijing National Olympic Stadium, famously known as the "Bird's Nest", was the first Public–Private Partnership (PPP) sports venue in China (Yuan et al., 2010). It was designed by Swiss architects Jacques Herzog and Pierre de Meuron, who are two prominent international architects, showing China's ambition and rise as a prominent economic and political power (Ren, 2008). The development started in October 2002. The pre-qualification for the financing, design, construction and operation of the stadium was initiated by the Beijing Municipal Government (BMG) (Liu et al., 2010).

There were two main parties involved in the construction of the stadium. The government representative was the Beijing State-Owned Assets Management Corporation, which contributed 58 percent of the total investment. The private partner was the China International Trust and Investment Corporation (CITIC), a consortium which comprises the CITIC Group, the Beijing Urban Construction Group, the Golden State Holding Group of the United States, and the CITIC Group affiliate Guoan Elstrong, which contributed 42 percent of the investment (Yuan et al., 2008). Two French companies provided project management advice to the project partners. They were: Vinci Construction Grands Projects and Bouygues Batiment (Liu et al., 2010).

Based on a Build Operate Transfer (BOT) PPP model, the initial intention was that the project partners had to bear the profit or loss relating to the project on their own. The government provided various incentives to support the project: BMG provided land at very low cost; while BMG contributed 58 percent of the total investment, it would not receive any dividend from the project revenue; BMG also provided the necessary infrastructure connections to the site and helped to create convenient conditions for the construction and operation of the stadium; during the Test Competitions/Events and the Olympic Games, the Beijing Organising Committee for the Olympic Games (BOCOG) agreed to pay fees to the project partners;

and during the 30-year concession period, BMG is not allowed to develop a new stadium to compete with the Bird's Nest, nor to expand any existing competitive stadium in the northern area of Beijing to be a competitor (Liu et al., 2010).

The responsibility of the project partners includes investing, financing, designing, constructing, operating, and maintaining the stadium throughout the concession period, and transferring the stadium at no cost to the government at the end of the concession period (Liu et al., 2010).

Project risks

From the beginning, various disputes arose among the project partners. All the partners wanted to gain profits from the construction work. Thus, the partners failed to establish good and independent control over the construction. There were also disputes between the project partners and BMG, especially relating to design changes, such as the reduction of parking areas, the reduction of the commercial areas in the stadium, and the cancellation of the retractable roof. These changes reduced significantly the project partners' projected sources of revenue (Liu et al., 2010).

There were also complaints from Chinese architects and academics concerning the major involvement of foreign architects in the project and the recent trend for bigger, newer, and foreign design that is expensive and can cause safety problems. Prominent petitioners criticised the design of the stadium by highlighting potential problems in safety and stability as well as creating unnecessary waste and risk (Ren, 2008). This aspect is discussed in further detail in the next section.

Cancellation of the retractable roof

Due to numerous complaints concerning the extravagant and complex design of the stadium, the government decided to halt the project and conduct a financial review. The government finally accepted the cancellation of the retractable roof that saved 15,000 tonnes of steel and US$50 million (Ren, 2008). This decision also helped to improve safety during the construction as the installation of the roof would be a difficult operation. On the other hand, the cancellation represented a major design change so the design consortium claimed almost one third of the primary design fee for redesigning. The evaluation and redesign process also took time and this caused a six-month delay, cost overruns, and further disputes on who should bear the blame (Liu et al., 2010).

The cancellation of the retractable roof affected the potential post-Olympic operations and revenue in three main ways. First, as opening and closing the roof should have carried an operation fee, the project partners would no longer have that source of income. Against this, though, related high maintenance costs could be saved. Second, without the ability to close the roof the stadium lost its attractiveness as an all-weather stadium. As only an open-air stadium, it became much less suitable for renting out to the exhibition, cultural and tourism markets. Third, the cancellation reduced the stadium's brand value. It no longer had any obvious unique characteristic relative to other large stadiums and the project partners began to worry about the brand value of the project (Liu et al., 2010).

Other project risks

There were many risks involved in the project that had to be managed vigilantly. The stadium was built for the Olympics, thus it had to be completed on time. The project had been halted due to the roof issue; this caused other problems as the schedule was already tight. Furthermore, the stadium also adopted new technologies, new materials, and an innovative design. The tight schedule and complexity of the design exposed the construction process to high risks (Yuan et al., 2010).

Another risk of the project is its long-term market potential after the 2008 Olympics. In 2009, there was only one event scheduled for an opera performance to celebrate the one-year anniversary of the Olympic opening ceremony. One of China's leading soccer clubs cancelled its intention to use the stadium because it would be an embarrassment to use a 91,000-seat stadium for matches that normally only attract 10,000 spectators. The stadium costs US$9 million a year to maintain and the project partners would suffer great loss if they could not market the stadium. There is a plan to turn the stadium into a shopping mall in several years' time (Demick, 2009). The project partners had never operated a stadium before, which is also a risk that should be considered. They had entered into an agreement with Stade de France to acquire relevant knowledge and experience, but the high consultancy fee led to the termination of that agreement. Thus, the project partners have to develop the necessary expertise by themselves (Liu et al., 2010).

Lessons learned

A study by Liu et al. (2010) summarised the lessons learned from this project. First, the government's support and commitment are important for this kind of project. Without this support most of such projects will not be viable. Second, a project's scope should be well defined before agreements or contracts are signed. The design and construction contracts should be formulated clearly to avoid future disputes. While short-term national prestige and pressure may press for speedy development, concerns about long-term public interest require that such projects are based on a sound contractual and commercial footing.

Third, all the parties concerned should share common project objectives in pursuing overall efficiency and cost control during the whole project lifecycle. If one partner is over-concerned with acquiring profits or revenue from some parts of the project only, then the viability and integrity of the project as a whole is likely to suffer. Yuan et al. (2010) suggested that an agreed level of performance should be decided upon as one of the objectives before the commencement of the project in order to reduce conflicts caused by different expectations of different stakeholders. Fourth, proper risk management is critical. In particular, the private sector's interests should be aligned with those of the public sector and there should be clear contractual arrangements as well as implementation of risk management arrangements.

Fifth, when disputes occur, renegotiation among the partners and especially with government is sometimes more efficient and effective than seeking mediation, arbitration, or pursuing a lawsuit. In the case of the Beijing National Olympic Stadium, renegotiation was essential to complete the project on time (Liu et al., 2010).

Sample population

A total of 823 sets of questionnaires were sent to design and construction companies across Australia. The firms in the sample were obtained from property developers and investors in the 2005 issue of Mahlab's *Who's Who in Property Directory*, 17th Edition. Of the 823 firms in the sample, 287 were architects/designers, 178 were engineering consultancy firms, and 358 were construction companies.

Forty-one responses were received, and 17 of the companies reported that they have business operations in China. The remaining 24 responses were from companies that did not have business operations in China. The respondents not doing business in China were only asked to complete specific sections in the survey questionnaire applicable to them. The company characteristics of the respondents are presented in Table 8.4. The respondents generally hold professional/technical and director/executive positions. The respondents from companies doing business in China most preferred business locations in Beijing and Shanghai. Forty percent of the respondent companies doing business in China have been operating in the country for an average of 8 years. Ninety-four percent of the companies doing business in China have less than 25 percent of their work from China.

Table 8.4 Characteristics of survey respondents

	Doing business in China		*Not doing business in China*	
Respondents' characteristics	Frequency	Percentage	Frequency	Percentage
Designation of respondents				
Professional/Technical	8	47.1	10	59.0
Director/Executive	6	35.3	12	71.0
Managerial	3	17.6	1	6.0
Clerical/Administrative	0	0	0	0
Others	0	0	0	0
Number of employees				
<10	3	17.6	3	12.5
10–100 employees	5	29.4	17	70.8
101–200 employees	1	5.9	1	4.2
201–500 employees	5	29.4	2	8.3
>500 employees	3	17.6	1	4.2
Value of projects (AUD)				
Up to $100,000	1	5.9	0	0
$100,000–$200,000	2	11.8	0	0
$200,000–$500,000	1	5.9	4	23.5
$500,000–$2million	2	11.8	5	29.4
$2million–$5million	1	5.9	1	5.9
Over $5million	10	58.8	13	76.5
Information not provided	1	5.9	1	5.9
Business nature				
Architectural/Design	13	76.5	14	82.4
Engineering	3	17.6	4	23.5
Construction	1	5.9	9	52.9
Project management	2	11.8	6	35.3
Urban planning	14	82.4	8	47.1
Real estate	1	5.9	3	17.6
Landscape design	8	47.1	1	5.9
Others	1	5.9	1	5.9

Results and discussion

Risks

The risks were classified into five categories: legal, political and economic; operational; financial/tax; human resource management; and others. The survey results are presented in Table 8.5. To the respondents doing business in China, the key risks were (in ranking order):

1 Lack of enforceable intellectual property rights;
2 Debt collection;
3 Knowledge of local business environment;
4 Restrictions on the transfer of capital and profits out of China;
5 Business profitability.

Moreover, there were seven medium-level risks and the remaining ten were deemed to be minor risks.

Risk perceptions

There are major differences in risk perceptions between the respondents whose firms have undertaken projects in China and those from companies without such experience. Generally, those whose firms have been doing business in China appeared to rate the risks at a much lower level than those without firsthand experience in China. For example, while only five risks (22.7 percent) were rated as high (H) by those whose companies have been doing business in China, 16 (72.7 percent) were rated as high (H) by those from firms which had not done any business in China. Furthermore, while 10 risks (45.5 percent) were rated as low (L) by those doing business in China, only one (4.5 percent) was rated as low by those whose firms have no experience in the China market. These results appear to imply that, once a firm has entered the market and gained practical experience through "learning by doing", the risks become less of an issue. They may also be interpreted to mean that, because some of the companies in the second group perceived the level of risks in China as being high, they did not enter the Chinese market.

Market entry modes

There are two types of entry modes: non-equity modes and equity modes. Non-equity modes reflect relatively small commitments to the overseas market. Examples are exporting and entering into contractual agreements. On the other hand, equity modes indicate relatively larger and harder-to-reverse commitments to the overseas markets. Some examples of these modes of entry are wholly foreign-owned enterprises and joint ventures

Table 8.5 Risks and their significance levels (levels: H–high, M–medium, L–low)

	Those doing business in China		*Those not doing business in China*	
Risk factors	*Risk index*	*Risk Level*	*Risk index*	*Risk Level*
Legal, political and economic				
Lack of enforceable intellectual property rights	0.484	H	0.458	H
Business profitability	0.360	H	0.446	H
Lack of transparency and consistency in the Chinese legal and political system	0.344	M	0.540	H
Legal protectionism (favouring local Chinese organisations)	0.328	M	0.395	H
Inconsistent application and interpretation of government's industry policies at a sub-national level	0.254	L	0.168	L
Environmental regulations	0.204	L	0.368	H
Operational Challenges				
Debt collection	0.444	H	0.532	H
Knowledge of local business environment	0.384	H	0.415	H
Guanxi (relationship) with government	0.337	M	0.466	H
Transparency of local management	0.302	M	0.587	H
Reliability of information provided by local management	0.296	M	0.329	M
Market potential and competitors	0.240	L	0.399	H
Control over operation of business (ownership restrictions)	0.214	L	0.454	H
Financial and Tax				
Restrictions in the transfer of capital and profits out of China	0.382	H	0.321	M
Complex tax regime/regulation	0.237	L	0.348	M
Financial reporting fraud	0.195	L	0.513	H
Human Resource Management				
Differences in culture, language and communication	0.325	M	0.337	M
Management quality and attitudes	0.262	L	0.450	H
Cooperating of the China partner	0.161	L	0.438	H

Others				
Licensing provisions for professional recognition in China	0.271	M	0.423	H
Bribery and corruption	0.230	L	0.321	M
Infrastructure	0.191	L	0.407	H

(Peng, 2006). There are three business structure options which a foreign investor can use to operate in China's market (Crowe Horwath, 2003) and two of these are particularly applicable to design and construction businesses: wholly foreign-owned enterprise, and joint venture. These business structures are now discussed in the context of the market in China.

A wholly foreign-owned enterprise is an entity that is a 100 percent owned by a foreign company (China–Britain Business Council, 2005). China's WTO commitments together with the relaxation of investment regulations have seen wholly foreign-owned enterprises outpace joint ventures (Lasserre and Schutte, 2006). A wholly foreign-owned enterprise provides the investor with complete control over the business direction and operation (China–Britain Business Council, 2005). A joint venture is a business structure where a foreign firm goes into business with a local Chinese partner (China–Britain Business Council, 2005). The primary objective for foreign companies to invest in China via joint ventures is to overcome potential trade barriers in penetrating China's market. However, until now, joint ventures in China have had a high failure rate.

As shown in Table 8.6, more than half of the respondents chose establishing a wholly foreign-owned enterprise as their market entry mode, and about 30 percent chose the joint venture. A survey conducted by PricewaterhouseCoopers (2004) on the issues encountered by foreign companies investing in China found that most respondents' foreign investments in China were in the form of wholly foreign-owned enterprises, which is consistent with the findings of this research. The growing popularity of wholly foreign-owned enterprises over joint ventures in the past few years took place as a direct result of China's accession to the WTO. This move has paved the way for changes in policy that have resulted in design and construction companies being permitted to establish wholly foreign-owned enterprises in China. However, a foreign firm should consider the higher level of scrutiny on the business scope of a wholly foreign-owned enterprise as compared to that of a joint venture when selecting the type of entry mode to adopt for the Chinese market.

As the design and construction business is project-based, the project-oriented joint venture market entry mode may be suitable for the enterprises to adopt when entering new international markets. This is partly because the specific political and macro-economic conditions in the host country may significantly impact project performance (Ng et al., 2007). Moreover, the unique characteristics of each project are highly associated with joint venture performance, and appropriate strategies should be developed to handle particular project risks (Ozorhon et al., 2007). Gale and Luo (2004) suggested that, on international construction projects in developing countries such as China, the five most important factors leading to joint venture success are: selection of partners; clear statement of joint-venture agreement; obtaining information about potential partners; partners' objectives; and control of the ownership of the capital. A successful example of a joint-venture project in China is the design of the Beijing Olympic Games National Swimming Centre (the "Water Cube") (Zou and Leslie-Carter, 2010), presented in Case Study 2.

Table 8.6 Respondents' preferred market entry modes into China

Rank	*Respondent's entry mode into China*	*Frequency*	*Percentage*
1	Wholly foreign-owned enterprise	9	52.9
2	Joint venture	5	29.4
3	Establishment	2	11.8
4	Others	1	5.9

Case Study 2: Beijing Olympic "Water Cube" international joint venture project

This case study is on a large-scale international project procured through a project-based joint venture strategy.

The Beijing 2008 Olympic Swimming Centre (popularly known as the "Water Cube") project was the result of an international design competition with 10 short-listed participants, judged by a panel of architects, engineers and eminent Chinese academics in 2003, and the winner was a Sydney-based joint-venture team consisting of ARUP, PTW Architects and China Construction Design International (CCDI). This team consisted of more than 100 engineers and specialists, spread across 20 disciplines and four countries. The project objectives included: create the best Olympic swimming venue; spend no more than US$100 million; create a green Games; create a hi-tech Games; and create a peoples' Games. ARUP Project Management was leading and coordinating the design process, and managing both the internal and external interfaces. It was a fast-track programme with the design delivered from competition stage through to a fully approved scheme and continued through to the official opening of the Water Cube. As well as delivering a fully coordinated scheme design, it also involved regular handing over of the design to the Chinese design partners for detailing, whilst ensuring that the technical approvals were all secured and that the innovative design was understood, accepted and then constructed safely. Ensuring the Water Cube became reality was achieved by establishing and maintaining clarity of the design vision, and full and transparent collaboration between the joint venture partners.

The key threads of the project implementation strategy covered everything from establishing a communication strategy, through to the dynamics of team leadership, a risk management strategy focused on the complex and dynamic nature of the Chinese market, and management of differences between Chinese and Australian stakeholders. It was a challenge to manage the contribution of 20 specialist engineering disciplines, and to ensure that they were properly coordinated and the complex interfaces of the Water Cube were properly understood and documented.

Forming an international joint venture partnership

The unusual thing about the Beijing Olympics is that international designers were invited to participate in the projects. This was not the case in Sydney and other previous Olympic host cities. One reason was that the challenge was of such a huge scale that Beijing recognised it needed solutions from home and abroad. This attitude set the tone for a genuine two-way collaboration on the Water Cube – where Western and Eastern perspectives worked together with mutual respect and openness.

The Water Cube team also came about after some deliberate relationship building by ARUP and PTW in the build-up to the international design competition. In 2003, Sydney had the halo effect of having just hosted "the best Olympic Games ever" and having built what was regarded as the fastest pool ever, which had been designed by ARUP and PTW. ARUP had also recently designed the Shenzhen Aquatic Centre from its Sydney office, so it understood some of the challenges of working in China as an international firm. Specifically, the opportunity to align with the Chinese design partners CCDI and their parent company China State Construction Engineering Corporation (CSCEC) (China's largest construction firm) came about from building up relationships and ARUP's reputation through a series of visits to China to make presentations on the firm's profile and the engineering behind "fast pools", and to discuss the opportunities for collaboration on projects for the Beijing Olympic Games.

The Water Cube was generated by equally integrating the requirements of ARUP's engineering, PTW's space planning, and Chinese cultural influences on the architecture from CCDI. It was not the result of any one single dominant party.

Establishing a legacy

As the cliché reads, there are only three things that matter when it comes to the Olympic Games, "Legacy, Legacy, Legacy". There were legacy-building opportunities that directly benefited the team relationship and the final outcome of the Water Cube. An ongoing challenge during the contract negotiations was the inclusion of standard clauses to protect intellectual property and copyright over design ideas and documentation. At the implementation planning workshop, the project managers presented the benefits of embracing a clear and simple policy that collaboration among all design partners be total and completely transparent. This was fundamental to establishing and maintaining trust and respect at the start of the project. In design terms, this involved accepting that the concepts and analytical approaches that were developed would become an important knowledge legacy to help the local partners to develop their capabilities.

The first legacy of the building is the ethylene tetrafluoroethylene (ETFE) façade design, construction and performance. The team members spent a week interviewing ETFE tenderers and being challenged by a panel of Chinese leaders, academics and engineers on various aspects of the ETFE façade design and performance. As an extension to the deliberate legacy-building approach, ARUP lobbied that the ETFE contractors and the people of Beijing would benefit by investing in local manufacturing and processing facilities in Beijing, which the winning tenderer accepted and implemented. This guaranteed local training and employment in the short term, but also led to a long-term local capability in producing an innovative material likely to feature heavily in Beijing's ongoing development programme.

Another often-debated legacy is the totally shared ownership of the Water Cube concept. The philosophy agreed at the implementation planning workshop was that the box of bubbles concept for the Water Cube was generated by integrating the requirements of ARUP's engineering, PTW's space planning, and Chinese cultural influences on architecture from CCDI. This was a powerful statement in terms of the success of the collaboration established among the three project partners.

Finally, for the team members involved, the relationships they have established, and the self-fulfilment they have achieved from being part of such a project, have provided a genuine legacy. As well as achieving the critical acclaim, the project has proved to be a successful investment in developing a project management approach to establishing and leading winning teams, managing relationships with stakeholders across cultures, project management processes required on major multidisciplinary projects, and technological improvements in the firms' technical capabilities.

Managing cultural risks and differences

The questions of how to manage communication both internally and externally, as well as how to handle the relationships among all parties involved in the project, were critical to the success of the Water Cube project. The Water Cube acted as a bridge for cultural exchanges, and deepened the understanding, trust and friendship among the project team members and stakeholders. This was achieved by establishing and maintaining clarity of the design vision, communicating that vision to project stakeholders with differing cultural expectations, and the collaboration between the joint venture parties.

On the Water Cube project, more challenging for the project management team than the technical aspects was learning and understanding the business culture and context in China. It was very difficult to read. To resolve this problem, the implementation plan workshops and follow-up sessions were held with all parties involved in design, particularly with Chinese team members, to agree the approach to the early "management of differences". These workshops partly focused on maintaining leverage over commercial arrangements, but mainly looked at how to minimise and manage the risks of the specific differences in norms, practices and expectations through the implementation of the project.

The complex and dynamic nature of the Chinese market, particularly in the context of the Olympics, meant that the risks associated with delivering the Water Cube project could not be underestimated. Beijing's lack of regulatory transparency, regional differences, as well as a relationship-based business culture, were among the factors identified as making China a challenging project environment.

The project management team identified a diverse range of risks, trying to understand and plan their approach to the project in the unfamiliar context of China's legal, social, cultural, economic and technological environment. Other than the commercial risk of delayed payment, the key risks identified were social – how China's business culture might affect the relationships and dynamics within the international Water Cube team, and those with the external stakeholders involved in approving the design concept.

Social risks such as cultural misunderstandings could have completely derailed or significantly delayed the project. Relationship-building is fundamental in Chinese business, so understanding *Guanxi* and how to authentically cultivate and manage it was vital to the project management team. Other important factors in the approach included emphasising the team's international reputation and the depth and diversity of its activities and locations. Senior engineers from ARUP's Beijing and Hong Kong offices were directly involved at key stages of the approvals process. Their influence and local knowledge of Chinese legislation, coupled with their involvement in other Olympic projects in Beijing, were leveraged to convince some conservative approving authorities to accept innovative approaches to the engineering design that did not follow the rules of the Chinese building codes. This was the number one risk in the early stages of the project, and the formal approval of the engineering design in early 2004 set a major precedent and direction for other Olympic projects.

Another example was the commercial negotiations. The project has been a financial success in that it made an acceptable profit despite the considerable risks of working with international partners and stakeholders, on a fast-track project involving such groundbreaking design techniques and materials. This is largely because the project managers were very specific during contract negotiations to clearly define their scope of services and the interfaces with the Chinese design partners, and were robust in contract negotiations that removed the project management company (i.e. ARUP) from some of the post-Olympic payment milestones that were unrelated to the project scope. By deliberately resolving any potential conflicts early, the project management team was able to sign a contract and facilitate a smooth and seamless handover to the Chinese partners with clearly understood and accepted interfaces.

Overall, the Beijing Olympic Water Cube joint venture project was successful from many perspectives and considerations, as discussed above. On reflection, one aspect that could have been improved was for ARUP to secure a role for the project management team during the construction phase, and also after the games for the conversion to legacy mode.

Source: Zou and Leslie-Carter (2010)

Business development strategies

Table 8.7 shows the results regarding the business strategies adopted by the respondents. Among the 16 possible business strategies, the following five were perceived to be effective in helping companies to achieve business success (in ranking order):

1. Provide speciality/niche services;
2. Build network and contacts in China;
3. Provide superior service;
4. Focus on customer/client satisfaction;
5. Post management staff from head office to manage projects.

These findings are similar to those identified by Zou and Wong (2008) who derived their conclusions through interviews with executives having experience in doing business in China. These findings also reflect China's unique market system and the relevant business operations.

According to Porter (1985), there are two strategies which companies can adopt in entering and sustaining a presence in a new market – *differentiation* and *cost leadership*. In a differentiation strategy, a firm seeks to be unique in its industry by selecting one or more attributes that many buyers in the industry perceive as important, and uniquely positions itself to meet those needs. The firm is rewarded for its uniqueness with a premium price for its product or service (Porter, 1985). Foreign design and construction businesses in China tend to dominate the high-end segments of the markets in which they operate. Under the cost leadership strategy, a firm sets out to become the low-cost producer in its industry. A low-cost producer must find and exploit all sources of cost advantage (Porter, 1985). Where a firm can achieve and sustain overall cost leadership, it will be an above average

Table 8.7 Strategies for business development in China

Rank	*Business strategies*	*Frequency*	*Mean*	*Standard deviation*
1	Provide specialty/niche services	12	6.65	2.98
2	Build network and contacts (Guanxi) in China	11	6.18	3.47
3	Provide superior service	9	6.10	3.18
4	Focus on customer/client satisfaction	7	6.00	3.07
5	Post management staff from head office to manage projects	6	5.12	2.37
5	Obtain political backing in China	6	5.12	2.37
6	Employ Chinese citizens to manage projects	4	4.71	1.85
6	Head office provides good support to office/s in China	4	4.71	1.85
7	Establish overall competitiveness	3	4.35	1.65
8	Reduce project time scale	2	3.94	1.53
8	Provide packaged/integrated services	2	3.94	1.53
8	Establish price competitiveness	2	3.94	1.53
8	Employ professional staff from China	2	3.94	1.53
9	Undertook in-depth studies of China before entry	1	2.68	1.50
9	Post professional staff from head office	1	2.68	1.50
10	Engage subcontractors from China	0	0	0

performer in its industry, provided it can command prices at, or near, the industry average. In the China market, differentiation by providing specialty and niche services or products or superior services is the only way forward while there is little scope for cost leadership. This is mainly because the Chinese firms have an abundant supply of cheap labourers and skilled workers. The local materials are also cheap. In contrast, they lack high-end technology and expertise that can provide high-level speciality and niche services or products.

Being "*customer focused*" is a universally recognised business concept. Customer satisfaction is a key business performance indicator. Good performance on the project of a major and influential client can lead to several other commissions and successful bids from the same client, or other clients. According to Zou and Wong (2008), securing personnel from the host country to work long-term in China is one of the major challenges. "*Post management staff from head office to manage projects*" would be relatively easy to accomplish if the personnel are to operate in China over a short period. However, it becomes a challenge in the long term because many people do not wish to work overseas for a very long period of time (Zou and Wong, 2008).

Building a network and contacts (Guanxi) in China and obtaining political backing in China could be a major challenge to overseas firms, due to the differences in political systems and culture between the home and host countries. It requires not only an understanding of the Chinese system but also excellent inter-personal and communication skills and an awareness of cultural differences. *Guanxi*, in broad terms, is about building lifelong relationships and is probably the most important single asset of any foreign business in China (China–Britain Business Council, 2005). To do business in China, one has to make the right connections by meeting the right people, developing long-term relationships with them, and making sure that favours and generosity are mutual (China–Britain Business Council, 2005).

One strategy that can be applied to tackle this challenge would be to employ those who have educational and working experience in both the company's home country and China to be the "bridge" – the communication channel. Employing such bi-cultural, bi-lingual and bi-technical and managerial people may also help to address the problems the firm may encounter in the posting of management personnel from the head office to China.

Implications of the findings of the study

The research findings presented in this chapter have several implications for the Chinese government's policies and those of foreign enterprises operating in (or planning to enter) the China market. For the Chinese government, there is a need to review its legal, political and economic systems in order to facilitate the establishment of businesses in the country by foreign companies, and to allow them to transfer capital and profit out of China. While it is difficult to change a nation's business and social culture, it is time for China to consider its ways of doing business such that they would be more closely aligned with the "international ways of doing business", which are less dependent on personal contacts (*guanxi*), but instead are based more on the company's competitive advantages and strengths as well as contract conditions (Zou and Wong, 2008).

For enterprises planning to enter an international construction market such as the one in China, it is necessary for them to assess the risks associated with the opportunities. The method and the inferences from the results presented in this chapter could be a useful starting point in identifying, assessing and responding to the opportunities and risks. It is important to point out that while the perception of risks in the market of a particular country may be high, the results of this research show that those firms that have entered an international construction market may perceive it to be less risky than those without actual experience in that market. Once the risks and opportunities have been identified and assessed, and appropriate response strategies have been put in place, it is necessary to decide which mode to use to enter the market. Each of the modes adopted by foreign construction firms in China, "wholly-owned" and "joint-venture", has its advantages and drawbacks.

Once the business is established in any market, it is important to put in place strategies for business positioning and development. For enterprises entering an international construction market in a developing country, a common strategy would be to provide a niche or speciality service. Focusing on client satisfaction is an essential strategy for today's businesses, regardless of the location and type of business and international construction is no exception in these regards. Developing a network of contacts is vital to the success of any business. While it is commonly referred to as *guanxi* in China, it is similar to the term 'relationship' in other contexts, but the basic reasons and fundamentals are similar. *Guanxi* is about "who you know" and includes personal emotional affections between the parties.

Conclusion

The aim of this chapter was to discuss the risks and opportunities in, and appropriate strategies for entering, an international construction market by using China as an example. Despite the attractive opportunities in the Chinese market, overseas design and construction management firms would face challenges in the attendant threats and risks. This research analysed the risks of the foreign firms when doing business in China. The key risks identified included the difficulties relating to the repatriation of capital and earnings; intellectual property protection; debt collection; and administrative and bureaucratic transparency. For the market entry strategies, wholly foreign-owned enterprise was found to be the most commonly preferred. It is also believed that providing a niche or superior product or service, the cultivation of a network of contacts in China, focusing on customer satisfaction, and being culturally adaptive were effective business strategies.

It should be pointed out that the response rate of the study was low. Nevertheless, the results provided indicative references for researchers and practitioners. Although the data were collected on the Chinese construction market, the research findings are applicable to other international markets. The methods and findings of this research may provide a starting point for the construction enterprises which plan to enter new international markets with respect to their decisions on issues including assessing the risks and comparing them with the opportunities, and deciding on the mode of entry into the particular new market as well as formulating the business development plans.

Acknowledgement

The authors wish to acknowledge that some contents of this chapter are based on the research undertaken by Ms Alison Wong for her dissertation under the supervision of the first author.

References

Aslam, N. (2002) Reform and growth in China: Prospects for foreign enterprises. *Thunderbird International Business Review*, 44(3): 437–43.

Austrade (2008) Green building to China. Retrieved 16 November 2008 from http://www.austrade.gov.au/Green-building-to-China/default.aspx/

Austrade (2009a) Country overview. Retrieved 20 April 2009 from http://www.austrade.gov.au/China-Market-Overview/default.aspx/

Austrade (2009b) Environmental technology to China. Retrieved 20 April 2009 from http://www.austrade.gov.au/Environmental-technologies-to-China/default.aspx/

Austrade (2009c) Building and Construction Overview. Retrieved 20 April 2009 from http://www.austrade.gov.au/Building-and-construction-overview/default.aspx/

Bongaarts, J. and Sinding, S.W. (2009) A response to critics of family planning programs. *International Perspectives on Sexual and Reproductive Health*, 35(1): 39–44.

Chen, C. and Lee, Y. (2008) *Leadership and management in China: Philosophies, theories and practices*. Cambridge: Cambridge University Press.

Chen, P. and Partington, D. (2004) An interpretive comparison of Chinese and Western conceptions of relationships in construction project management work. *International Journal of Project Management*, 22(5): 397–406.

Chevrier, S. (2003) Cross-cultural management in multinational project groups. *Journal of World Business*, 38(2): 141–9.

China–Britain Business Council (2005) China-Britain Business Council, Beijing. Retrieved 28 November 2005 from http://cbbc.org/the_review/review_archive/sectors/21.html/

Crowe Horwath (2003) *International tax planning manual*. New York: Crowe Horwath.

Demick, B. (2009) Beijing's Olympics building boom becomes a bust. *Los Angeles Times*, 22 February. Retrieved 14 September 2010 from http://articles.latimes.com/2009/feb/22/world/fg-beijing-bust22/2/

Fang, D., Li, M., Fong, P.S. and Shen, L. (2004) Risks in Chinese construction market – Contractors' perspective. *Journal of Construction Engineering and Management*, 130(6): 853–61.

Firat, C.E. and Huovinen, P. (2005) Entering regional construction markets in Russia. In Kahkonen, K. and Porkka, J. (eds) *Global Perspectives on Management and Economics in the AEC Sector*. Technical Research Centre of Finland and Association of Finnish Civil Engineers, Helsinki, pp. 57–68.

Flanagan, R. (2005) The synergy between business and global drivers in futures planning for construction enterprises, In Kähkönen, K. and Sexton, M. (Eds) *Understanding the construction business and companies in the new millennium*. VTT Helsinki pp. 42–53.

Gale A. and Luo J. (2004) Factors affecting construction joint ventures in China. *International Journal of Project Management*, 22(1): 33–42.

Grisham T.W. (2010) *International project management: Leadership in complete environment*. Hoboken, NJ: John Wiley and Sons.

Gunhan, S. and Arditi, D. (2005a) Factors affecting international construction. *Journal of Construction Engineering and Management*, 131(3): 273–82.

Gunhan, S. and Arditi, D. (2005b) International expansion decision for construction companies. *Journal of Construction Engineering and Management*, 131(8): 928–37.

Hill, C.W.L. (2005) *International business: Competing in the global marketplace*. 5th edn. New York: McGraw Hill/Irwin.

Javernick-Will, A.N. and Scott, W.R. (2010) Who needs to know what? Institutional knowledge and global projects. *Journal of Construction Engineering and Management*, 136(5): 546–57.

Jin, X. and Ling, F.Y.Y. (2005a) Constructing a framework for building relationships and trust in project organizations: two case studies of building projects in China. *Construction Management and Economics*, 23(7): 685–96.

Jin, X. and Ling, F.Y.Y. (2005b) Model for fostering trust and building relationships in China's construction industry. *Journal of Construction Engineering and Management*, 131(11): 1224–32.

Lasserre, P. and Schutte, H. (2006) *Strategies for Asia Pacific: Meeting new challenges*. New York: Palgrave Macmillan.

Ling, F. and Lim, H.L. (2007) Foreign firms' financial and economic risk in China. *Engineering, Construction and Architectural Management*, 14(4): 346–62.

Ling, F.Y.Y. (2005) Benefits that foreign AEC firms derive when undertaking construction project in China. *Management Decision*, 43(4): 510–15.

Ling, F.Y.Y. and Gui, Y. (2009) Strengths, weaknesses, opportunities, and threats: Case study of consulting firms in Shenzhen, China. *Journal of Construction Engineering and Management*, 135(7): 628–36.

Ling, F.Y.Y., Ibbs, C.W. and Chew, E.W. (2008) Strategies adopted by international architectural, engineering, and construction firms in Southeast Asia. *Journal of Professional Issues in Engineering Education and Practice*, 134(3): 248–56.

Ling, F.Y.Y., Ibbs, C.W. and Cuervo, J.C. (2005) Entry and business strategies used by international architectural, engineering and construction firms in China. *Construction Management and Economics*, 23(5): 509–20.

Ling, F.Y.Y., Ibbs, C.W. and Hoo, W.Y. (2006) Determinants of international architectural, engineering, and construction firms' project success in China. *Journal of Construction Engineering and Management*, 132(2): 206–14.
Ling, F.Y.Y., Low, S.P., Wang, S.Q. and Egbelakin, T.K. (2007a) Foreign firms' strategic and project management practices in China. International construction management – China. In W. Hughes (ed.) *Proceedings of Construction Management and Economics, Past, Present, and Future*, Reading, University of Reading, CME 25 Conference, 16–18 July.
Ling, F.Y.Y., Low, S.P., Wang, S.Q. and Lim, H.H. (2007b) Key project management practices affecting Singaporean firms' project performance in China. *International Journal of Project Management*, 27(1): 59–71.
Liu, Y.W., Zhao, G.F. and Wang, S.Q. (2010) Many hands, much politics, multiple risks – The case of the 2008 Beijing Olympics Stadium. *The Australian Journal of Public Administration*, 69(S1): S85–S98.
Low, L.S. and Leong, C.H.Y. (2000) Cross-cultural project management for international construction in China. *International Journal of Project Management*, 18(5): 307–16.
Lu, S. and Yan, H. (2007) A model for evaluating the applicability of partnering in construction. *International Journal of Project Management*, 25(2): 164–70.
MOC (Ministry of Construction) Policy Research Centre (2007a) *Reform and development: A report on construction industry and market in China*. Beijing: China Architecture and Building Press (in Chinese).
MOC (Ministry of Construction) (2007b) *Decree 114*. Issued by the Ministry of Construction on 5 January 2007.
Mohamed, S. (2003) Performance in international construction joint ventures: Modeling perspective. *Journal of Construction Engineering and Management*, 129(6): 619–26.
Navaratnam, P. (2002) *Building construction and engineering*. Singapore: IE Singapore.
NBSC (National Bureau of Statistics of China) (2007) *2007 year book*. Beijing.
Ng, P.W.K., Lau, C.M. and Nyaw, M.K. (2007) The effect of trust on international joint venture performance in China. *Journal of International Management*, 13(4): 430–48.
Ofori, G. (2003) Frameworks for analysing international construction. *Construction Management and Economics*, 21(4): 379–91.
Ofori, G. (2006) Chinese contractors and international construction: Tentative analytical models and research agenda. *Proceedings of the CRIOCM 2006 International Symposium on Advancement of Construction Management and Real Estate*, Beijing, China, 3–5 November.
Ozorhon, B., Arditi, D., Dikmen, I. and Birgonul, M.T. (2007) Effect of host country and project conditions in international construction joint venture. *International Journal of Project Management*, 25(8): 799–806.
Peng, M.W. (2006) *Global strategy*. Cincinnati: Thomson South-Western.
Porter, M. (1985) *Competitive advantage*. New York: The Free Press.
PricewaterhouseCoopers (2004) *Doing business in China: A survey of issues encountered by foreign companies doing business or investing in China*. Hong Kong: PricewaterhouseCoopers.

Ren, X. (2008) Architecture and nation building in the age of globalization: Construction of the National Stadium of Beijing for the 2008 Olympics. *Journal of Urban Affairs*, 30(2): 175–90.

So, Y.L. and Walker, A. (2006) *Explaining guanxi: The Chinese business network*. London: Routledge.

Wang, C.L. (2007) *Guanxi* vs. relationship marketing: Exploring underlying differences. *Industrial Marketing Management*, 36(1): 81–6.

Wang, X.W., Pham, L. and Fang, D.P. (2007) Risk assessment of Australian construction and engineering firms in Chinese market. *The CRIOCM2007 International Symposium on Advancement of Construction Management and Real Estate*, 8–13 August, Sydney, Australia.

Wilfred, V. (1997) Entering China: An unconventional approach. *Harvard Business Review*, 75(2): Mar-Apr, 130–1, 13406, 138–40.

Xu, T. and Greenwood, D. (2006) Using design-and-build as an entry strategy to the Chinese construction market. *International Journal of Project Management*, 24(5): 438–45.

Xu, T., Smith, N.J., Bower, D.A. and Chew, A.S. (2004) Development strategies for Chinese design institutes. *Journal of Management in Engineering*, 20(2): 62–9.

Yuan, J., Skibniewski, M.J., Li, Q. and Zheng, L. (2010) Performance objectives selection model in Public–Private Partnership projects based on the perspective of stakeholders. *Journal of Management in Engineering*, 26(2): 89–104.

Zou, P.X.W. and Leslie-Carter, R. (2010) Lessons learned from managing the design of the "Water Cube" national swimming centre for the Beijing 2008 Olympics Games. *Architectural Engineering and Design Management*, 6(3): 175–88.

Zou, P.X.W. and Wong A. (2008) Breaking into China's design and construction market. *Journal of Technology Management in China*, 3(3): 279–91.

Zou, P.X.W. and Zhang, G.M. (2009) Managing risks in construction projects: Life cycle and stakeholder perspectives. *International Journal of Construction Management*, 9(1): 61–77.

Zou, P.X.W., Zhang, G.M. and Wang, J.Y., (2007) Understanding the key risks in construction projects in China. *International Journal of Project Management*, 25(6): 601–14.

9 Leadership and its development in construction in developing countries

Shamas-ur-Rehman Toor
Islamic Development Bank, Kingdom of Saudi Arabia

Stephen Ogunlana
Heriot-Watt University, United Kingdom

and

George Ofori
National University of Singapore

Introduction

Leadership has attracted the attention of researchers in a wide range of fields, from history to sociology, from military studies to political science, and from business to education (Vroom and Jago, 2007). The study of leadership dates back many centuries, and a large volume of literature has been built up on it. Nevertheless, there are still many gaps in the knowledge on leadership, and many of its aspects have yet to be explained adequately (Conger, 1998; Avolio, 2007). Indeed, there is still no single agreed definition of leadership (Vroom and Jago, 2007; Bennis, 2007), and there continues to be a debate about many aspects of leadership including the best strategies for developing and exercising leadership (Hackman and Wageman, 2007).

Today, the combination of the information age and globalisation, and the implications of this congruence of phenomena, are posing new challenges to leaders of construction projects. Among the key challenges are human resource issues such as the need for effective leadership of multi-cultural and multi-disciplinary teams; the rapidly changing operating environment of the industry; the growing size and complexity of projects; and the large numbers of stakeholders who have a legitimate interest in most construction projects. Other researchers on influences on, and developments in, the construction industry and its projects note challenges such as: the gap between research and practice (Ofori, 1993; Chemillier, 1988), and the need to attain the highest client value as well as its creation and distribution (Huovinen, 2006). It has also been suggested that the industry needs to become more client- and market-oriented (Pries et al., 2004). Songer et al. (2006) note that the construction industry faces major leadership challenges such as those relating to the workforce, including: lack of quality people owing to difficulty in attracting talent; ageing workforce; and the need to deal with issues such as teamwork and communication, and training and education (Songer et al., 2006).

These challenges, together with rapid changes in the business culture and complexity of the professions and trades, call for a change in the conventional paradigm of leadership in construction that is largely technology- and project-oriented (Pries et al., 2004). This call for change underpins the need for more leaders who can drive the future of construction.

These challenges are also coupled with ethical and moral crises which continue to emerge even now in business, and specifically in the construction industry, and which are eroding the trust of the followers and the general public in leaders (George, 2003; George et al., 2007). Construction has been rated in several international reports as the most corrupt sector. This concern about the construction industry is all the more grave in the developing countries, which are characterised by lack of professionalism and poor implementation of laws and regulations. It is imperative that the construction industry in developing countries is given a new leadership vision to enable it to contribute to national, social and economic development. To achieve this, a new generation of leaders should be developed who are not only fully aware of the gravity of the challenges but are also capable of finding innovative solutions to the old problems in construction. Such leaders need to be competent professionals. They should also be ethical and moral in the way they approach their leadership functions. Such leaders are driven by values, purpose, vision and positive behaviours rather than by charisma, style and self-interest.

Therefore, there is an urgent need to find ways (or formulate interventions) that can help to develop what many researchers now call "authentic leaders" (Avolio and Gardner, 2005; Luthans and Avolio, 2003). Ofori (2009) suggests that issues which must be explored in relation to leadership in the developing countries include: the need for leaders; the difference leaders can make in their organisations, industries and societies; the best approach to the exercising of the leadership function; how leaders are developed; and how leaders influence their organisations (in this context, projects, companies and industries), and followers.

Objectives of chapter

The objectives of this chapter are to: review the literature on the subject of leadership, focusing on recent developments in the area; assess the importance of leadership in the construction industry in general, and in the developing countries in particular; consider the state of leadership research and leadership development in construction, and consider its relevance to the developing countries; discuss ways and means of developing leaders for the construction industries in the developing countries; and present a research agenda for the development of leaders for the construction industries in the developing countries.

Leadership: a review

What is leadership? Many works on leadership start with this question. Many have observed that the conceptualisation of leadership has no common and established definition by which it can be evaluated, no dominant paradigms for studying it, and little agreement about the best strategies for developing and exercising it (Hackman and Wageman, 2007; Barker, 1997; Higgs, 2003). Bennis (2007) notes that it has almost become a cliché that there is no single definition of leadership. Bass (1990) observes that there are as many definitions of leadership as there are authors who have attempted to define it. Goethals et al. (2004) suggest that the large volume of literature on leadership which is available makes it difficult to discuss the concept of leadership under a single definition. Some definitions of leadership in the literature are now presented.

> Leadership may be defined as the behavior of an individual while he is involved in directing group activities (Hemphill, 1949: 4).

> Leadership behavior means particular acts in which a leader engages in the course of directing and coordinating the work to his group members (Fiedler, 1967: 36)

> Leadership is the reciprocal process of mobilizing by persons with certain motives and values, various economic, political and other resources, in context of competition and conflict, in order to realize goals independently or mutually held by both leaders and followers (Burns, 1978: 425).

> [Leadership is the] capacity to create a compelling vision and translate it into action and sustain it (Bennis, 1989: 65)

> Leadership involves influencing task objectives and strategies, influencing commitment and compliance in task behavior to achieve these objectives, influencing group maintenance and identification and influencing the culture of an organization (Yukl, 1989: 253).

> [Leadership is the] principal dynamic force that motivates and coordinates the organization in the accomplishment of its objectives (Bass, 1990).

> Leadership is the process of persuasion or example by which an individual (or leadership team) induces a group to pursue objectives held by the leader and his or her followers (Gardner, 1990: 1).

> Leadership is an influence relationship among leaders and followers who intend real changes that reflect their mutual purposes (Rost, 1991: 102).

> Leadership is a process of social influence in which one person is able to enlist the aid and support of others in the accomplishment of a common task. The main points of this definition are that leadership is a group activity, is based on social influence, and revolves around a common task (Chemers, 1997: 1).

> Leadership is a process—a dynamic process in which the leader(s) and followers interact in such a way as to generate change (Kellerman and Webster, 2001: 487).

> [Leadership is a] process of motivating people to work together collaboratively to accomplish great things (Vroom and Jago, 2007: 18).

These definitions, proposed at different points over a period of half a century, show that the understanding of leadership has travelled from a focus on "behaviours" to concentration on "actions" to eventually a consideration of leadership as a "social process" which involves the leader, followers and situations. The abundance of different definitions of leadership, and lack of agreement on its real nature and essential features, led Burns (1978) to argue that leadership is the most observed but least understood phenomenon. It is a field which has both fascinated and perplexed researchers and practitioners, creating a significant amount of research and theories to conceptualise and explain the phenomenon (see, for example, Ayaman, 2000, cited in Goethals et al., 2004). Other researchers have also alluded to the complexity and elusiveness of leadership (see, for example, Chemers, 1997).

The abundance of literature on leadership is reflected by the increase in the number of articles in its key reference, *Stogdill's Handbook of Leadership*. Only 3000 studies were listed in the earlier publication, and the number increased to 5000 within seven years. However, despite this growth in volume, the authors of *Stogdill's Handbook of Leadership* concluded that "the endless accumulation of empirical data has not produced an integrated understanding of leadership" (Bass, 1990: vii). DuBrin (1995) claims that about 30,000 research articles, books and articles in magazines had been written on the subject of leadership by the mid-1990s. Goffee and Jones (2000) observe that nearly 2000 books were published on leadership in the year 1999. These observations depict the volume and pace of growth of publications on the subject of leadership. At the same time, it is evident that the outcomes from this increasing volume have been limited. The new reader or researcher of leadership is often confused by the many theories and concepts. Nevertheless, new topics, arenas, horizons and conceptualisations keep emerging in the study of leadership. Researchers are now breaking out from the confinement of social science laboratory experiments to observe real leaders in action. This is essentially because leadership has become much more relevant and even more complex in the globalised world.

Researchers from various fields of knowledge, such as political science, psychology, education, history, agriculture, public administration, management, anthropology, medicine, military sciences, philosophy and sociology, have undertaken studies on leadership which have contributed to the understanding of the subject. With better availability of research facilities, resources and infrastructure, leadership researchers are more interested in integrating various concepts of leadership rather than studying them in isolation in the domain of a single subject.

Various forms of leadership

Leadership research can be divided into seven categories. The first group comprises the ancient approaches and those prior to the 20th century, such as *The Republic* by Plato, *Politics* by Aristotle, *The Art of War* by Sun Tzu and *The Treatise on General Sociology* by Vilfredo Pareto. The second group may be termed classical approaches. Examples are: Scientific Management by Frederick Winslow Taylor, Elton Mayo's human relations approach, McGregor's Theory X and Theory Y, William Ouchi's Theory Z, Abraham Maslow's Hierarchy of Needs Theory, Frederick Herzberg's Hygiene Theory, McClelland's Achievement Motivation Theory, and various Trait Theories. The third classification comprises transactional approaches (Behavioral Theories originating from Ohio and Michigan; Contingency Theories such as the Vroom and Jago Model, Path–Goal Theory, Leadership Substitute Theory, Leader–Member Exchange (LMX) theory, and The Multiple Linkage Model). The fourth group are transformational approaches (examples are Burns's Transforming Leadership Theory, Bass's Transformational Leadership Theory). The fifth classification of works would be the Charismatic Leadership Theories propounded by various authors. In the sixth group would be various perspectives on Cross-cultural Leadership that have also contributed to the wealth of knowledge on leadership. Finally, there are several miscellaneous approaches to leadership that have also been presented. They include: Self-Leadership Theory, Servant Leadership Theory, Collins's Model of Level-5 Leadership, and Shared Leadership Theory.

Recent works in leadership research have also paid much attention to negative forms of leadership (such as laissez-faire leadership, derailed leadership and toxic leadership). An important reason for scholars to explore negative or ineffective forms of leadership is the growing concern about leadership malfeasances and ethical problems that have emerged in the corporate world. The wrongful use of power, exploitation of organisational procedures and systems, and poor corporate governance are some subjects which have been discussed in the literature under this stream of research.

In parallel, there have been significant developments on topics such as ethical or moral leadership, servant leadership and spiritual leadership. Although characteristics such as honesty, integrity, altruism, trustworthiness, collective motivation, encouragement and justice have often been considered to be attributes of good leaders, research has shown that there is more to ethical leadership than these. Ethical leadership, in a true sense, practices and also consciously manages ethics, and holds everyone in the organisation accountable for it (Brown et al., 2005). In other words, as Brown et al. (2005) suggest, ethical leadership has two dimensions: moral persons and moral managers. "Moral persons" are those who model "normatively appropriate" conduct such that they appear honest, trustworthy and credible to others. Moral persons are perceived as fair and just decision-makers, ethically principled, caring and altruistic. "Moral person" has to do with how others perceive the leader's character, traits, attributes and personal characteristics (Brown and Treviño, 2006). On the other hand, "Moral Manager" is concerned with the promotion of moral and ethical conduct to followers through two-way communication, reinforcement, and decision-making.

The philosophy of servant leadership, originally proposed by Robert Greenleaf in 1977, suggests that leaders must place the needs of their subordinates, the organisation's customers and the community ahead of their own interests in order to be effective. Servant leaders choose to serve first, and then lead as a way of expanding service to individuals and institutions. Servant leaders may or may not hold formal leadership positions. Servant leadership encourages collaboration, trust, foresight, listening and the ethical use of power and empowerment. Leaders transcend their self-interests to serve others and the organisations. Characteristics of servant leaders include empathy, stewardship and commitment to the personal and professional growth of their subordinates. Extending the philosophy of servant leadership, Liden et al. (2008) offer the following dimensions of such leadership: conceptual skills, empowering, behaving ethically, creating value for the community, helping subordinates to grow and succeed, putting subordinates first, and emotional healing. Liden *et al.* (2008), in their empirical study, found that servant leadership is a multi-dimensional construct and at the individual level makes a unique contribution in explaining community citizenship behaviour, in-role performance, and organisational commitment.

The concept of spirituality in modern leadership theories is quite recent. Most leadership theories had not attended to the role of religion or spirituality in political or workplace institutions (Hicks, 2002). Fry (2003) proposed a causal theory of spiritual leadership that incorporates vision, hope/faith and altruistic love, theories of workplace spirituality, and spiritual survival. He argues further that spiritual leadership theory includes other major existing motivation-based theories of leadership, and is also more conceptually distinct, parsimonious and less conceptually confounded. As it incorporates calling and membership as two key follower needs for spiritual survival, spiritual leadership theory is inclusive of the religious-, ethics- and values-based approaches to leadership (Fry, 2003).

Finally, one of the emerging forms of leadership in the recent literature is authentic leadership. "Authenticity" has been discussed in philosophical and psychological terms (see, for example, Harter, 2002; Erickson, 1995). In relation to leadership, "authenticity" was popularised by George (20007). His best-selling books, *Authentic Leadership: Rediscovering the Secrets to Creating Lasting Value* and *True North: Discover Your Authentic Leadership* focus on the role of authenticity in leadership and leadership development. Following George's work, other researchers on leadership have realised that leadership is not merely a matter of style, charisma, motivation, inspiration or strategy. It should be considered as embracing character, positive behaviour and authenticity.

Authentic leaders are described as possessing the highest level of integrity, a deep sense of purpose, courage to move forward, genuine passion, and skilfulness in leadership (George, 2003; George et al., 2007). They are ready to embrace the challenges of a constantly changing operating environment by continuously growing themselves as authentic leaders, and their followers as authentic followers. Extracted from positive psychology, ethical leadership and positive organisational behaviour, the authentic leadership construct focuses on authenticity of character, awareness of the self, regulation of self, faithfulness of individuality, genuineness in beliefs, truthfulness of convictions, practicality of ideas, veracity of vision, sincerity in actions, and openness to feedback (see, for example, Avolio and Luthans, 2006; George and Sims, 2007; Toor and Ofori, 2008a; Walumbwa et al., 2008). These characteristics may portray some of the features of transformational, charismatic, servant, spiritual and ethical leaderships. However, proponents of authentic leadership contend that it is distinct from other forms of leaderships in many respects (see, for example, Avolio and Gardner, 2005) and its proponents suggest that it lies at the root of all forms of positive leadership.

Leadership and construction

There is a growing recognition that leadership plays a vital role in the success of projects (Ofori and Toor, 2009b). In construction, the need for leadership is even more pressing. Many researchers report that effective performance of the project manager is the single most critical factor affecting success on construction projects (Hartman, 2000; Powl and Skitmore, 2005). More recently, some researchers have noted the importance of leadership in the implementation of construction projects. For example, Thamhain (2003) highlights the importance of the leader in creating a supportive working environment for the project team members. Munns and Bjeirmi (1996) emphasise that the success or failure of project management is highly dependent on the choice of the project leader. Chinyio and Vogwell (2007) found that effective leadership of the many stakeholders on a construction project can help to harmonise their goals and prevent conflict.

Despite the realisation of the importance of leadership at all levels of the construction industry, emphasis is usually put on the technical aspects and on management, and, as a result, inadequate attention is devoted to leadership research (Skipper and Bell, 2006a). Many authors suggest that this situation needs to be rectified (see, for example, Keegan and Den Hartog, 2004; Chan and Chan, 2005). Several reasons can account for the relatively low volume of work on leadership in construction, of which three may be outlined here. First, social scientists undertaking research on leadership do not have much knowledge of the construction industry (Langford et al., 1995), and the rich opportunities it offers for significant studies on the subject. Second, as the construction industry has tended to focus on management in relation to projects and organisations, it has not stressed the importance of leadership; thus the subject has not emerged strongly as one worth studying. Finally, there are few researchers who have the necessary skills to undertake the detailed studies of the complex psychological, social and cultural construct that is leadership, and, in most countries, there is also inadequate funding for such work (Chinowsky and Diekmann, 2004).

Development of leadership research in construction

Bresnen (1986) and Bryman et al. (1987) undertook some of the earliest works on leadership in construction when they examined the leadership orientation of construction professionals. Over the next quarter of a century of empirical research on leadership in construction, the main trends in terms of areas of emphasis have changed several times. The initial works examined the motivational factors, personal characteristics, traits and behaviours (see Muir and Langford, 1994; Ogunlana et al., 2002; Odusami, 2002), and leadership style (Chan and Chan, 2005). In recent years, the focus has been on leadership development (Skipper and Bell, 2006a, 2006b; Toor and Ofori, 2008a; Skipper and Bell, 2008), cross-cultural leadership (Grisham, 2006; Toor and Ogunlana, 2008b; Ofori and Toor, 2009a), and leadership succession (Skipper and Bell, 2008). Other recent works have considered the new constructs such as ethical leadership (Toor and Ofori, 2009b), authentic leadership and transformational leadership (Limsila and Ogunlana, 2008a, 2008b) and their relevance to construction. Yet other recent studies have considered more complex issues relating to leadership such as its relationship with psychological wellbeing and contingent self-esteem (Toor and Ofori, 2009a), and psychological capital (Toor and Ofori, 2010). This account suggests that more researchers on the construction industry are developing interest in leadership and breaking new grounds in terms of the theories and aspects of leadership they study and methodologies they apply.

Despite its vital relevance for construction projects, leadership has not been a popular topic of research until recently. The volume of the literature on leadership research in construction is both thin and shallow (Toor and Ofori, 2008e). There is also inadequate understanding of leadership and heavy dependence on managerial functionalism (Chan, 2008). There is little appreciation of the nature of leadership as a process (Toor and Ofori, 2008e). Researchers on construction have also tended to focus on the motivational factors and the personal characteristics of project managers (Dulaimi and Langford, 1999). In effect, leadership research in construction has been mostly a replication of what has already been done and found in the mainstream management studies, with little innovation and lack of focus on practical implications. Another feature of leadership research in construction is that there is an excessive inclination towards the application of quantitative methods (Toor and Ofori, 2008d).

Nevertheless, there appears to have been a recent boost in the level of interest in, and volume of works on, leadership in construction. Several developments point to a new era for research on leadership in construction. Some of these are now discussed.

First, the CIB Task Group 64 on Leadership in Construction was formed in 2006. Its objectives are: to establish an international group to identify the issues facing construction leadership and to research the state of construction leadership and barriers for development; to broaden the current research of construction leadership issues by involving industry representatives and experts from different regional backgrounds; and to raise the awareness of leadership issues within the construction industry and of the need for further research in this area (CIB, 2006). The task group was closed at the end of its tenure in November 2010.

Second, two journals dedicated to the subject of leadership in construction have been launched recently. They are: *Leadership and Management in Engineering*, published by the American Society of Civil Engineers, and *Engineering and Construction Leadership*. Third, leadership has been the subject of special issues of many construction management journals. These include *Construction Information Quarterly* (Volume 10, Issue 2, 2009). Fourth, conferences focused on leadership have been successfully organised (examples are LEAD2008 and LEAD2009); many of these conferences have had special streams for construction. Finally, research centres dedicated to the study of leadership in construction have been established in some institutions around the world. An example is the Center for Project Leadership at Columbia University in the US.

Leadership research in developing countries

Much of the present literature on leadership results from research carried out in the developed countries. Toor and Ofori (2008d) reviewed a total of 49 empirical works on leadership in construction, published between 1986 and 2008. Their findings show that the majority of the studies were done in the UK, USA, Hong Kong and Singapore. Only 11 (or just above 20 percent) of the empirical works were carried out in the developing countries. Studies undertaken in the developing countries were done in: Thailand (see Low and Chuvessiriporn, 1997; Ogunlana et al., 2002; Limsila and Ogunlana, 2008a, 2008b; Toor and Ogunlana, 2009); Nigeria (Odusami, 2002; Odusami et al., 2003); Egypt (Seymour and Elhaleem, 1991); Tanzania (Debrah and Ofori, 2005); and China (Liu and Fang, 2006; Wong et al., 2007).

The review of the literature for this chapter revealed that there has not been much research on leadership in construction in the developing countries. The inadequacy of studies is not only in terms of number of works but also in terms of the range of topics that have been explored and methodologies that have been employed. There are several apparent reasons for the scarcity of such studies in the developing countries (see Ofori and Toor, 2008). They include excessive focus on technical and management aspects of projects at the cost of leadership, lack of focus on leadership in the curricula of academic programmes leading to lack of knowledge or interest among students and the faculty, lack of funding for research resulting in too little research with too little impact, lack of access of researchers in developing countries to various research platforms and outlets such as international journals and conferences, and lack of coordination between these researchers and their counterparts in the developed countries. There is also lack of collaboration between industry and academic institutions in these countries. Owing to these reasons, research on leadership in construction has failed to flourish in the developing countries, leading to low awareness and appreciation of leadership among practitioners. As a consequence, the professionals from such systems invariably have difficulties when they face the challenges of today's complex business environment.

The current practices and general orientation of the construction industry is responsible for the lack of leadership research. In construction, the professionals usually focus on management, to the exclusion of leadership. In many instances, management is confused with leadership although they are completely different concepts (Toor and Ofori, 2008e). That is probably the reason that those who lead construction projects are not perceived as leaders, and are mostly termed as managers (Russell and Stouffer, 2003). Their day-to-day work involves management of activities and achievement of the short-term goals of the project (Toor and Ofori, 2008a). This mindset of construction project management results in managers who mostly end up managing their teams rather than leading them. Due to excessive focus on management and lack of cognisance of the vital role that leadership plays in the achievement of success on construction projects, the industry has paid little attention to understanding leadership and its implications. This may be one of the reasons that are responsible for the incidence of poor performance on construction projects that is a common feature of the construction industries in most countries, and especially in developing countries.

Features of construction and implications for leadership

There is greater need for leadership in the construction industry than, arguably, in any other field of human endeavour. There are several reasons for this, and many of them are evident in the nature of construction projects, the industry and the constructed product (Hillebrandt, 2000).

First, construction projects are technically complex and large. They require a combination of specialist skills to undertake them, and the ability to manage a wide range of risks. Second, construction projects typically involve relatively high expense, and the stock of buildings in any country represents a large proportion of a nation's savings. Thus, the quality of the built product is of the essence. Third, the projects take quite a long time to complete, and involve a very large number of discrete activities. This increases the probability that at least some of the time-related risks would occur, and also exacerbates the communication and coordination problems. Fourth, the teams involved in construction projects are not only large, but are also multi-disciplinary, and comprise members from several different organisations. On today's large projects, construction teams are also multi-cultural. Fifth, the projects and the constructed products have serious implications for the health and safety of the workers involved, as well as those of the general public. Thus, due care, diligence and expertise are needed to safeguard the safety and health of the workers and the community. Finally, construction projects have many stakeholders whose interests are diverse and often in conflict with each other.

There is even greater need for leadership in construction in the developing countries. Again, the reasons can be outlined. First, there are more reports of project performance deficiencies in developing countries than in the developed nations. Some of the commonly highlighted deficiencies include cost and time overrun, poor quality of work, technical defects, poor durability, and inadequate attention to safety, health and environmental issues (see, for example, Ofori, 2007). Second, the management of projects in the developing countries is fraught with many problems because of the nature of the industries and their operating environments. The importance of the effective management of the stakeholders of construction projects in the developing countries is most clearly evident on the international projects that are common at the large and complex project end of the markets in these countries (Ofori, 2003). On such projects, the teams are invariably multi-cultural; thus, they require skilful integration of the efforts of workers from diverse national as well as corporate and professional backgrounds. These large, international projects also tend to have a wide range of beneficiaries who derive different benefits from the project itself, and the constructed items. At the same time, they also have an impact on large numbers of people whose interests need to be addressed. This underscores the need for leadership skills.

Third, there are even greater adverse implications of poor performance on construction projects for long-term national socio-economic development in the developing countries as the constructed product is critical to this process. It is important that the projects in which the limited funds available to these countries are invested do not only give value for clients' money upon completion, but also continue to provide a flow of benefits in the course of their lives. Finally, the clients, end purchasers, users and other stakeholders of the construction project have no knowledge of aspects of construction. This implies a need for professionalism among the participants in the construction project, and dedication to meet the objectives and aspirations of the stakeholders in the most innovative, imaginative and value-adding manner for the benefit of the client and all the stakeholders. These points put the need for leadership in construction in developing countries into sharp relief.

In the next section, the critical need for effective leadership in the construction industries in the developing countries is discussed further.

Need for leadership in construction industries of developing countries

As stated in the preceding sections, the construction industries in the developing countries face formidable challenges. In a study on the competitiveness of the industry in Indonesia, Budiwobowo et al. (2009) found that the internal features of the construction industry that they termed "factor conditions" of the construction cluster were not highly rated. These 17 conditions were: geographical condition; labour productivity; level of technology for product development; level of technology for business processes; level of technology for construction plant and equipment; quality of education; flow of technology from higher educational institutions; institution for transfer of technology; collaboration for technology development; codes and standards for the construction industry; implementation of policies; role of professional associations; role of construction associations; role of construction services board; role of government; role of higher educational institutions; and availability of basic infrastructure. It is pertinent to note that all the factors were rated 3.0 (on a scale of 1 to 5, where 1 = "very poor" and 5 = "very good") except "flow of technology from higher educational institutions"; and "role of higher educational institutions" which were rated at an even poorer 2.0. Thus, research and its flow into the construction industry to be applied are viewed by practitioners in Indonesia as having little relevance and impact.

The study by Budiwobowo et al. (2009) also found that the operating environment of the construction industry in Indonesia is rather unfavourable. The three attributes of "related and supporting industries" were all rated 3.0. They were: completeness of the supporting industries; quantity of the supporting industries; and competitiveness of the supporting industries. Therefore, the participants in the construction industry in Indonesia perceive it as having major weaknesses, as a result of which the local enterprises and professionals are unable to compete with their foreign counterparts. Thus, the study showed that indigenous firms are not able to take advantage of the favourable market conditions in the industry. It is striking that leadership was not highlighted in the study as one of the features of Indonesia's construction industry.

Alinaitwe (2009) found the following as the top ten barriers to the implementation of lean construction in Uganda: inputs exactly when required; infrastructure in transportation and communication; capability of teams to maintain alignment with other teams; certainty in the supply chain; steady prices of commodities; reward systems based on team goals; buildable designs; participative management style for the workforce; parallel execution of development tasks in multi-disciplinary teams; and accurate pre-planning. It is evident that most of these areas can be effectively addressed if there is good leadership of projects in Uganda.

The low level of importance accorded to leadership emerges even where practitioners are asked questions on it in unambiguous terms. For example, in a study in Palestine, "leadership and motivation" was ranked at as low as 17th in a list of 20 "skills important to contractors" (Enshassi et al., 2009). The top-ranked skills were: decision making; problem solving; financial management; project management; administration; risk taking; organisation; creativity; planning and goal setting; and delegation. Among clients, "leadership and motivation" was ranked equal 12th among 21 important skills. The skills ranked above leadership were: planning and goal setting; decision making; problem solving; project management; financial management; attitude; administration; organisation; creativity; communication; negotiation; and risk taking. This confirms the points made earlier in the chapter about the tendency in the construction industry to focus on management issues, rather than leadership.

From the discussion so far, it is clear that there is a need for leadership at many of the levels of the construction industry, and especially in the developing countries. First, at the level of the industry, there is the need for strategic leadership and championing of the continuous development and improvement of the industry. Second, each of the professional institutions and trade associations in the construction industry requires effective leadership to ensure the development of the expertise and professionalism of its members. Third, the construction companies must be led with vision, competence and innovation, considering the formidable challenges within the construction industries and in their operating environments. Finally, leadership is the key at the project level owing to the many peculiar features of construction projects that are highlighted above.

Leadership development

There has been a long debate on whether leaders are born or can be developed. However, the current consensus appears to be that whereas certain traits considered in some contexts to be desirable in leaders may be naturally endowed, one can develop some, if not most, of the attributes and capabilities of leadership through appropriate structured interventions. Thus, there have been attempts to study how leaders develop, in order to help formulate these interventions. Avolio (2007) noted that leadership can be better understood by researching when, where and how it is activated and how it makes a difference in team performance and process effectiveness. Luthans and Avolio (2003), in their initial conceptualisation of authentic leadership development, stress the need to construct "taxonomies of trigger events" that promote positive leadership development.

These trigger events may include influential role models and various significant others in one's life, events and experiences, and social institutions which influence one's behaviour. Similarly, others, such as Rothstein et al. (1990) suggest that a leader's personal history, trigger events, experiences at work, and personal and organisational factors may be potential antecedents to the person's emergence as a leader and the person's effectiveness in this role. The biographies of several political and business leaders have shown that challenges, struggles, obstacles, crises, and dilemmas also help to hone a potential leader's talents (see Shamir et al., 2005). Exploring the antecedents also clarifies the contextual variables that play a mediating role in leadership development and emergence.

The study of leadership development is still a fledgling field. There is a lack of a clear theoretical framework, and the work is of a retrospective nature, with some recent works based on narratives (Sparrowe, 2005) and life stories (Shamir and Eilam, 2005; George and Sims, 2007; George et al., 2007) as important lenses of leadership emergence and influence.

Case study: Authentic leadership development in Singapore

A study on authentic leadership development in the construction industry in Singapore was undertaken during 2007–8. The principal aim of the study was to explore how authentic leaders develop, and how they remain authentic over their lives. Table 9.1 presents a summary of the objectives of the study, set against the approach taken to achieve each of them, and the main results attained. A grounded theory approach was adopted, and the field study comprised interviews with 45 leaders of the industry who fitted the description of authentic leaders suggested by George (2003). The interviewees were leaders of organisations and ranged from clients, through architects, various types of engineers and contractors, to project managers and quantity surveyors. The initial group of interviewees were current or former (elected) presidents of professional institutions and trade associations. The others were identified through a snowballing process.

It was found that authentic leadership development and influence are not two discrete aspects of authentic leadership. Instead, how authentic leaders manifest their leadership and how they influence others is inculcated in their life stories, values, purpose, goals, and motivations. The study found that authentic leaders develop in four phases: preparing, polishing and practising, performing, and passing, through experiential learning under the influence of significant individuals (such as mentors, parents, siblings and teachers) and significant social institutions (such as school, university and professional organisation). Over the course of their development, authentic leaders learn from both their successes and their failures. They discover their authentic selves through transformative events (such as a difficult childhood, the death of a loved one, or a leadership opportunity during their youth). The discovery of the authentic self leads to further conscious development (self-regulation) on the part of leaders. Self-transcendence,

Table 9.1 Objectives, approach and achievements of study on leadership in Singapore construction

Objective of study	*Approach to, and extent of its achievement*
To develop theoretical frameworks to explain authentic leadership development.	A review of the literature resulted in a theoretical model: Integrated Antecedental Model of Leadership Development.
To explore how authentic leaders develop and remain authentic during their lives.	Based on the information collected in the study, a four-phase authentic leadership development model was proposed.
To examine the leadership processes at the personal level by which authentic leaders influence their followers and organisations.	The two main processes by which leaders influence their followers and organisations at the personal level were found to be: self-leadership and self-transcendent leadership.
To examine the leadership processes at the group, team and organisational levels by which authentic leaders influence their followers and organisations.	Sustainable leadership emerged as a process at the group, team and organisational levels, through which authentic leaders influence their followers and organisations.
To make recommendations for authentic leadership development.	Apart from recommendations on authentic leadership development, an agenda for future research on authentic leadership in construction was prepared.

positive reframing, systemic perspective, meta-reflection, and role modelling for followers are key processes which leaders engage in during the "performing" and "passing" phases of leadership development.

In the course of the study, as the conceptual categories were emerging, it was found that authentic leaders are primarily driven by self-leadership, self-transcendent leadership, and sustainable leadership. However, what makes them effective and successful is their ability to dynamically and creatively reconcile their self and social realities – a core social process of authentic leadership. In doing so, authentic leaders are able to harmonise their self-related knowledge by being cognisant of their values, ambitions, aspirations, goals, motivations and purpose. They then seek an active alignment of their purpose with various social agents around them. They are able to garner the support of others by inspiring mutual trust, commitment and singularity of goal. They are adept at harnessing followership due to their ability to creatively reconcile the social realities – dilemmas, paradoxes, uncertainties and ambiguities. This dynamic and creative reconciliation of self and social realities also makes them authentic, and makes them appear authentic to other people. This congruence provides them with the attributes of authenticity of leadership, for which others are willing to render their support and followership.

The study also showed that leadership enactment, manifestation or influence cannot be seen in isolation from leadership development. Authentic leadership is influenced by various trigger events that result in a progressive development of the leader's purpose and vision that is grounded in the leader's self-awareness that largely drives the leader's subsequent actions – or self-regulation. Authentic leaders consciously engage in several cognitive and social processes over their lifespan. Experiential learning helps them to develop the competencies, higher-order attributes, skills and dispositions to engage in self-leadership, self-transcendent leadership and sustainable leadership – three dimensions of authentic leadership which were found in this study.

Fully aware of their strengths, weaknesses, motivations and values, authentic leaders actively establish social networks to gain support from various social agents around them – including followers, clients, external stakeholders – to achieve their purpose and vision. This view also confirms that authentic leaders are not "perfect", "heroic", or "great men or women" leaders. Their success lies in their ability to look at the bigger picture and balance the conflicting voices within their inner selves as well as in the environment of which they are part. Their ability to creatively reconcile the context, social agents, complexities, paradoxes, dilemmas, uncertainties, ambiguities and other challenges makes them stand out as effective and successful leaders. They do not have to be charismatic or flamboyant. They are leaders who lead themselves first, then go beyond themselves to lead others around them through self-transcendence, and finally they take appropriate actions to sustain the effect of their leadership.

In this process of dynamic and creative reconciliation, authentic leaders resolve inner and environmental paradoxes and dilemmas; comprehend, manage and shape their inner and social states by reducing the socio-cognitive dissonance; frame and reframe their life stories to reconcile them with their inner selves and shape their social contexts; manage the organisational meanings by sense-making and sense-giving; enrich working environment through positive and constructive politics; and are driven to interplay between order and chaos within their inner selves as well as in their surroundings.

Relating the case study to the literature: Fallacies and tipping points

As many scholars assert, authentic leadership is a lifelong journey (Avolio and Luthans, 2006; George and Sims, 2007), and not the destination. Authentic leaders realise it and therefore continuously strive for it. Their values and purpose provide them with the moral compass (George et al., 2007) to keep them on track without letting them become trapped in the six well-known fallacies (Sternberg, 2002; 2007): the unrealistic-optimism fallacy – that they are so smart and effective that they can do whatever they want; the egocentrism fallacy – that they are the only ones that matter, and not the people; the omniscience fallacy – that they know everything and therefore they lose sight of the limitations of their own knowledge; the omnipotence fallacy – that they are all-powerful and can do whatever they like; the invulnerability fallacy – that they can get away with anything because they are too clever to be caught; and finally the moral disengagement fallacy – that they cease to view their leadership in moral terms but rather only in terms of what is expedient. Authentic leaders are not only driven to be true to themselves and others (Luthans and Avolio, 2003), but also they manage their authenticity within their social contexts by treating different people differently and making adequate sense of varied situations and circumstances.

The most central quality of authentic leaders that emerged as the core category in the study in Singapore is their ability to reconcile all three dimensions of authentic leadership: self leadership, self-transcendent leadership and sustainable leadership. They do not compartmentalise their leadership into “self” and “others”. Instead, they develop and influence others by simultaneously being self-leaders as well as self-transcendent leaders. Appreciating these two crucial aspects, authentic leaders are able to sustain their leadership strategies and outcomes.

The views and experiences of the leaders in the study in Singapore supplemented the existing literature in explaining the nature and essence of authentic leadership. The study shows that the values, motivations and purpose of leadership of authentic leaders are a reflection of their past experiences, indicating that leadership influence cannot be seen in isolation from leadership development. The tipping points of the development of the leaders interviewed, as well as the main factors which influence them, provided ideas on how leadership development in construction could be approached. The aspirations of the leaders for their own firms and the construction industry gave indications of possible initiatives that could be adopted to develop the industry. The study also highlighted the importance of "followership". The study led to some answers to some of the perennial questions in leadership research such as: the difference between "leadership" and "management"; whether leaders learn more from mistakes or successes; and whether leaders are born or made.

Such studies need to be replicated in the developing countries as well. There is strong evidence that cultural differences account for the variations in the way management and leadership approaches are perceived and operationalised in different societies. Cultural differences are also likely to influence the way people use their own life stories to shape their leadership approaches, and various tactics that leaders use to lead their teams and organisations, and to influence their followers. Attempts to understand leadership without understanding culture are not likely to succeed. Hence, in order to gain in-depth understanding of leadership development and leadership influence, a cross-cultural view is necessary. Also, opportunities should be explored to undertake a large leadership study, spread over several developing countries, under a consortium of interested researchers. Whereas such a research endeavour will face many complex conceptual, methodological, operational and coordination issues, existing networks of researchers such as CIB W112 on Culture in Construction and W107 on Construction in Developing Countries can provide the platform for the conduct of such research studies. The findings of such studies would make a significant impact on knowledge on, and practice of, leadership in the developing countries.

Suggestions for developing future leaders

Toor and Ofori (2008b) present a set of suggestions for developing leaders in construction in the poorer countries. They put forward a proposal for a partnership among the industry (through professional institutions and trade associations), tertiary academic institutions and government agencies. They describe the various roles of each of the partners during each stage of the development of both the technical competencies and the leadership skills of construction professionals, from their university education to professional practice. For example, the construction industry, through its professional institutions and trade associations, should establish internship and mentorship programmes, and coaching circles for new graduates. They should also set up scholarship and fellowship schemes for deserving students. The universities should actively collaborate with industry in order to develop curricula that fulfil the future needs of the various professions in the construction industries, and identify and undertake research projects which explore real issues in the industry, and make proposals which are of direct benefit to practitioners.

Educational programmes for construction professionals should include: a reasonable proportion of management-related subjects in the syllabuses of predominantly technical programmes; setting of appropriate role-play exercises in which students work in cross-disciplinary teams, to enable them to develop skills necessary for decision making, communication, and collaboration; institution of co-curricular activities for developing personal and interpersonal skills among students; formulation of leadership development programmes customised for students of different levels; and administration of outreach programmes such as those involving students in volunteering to participate in social development projects on which they make practical use of the knowledge they have acquired.

Some scholars also suggest that some courses on human resource management should be offered to students in construction programmes so that they can understand the unique challenges of human resources in the construction industry (Laufer, 1987) and thereby learn to function effectively as leaders. Russell and Stouffer (2003) are of the view that training in leadership and management must continue from the engineers' entry level to the person's senior executive level. Similarly, Toor and Ofori (2008b) note that the construction industry should not rely on the organic growth of leaders who develop through experience. There is a need to promote and support the authentic leadership development of capable individuals at all levels of organisations.

Researchers should direct their endeavours towards developing effective leadership interventions that can help to accelerate authentic leadership development. Future studies should focus on the development of customised, action learning-based interventions that are culturally-specific, involve complex challenges, and are structured in order to develop leaders who can effectively function in the complex and dynamic business world of the construction industry.

Agenda for research on leadership in construction in developing countries

It is evident from the discussion so far that more work is required on leadership in construction in the developing countries. Some possible subjects for these research studies are now considered.

1 The relevance of leadership to the particular problems of the developing countries such as poverty alleviation, provision of work opportunities and small enterprise development;
2 Leadership in the operating environment of construction industries in the developing countries, the features of which include a vacuum of

enforcement of regulations, lack of appropriate and enabling policies, and poor level of development of physical infrastructure and supporting industries;

3 Leadership development in construction in the developing countries. What are the most effective approaches? What are the roles of government, industry and educational institutions?;
4 Consideration of the various forms of leadership, and assessment of the suitability of the leadership approaches to the situational context of construction in the developing countries. What kind of leadership is suitable for construction projects, in particular, in those countries?

Conclusions

Leadership has been studied for centuries by philosophers, historians, political scientists, sociologists and authors of works on business management, and there is a large and still growing body of literature on the subject. Many constructs of leadership have been developed. Moreover, some frameworks for formulating interventions that can be used to develop leaders have been proposed. Authentic leadership appears to be an appropriate construct for the developing countries owing to its stress on hope, perseverance, leading with the heart and self-transcendence, including commitment to the development of followers as leaders.

Leadership is critical to the undertaking of construction projects, the management of the enterprises involved in the construction industry, and the development of the industry as a whole. The recent surge of interest in research on leadership in construction is encouraging, since there is scope for more work to be undertaken on leadership and its development with respect to construction projects, enterprises and industries, especially in the developing countries.

Leadership development in the construction industry should start at the universities and continue through the practitioners' professional careers. A partnership of universities, government agencies and the construction industry can spearhead the development of authentic leaders in each country. The construction enterprises can help in the development of leaders through the provision of structured internship and mentorship programmes, and coaching circles.

Authentic leaders and authentic followers are needed in the construction industries in the developing countries. They can make a difference at many levels: at the strategic level, in the continuous development of the construction industry; at the corporate level in ensuring growth in competitiveness and profitability; and on projects to realise improved performance across all the relevant indicators.

References

Alinaitwe, H.M. (2009) Prioritising lean construction barriers in Uganda's construction industry. *Journal of Construction in Developing Countries*, 14(1): 15–30.

Avolio, B.J. (2007) Promoting more integrative strategies for leadership theory building. *American Psychologist*, 62, 25–33.

Avolio, B.J. and Gardner, W.L. (2005) Authentic leadership development: Getting to the root of positive forms of leadership. *The Leadership Quarterly*, 16(3): 315–38.

Avolio, B.J. and Luthans, F.L. (2006) T*he high impact leader: Moments matter in authentic leadership development*, New York: McGraw-Hill.

Ayman, R. (2000) Leadership. In E.E. Borgatta and R.J.V. Montgomery (Eds.), *Encyclopedia of sociology*, 2nd ed., (1563–1575), New York: Macmillan Reference.

Bass, B.M. (1990) *Bass & Stogdill's handbook of leadership*, 3rd ed. New York: The Free Press.

Barker, R.A. (1997) How can we train leaders if we do not know what leadership is? *Human Relations*, 50, 343–62.

Bass, B.M. (1990) *Bass & Stogdill's handbook of leadership*, 3rd ed. New York: The Free Press.

Bennis, W.G. (1989) *On Becoming a Leader*, Reading, MA: Addison-Wesley.

Bennis, W.G. (2007) The challenges of leadership in the modern world. *American Psychologist*, 62, 2–5.

Bresnen, M.J. (1986) The leader orientation of construction site managers. *Construction Engineering and Management*, 112(3): 370–86.

Brown, M.E., Treviño, L.K., and Harrison, D. (2005) Ethical leadership: A social learning perspective for construct development and testing. *Organizational Behavior and Human Decision Processes*, 97, 117–34.

Brown, M.E. and Treviño, L.K. (2006) Ethical leadership: A review and future directions. *The Leadership Quarterly*, 17, 595–616.

Bryman, A., Bresnen, M., Ford, J., Beardsworth, A. and Keil, T. (1987) Leader orientation and organizational transience: An investigation using Fiedler's LPC Scale. *Journal of Occupational Psychology*, 60, 13–19.

Bryman, A. (2004) Qualitative research on leadership: A critical but appreciative review. *The Leadership Quarterly*, 15(6): 729–69.

Budiwobowo, A., Trigunarsyah, B., Abidin, I.S. and Soeparto, H.G. (2009) Competitiveness of the Indonesian construction industry. *Journal of Construction in Developing Countries*, 14(1): 51–68.

Burns, J.M. (1978) *Leadership*, New York: Harper & Row.

Chan, P. (2008) Leaders in UK construction: the importance of leadership as an emergent process. *Construction Information Quarterly*, 10(2): 53–8.

Chan, A.T. and Chan, E.H. (2005) Impact of perceived leadership styles on work outcomes: Case of building professionals. *Construction Engineering and Management*, 131(4): 413–22.

Chemers, M.M. (1997) *An integrative theory of leadership*, Mahwah, NJ: Lawrence Earlbaum Publishers.

Chemillier, P. (1988) Facing the urban housing crisis: Confused responses to growing problems. *Building Research and Practice*, 16(2), 99–103.

Chinowsky, P.S. and Diekmann, J.E. (2004) Construction engineering management educators: History and deteriorating community. *Journal of Construction Engineering and Management,* 130(5): 751–8.

Chinyio, E. and Vogwell, D. (2007) Towards effective leadership in construction stakeholder management. Proceedings, CME 25 Conference, Reading, 16–18 July.

International Council for Research and Innovation in Building and Construction (CIB) (2006) TG64 – Leadership in Construction: Introducing new CIB Task Group. *CIB Newsletter*, December Issue. Available online at: http://heyblom.websites.xs4all.nl/website/newsletter/0608/tg64.pdf.

Conger, J.A. (1998) Qualitative research as the cornerstone methodology for understanding leadership. *The Leadership Quarterly*, 9(1): 107–21.

Debrah, Y.A. and Ofori, G. (2005) Emerging managerial competencies of professionals in the Tanzanian construction industry. *Human Resource Management*, 16(8), 1399–1414.

DuBrin, A.J. (1995) *Leadership: Research findings, practice, and skills*, Boston: Houghton Mifflin.

Dulaimi, M.F. and Langford, D.A. (1999) Job behaviour of construction project managers: Determinants and assessment. *Construction Engineering and Management*, 125(4): 256–64.

Enshassi, A., Mohamed, S. and Ekarriri, A. (2009) Essential skills and training provisions for building project stakeholders in Palestine. *Journal of Construction in Developing Countries*, 14(1): 31–50.

Erickson, R.J. (1995) The importance of authenticity for self and society. *Symbolic Interaction*, 18(2), 121–44.

Fiedler, F.E. (1967) *A theory of leadership effectiveness*, New York: McGraw-Hill.

Fry, L.W. (2003) Toward a theory of spiritual leadership. *The Leadership Quarterly,* 14, 693–727.

Gardner, J.W. (1990) *On leadership*, New York: Free Press.

George, B. (2003) *Authentic leadership: Rediscovering the secrets to creating lasting value,* San Francisco: Jossey-Bass.

George, B. and Sims, P. (2007) *True North: Discover your authentic leadership*. San Francisco: Wiley.

George, B., Sims, P., McLean, A.N. and Mayer, D. (2007) Discovering your authentic leadership. *Harvard Business Review*, 85, 129–38.

Goethals, G.R., Sorenson, G.J. and Burns, J.M. (Eds.) (2004) *Encyclopedia of Leadership,* Volumes 1–4. Thousand Oaks, Calif.: Sage.

Goffee, R. and Jones, G. (2000) *Why should anyone be led by you? Harvard Business Review*, September/October, 63–70. Thousand Oaks, CA: Sage Publications, 2004.

Grisham, T. (2006) Cross-cultural leadership in construction, in Proceedings of the International Conference on Construction Culture, Innovation, and Management, Dubai, UAE, 26–29 November.

Hackman, J.R. and Wageman, R. (2007) Asking the right questions about leadership. *American Psychologist*, 62, 43–7.

Harter, S. (2002) Authenticity. In: C.R. Snyder and S.J. Lopez (Eds.), *Handbook of positive psychology* (382–94), Oxford, UK: Oxford University Press.

Hartman, S.J., and Harris, O.J. (1992) The role of parental influence on leadership. *Journal of Social Psychology,* 132(2), 153–67.

Hemphill, J.K. (1949) *Situational factors in leadership*. Columbus, Ohio: The Ohio State University Press.

Higgs, M. (2003) How can we make sense of leadership in the 21 st. century? *Leadership and Organizational Development Journal*, 24(5), 273–284.

Hicks, D.A. (2002) Spiritual and religious diversity in the workplace: implications for leadership. *The Leadership Quarterly*, 13, 379–96.

Hillebrandt, P.M. (2000) *Economic theory and the construction industry*, 3rd edn. Basingstoke: Macmillan.

Huovinen, P. (2006a) Contextual platform for advancing the management of construction and engineering businesses: 52 concepts published between the years 1990–2005. In Songer, A., Chinowsky, P. and Carrillo, P., eds., Proceedings of the 2nd Specialty Conference on Leadership and Management in Construction and Engineering – International Perspectives. Grand Bahama Island, 381–88.

Irving, J.A. and Longbotham, G.J. (2007) Team effectiveness and six essential servant leadership themes: A regression model based on items in the organizational leadership assessment. *International Journal of Leadership Studies*, 2(2): 98–113.

Keegan, A.E. and Den Hartog, D.N. (2004) Transformational leadership in a project based environment: A comparative study. *International Journal of Project Management*, 22(8): 609–17.

Kellerman, B., Webster, S.W. (2001) The recent literature on public leadership reviewed and considered. *The Leadership Quarterly*, 12, 485–514.

Langford, D.A., Fellows, R., Hancock, M. and Gale, A. (1995) *Human resource management in construction*. London: Longman.

Laufer, A. (1987) Aptitude development of civil engineers for the management of human resources in construction projects. *International Journal of Project Management*, 5(4): 209–16.

Liden, R.C., Wayne, S.J., Zhao, H., Henderson, D. (2008) Servant leadership: Development of a multidimensional measure and multi-level assessment. *The Leadership Quarterly*, 19, 161–77.

Limsila, K. and Ogunlana, S.O. (2008a) Linking personal competencies with transformational leadership style: Evidence from the construction industry in Thailand. *Journal of Construction in Developing Countries*, 13(1): 27–50.

Limsila, K. and Ogunlana, S.O. (2008b) Performance and leadership outcome correlates of leadership styles and subordinate commitment. *Engineering, Construction and Architectural Management,* 15(2): 164–84.

Liu, A.M. and Fang, Z. (2006) A power-based leadership approach to project management. *Construction Management and Economics*, 24, 497–507.

Low, S.P. and Chuvessiriporn, C. (1997) Ancient Thai battlefield strategic principles: lessons for leadership qualities in construction project management. *International Journal of Project Management*, 15(3), 133–40.

Luthans, F. and Avolio, B.J. (2003) Authentic leadership development. In K.S. Cameron, J.E. Dutton and R.E. Quinn (eds) *Positive Organizational Scholarship: Foundations of a new discipline*. San Francisco: Berrett-Koehler; 241–58.

McCaffer, R. (2008) Editorial. *Engineering Construction and Architectural Management*, Vol. 15(4).

Muir, I. and Langford, D. (1994) Managerial behaviour in two small construction organizations. *International Journal of Project Management*, 12(4): 244–53.

Munns, A.K. and Bjeirmi, B.F. (1996) The role of project management in achieving project success. *International Journal of Project Management,* 14(2): 81–7.

Odusami, K.T. (2002) Perceptions of construction professionals concerning important skills of effective project leaders. *Journal of Management in Engineering*, 18(2): 61–7.

Odusami, K.T., Iyagba, R.R. and Omirin, M.M. (2003) The relationship between project leadership, team composition and construction project performance in Nigeria. *International Journal of Project Management*, 21, 519–27.

Ofori, G. (1993) Research on construction industry development at the crossroads. *Construction Management and Economics*, 12, 295–306.

Ofori, G. (2003) Frameworks for analysing international construction. *Construction Management and Economics*, 21(4): 379–91.

Ofori, G. (2007) Construction in developing countries. *Construction Management and Economics*, 25(1): 1–6.

Ofori, G. (2009) Leadership and Construction Industry Development in Developing Countries, CIBW 107 International Symposium on Construction in Developing Economies: Commonalities Among Diversities. Bayview Hotel Georgetown, Penang, Malaysia, 5th–7th October 2009

Ofori, G. and Toor, S.R. (2008) Leadership: A pivotal factor for sustainable development. *Construction Information Quarterly (CIQ)*, Special Issue on Leadership, 10(2), 67–72.

Ofori, G. and Toor, S.R. (2009a) Research on cross-cultural leadership and management in construction: A review and directions for future research (Extended Paper). *Construction Management and Economics*, 27(2): 119–33.

Ofori, G. and Toor, S.R. (2009b) Project leadership: A global study of new trends. Keynote Paper, Proceedings of the 3rd International Conference on Construction Engineering and Management and the 5th International Conference on Construction Project Management, May 27–30, Jeju, South Korea.

Ogunlana, S.O., Siddiqui, Z., Yisa, S. and Olomolaiye, P. (2002) Factors and procedures used in matching project managers to construction projects in Bangkok. *International Journal of Project Management*, 20, 385–400.

Powl, A. and Skitmore, M. (2005) Factors hindering the performance of construction project managers. *Journal of Construction Information*, 5, 41–51.

Pries, F., Doree, A., Veen, B.V.D. and Vrijhoef, R., (2004) The role of leaders' paradigm in construction industry change. *Construction Management & Economics*, 22(1), 7–10.

Rost, J.C. (1991) *Leadership for the twenty-first century*, New York: Praeger.

Rothstein, H.R., Schmidt, F.L., Erwin, F.W., Owens, W.A. and Sparks, C.R. (1990) Biographical data in employment selection: Can validities be made generalizable? *Journal of Applied Psychology,* 75, 175–84.

Russell, J.S. and Stouffer, B. (2003) Leadership: Is it time for an educational change? *Leadership and Management in Engineering*, 3(1): 2–3.

Seymour, D. and Elhaleem, T.A. (1991) *Horses for courses* – effective leadership in construction. *International Journal of Project Management*, 9(4), 228–32.

Shamir, B., Dayan-Horesh, H. and Adler, D. (2005) Leading by biography: Towards a life-story approach to the study of leadership. *Leadership*, 1(1): 13–29.

Shamir, B. and Eilam, G. (2005) What's your story? A life-stories approach to authentic leadership development. *The Leadership Quarterly*, 16(3): 395–417.

Skipper, C.O. and Bell, L.C. (2006a) Assessment with 360° evaluations of leadership behaviour in construction project managers. *Journal of Management in Engineering*, 22(2): 75–80.

Skipper, C.O. and Bell, L.C. (2006b) Influences impacting leadership development. *Journal of Management in Engineering*, 22(2): 68–74.

Skipper, C.O and Bell, L.C. (2008) Leadership development and succession planning. *Leadership and Management in Engineering*, 8(2), 77–84.

Songer, A., Chinowsky, P. and Butler, C. (2006) Emotional intelligence and leadership behavior in Construction executives. In *Proceedings of 2nd Specialty Conference on Leadership and Management in Construction*, May 4–6, Grand Bahama Island, Bahamas, pp. 248–58.

Sparrowe, R.N. (2005) Authenticity and the narrative self, *The Leadership Quarterly*, 16(3), 419–39.

Sternberg, R. J. (2002). Smart people are not stupid, but they sure can be foolish: The imbalance theory of foolishness. In R. J. Sternberg (Ed.), *Why smart people can be so stupid* (pp. 232–42). New Haven: Yale University Press.

Sternberg, R. J. (2007). A systems model of leadership: WICS. *American Psychologist*, 62 (1), 34–42.

Thamhain, H.J. (2003) Team leadership effectiveness in technology based project environments. *Project Management Journal*, 35(4): 35–46.

Toor, S.R. and Ofori, G. (2008a) Tipping points that inspire leadership: An exploratory study of emergent project leaders. *Engineering, Construction and Architectural Management*, 15(3): 212–29.

Toor, S.R. and Ofori, G. (2008b) Developing construction professionals of the 21st century: Renewed vision for leadership. *Journal of Professional Issues in Engineering Education and Practice*, 134(3): 279–86.

Toor, S.R. and Ofori, G. (2008c) Leadership in the construction industry: Agenda for authentic leadership. *International Journal of Project Management*, 26(6): 620–30g

Toor, S.R. and Ofori, G. (200d) Taking leadership research into future: A review of empirical studies and new directions for research. *Engineering, Construction and Architectural Management*, 15(4): 352–71.

Toor, S.R. and Ofori, G. (2008e) Leadership vs. Management: How they are different, and why! *Journal of Leadership and Management in Engineering*, 8(2), 61–71.

Toor, S.R. and Ofori, G. (2009a) Authenticity and its influence on psychological well-being and contingent self-esteem of leaders in Singapore construction sector. *Construction Management and Economics*, 27(3): 299–313.

Toor, S.R. and Ofori, G. (2009b) Ethical leadership: Examining the relationships with Full Range Leadership Model, employee outcomes, and organizational culture. *Journal of Business Ethics*, 90(4): 533–47.

Toor, S.R. and Ofori, G. (2010) Positive psychological capital (PsyCap) as a source of sustainable competitive advantage for organizations. *Journal of Construction Engineering and Management*, 136(3): 341–52.

Toor, S.R. and Ogunlana, S.O. (2008a) Critical COMs of success in large-scale construction projects: Evidence from Thailand construction industry, *International Journal of Project Management*, 26(4): 420–30.

Toor, S.R. and Ogunlana, S.O. (2008b) Leadership skills and competencies for cross-cultural construction projects. *International Journal of Human Resources Development and Management*, 8(3): 192–215.

Toor, S.R and Ogunlana, S. O. (2009) Ineffective leadership: Investigating the negative attributes of leaders and organizational neutralizers. *Engineering, Construction and Architectural Management*, 16(3): 254–72.

Vroom, V.H. and Jago, A.G. (2007) The role of situation in leadership. *American Psychologist*, 62, 17–24.

Walumbwa, F.O., Avolio, B.J., Gardner, W.L., Wernsing, T. and Peterson, S.J. (2008). Authentic leadership: Development and validation of a theory-based measure. *Journal of Management*, 34(1), 89–126.

Wong, J., Wong, P.N. and Li, H. (2007) An investigation of leadership styles and relationship cultures of Chinese and expatriate managers in multinational construction companies in Hong Kong. *Construction Management and Economics*, 25, 95–106.

Yukl, G. (1989) *Leadership in organizations*, Englewood Cliffs, NJ: Prentice-Hall.

Part IV

Reconstruction programmes

10 Sustainable low-cost housing in developing countries

The role of stakeholders "after" the project

Gonzalo Lizarralde
University of Montreal, Canada

Introduction

It is widely known that the participation of stakeholders greatly influences both the construction project process and its outcome (Chinyio and Olomolaiye, 2010; Friedman and Miles, 2006). However, little evidence exists on *how* the different roles of project participants in developing countries influence low-cost housing projects during the pre-occupancy phase and the post-occupancy phase. This chapter discusses research results that analyse and compare the role of stakeholders during these two phases.

The chapter begins with a review of the literature on low-cost housing in developing countries. This introduction serves to explain the complexity of the project team that is required to conduct the projects: the so-called Temporary Multi-Organisation (TMO), in construction management jargon. It then explains the characteristics of the study (including a description of the methods used) and it finally presents the results of the analysis.

Housing shortages and vulnerabilities

Three factors are often associated with the "housing problem" in developing countries. First, land market failures in developing countries often limit the conversion of agricultural land into urban land and, thus, into land for residential development (Ferguson and Navarrete, 2008). Second, capital market failures reduce the capacity of low-income residents to obtain the financial services (mortgages and low-interest loans) that are required to purchase a housing unit (Datta and Jones, 2001). Third, some inherent characteristics of the building industry also affect housing market failures (Lizarralde and Root, 2008). For instance, in hostile economic environments, builders and developers in the low-cost housing sector are required to accept slim profits per unit to maximise affordability. Thus, land development for residential use is only profitable if housing units are delivered in large numbers, taking advantage of economies of scale (Keivani and Werna, 2001a). Nevertheless, this development is often constrained by the absence of large pieces of affordable urban land in desirable locations (that is, close to means of transport, services and jobs); which ultimately leads back again to the first factor: the constraints of urban land development.

The most common consequence of residential market failure is housing deficits, which manifest themselves in the lack of formal housing solutions that poor families can realistically afford (Choguill, 2007). Families that cannot gain access to formal financial solutions, welfare and/or employment in order to purchase or rent formal housing frequently resort to informal housing products. Informal housing solutions often rely on self-help construction and planning, design and construction conducted by informal companies, paid and unpaid independent construction workers and self-help (Choguill, 2007; Datta and Jones, 2001; Keivani and Werna, 2001a, 2001b). They are usually the result of a progressive system of housing production that starts with illegal occupation of land (sometimes through invasion of vacant land located in risk-prone areas) and that continues with incremental construction over time until both the housing units and the urban settlements – often called slums – approach formality (Gough and Kellett, 2001). It is estimated that the percentage of the urban population living in slums is about 70 percent in Bangladesh, 65 percent in Nigeria, 50 percent in Iraq, 30 percent in South Africa, and 18 percent in Colombia (*Economist*, 2010). Through a process of progressive adaptation, informal housing solutions respond very well to the needs and priorities of individual users (Bhatt and Rybczynski, 2003). Contrary to common belief, informal housing is not produced in a spontaneous or random manner; instead, it responds to rational decisions that anticipate long-term development and economic and physical upgrading (Lizarralde and Davidson, 2006; Lizarralde et al., 2009; Lizarralde and Root, 2007).

Incremental construction (through later enlargement of the units and self-help improvements) is a natural strategy to build housing in the informal sector (*Economist*, 2010; Ferguson and Smets, 2009; Lizarralde and Boucher, 2004; Lizarralde, et al., 2009; Morado Nascimento, 2009). Incremental construction not only accounts for 70 percent of housing investment in developing countries (Ferguson and Navarrete, 2003; Gilbert, 2004), but also allows poor residents to gradually "climb" the housing ladder through improvements that follow the pace of their own economic possibilities (Strassmann, 1984). This process of evolution often includes horizontal and vertical expansion of the units and the integration of space or facilities for income generating activities (Kellett and Tipple, 2000).

Therefore, informal housing solutions are the manifestation of social, economic and political vulnerabilities, which together increase the risks of exposure to natural hazards such as earthquakes, floods and tsunamis (Blaikie et al., 1994). If increased physical and non-physical vulnerabilities clash with natural hazards, the result is a "natural" disaster, which not only suddenly increases housing deficits but also causes important pressure on the political and administrative systems that are responsible for alleviating them (Lizarralde et al., 2008; Lizarralde et al., 2009).

Housing solutions

During the last 30 years, responses to housing deficits in developing countries have mostly relied on three strategies: slum upgrading, subsidised new housing, and reactive measures to post-disaster reconstruction. The latter often include the development of temporary housing, housing relocation and delivery of subsidised solutions for permanent housing. In fact, subsidised housing is one of the most important short-term solutions used now in developing countries to alleviate low-cost housing shortages (Gilbert, 2004). However, subsidised projects often require that governments play an important role in reducing land market distortions, particularly by reducing requirements and streamlining the land development process in order to stimulate private developers and builders to act (Ferguson and Navarrete, 2003, 2008).

Local governments must also play an important role in promoting the projects and in recognising the influence of informal markets (Kombe and Kreibich, 2000). However, small municipalities often do not have the legal, financial and administrative means to initiate, manage and transfer the projects. Facilitating the provision of safe land for urban development, matching public and private resources, encouraging individual savings, selecting, evaluating and approving beneficiaries and choosing minimum standards are some of the major barriers in this process (Harpham and Boateng, 1997; Lizarralde, 2008).

Subsidised housing can also have two secondary effects that exacerbate the housing problems. First, the solutions proposed by governments might distort the market even further, particularly if public housing standards do not bridge the gap between the houses available in the informal market and non-subsidised housing products (Lizarralde and Root, 2008). Second, the participation of local organisations and beneficiaries in subsidised housing does not often guarantee appropriate levels of community participation and significant decision-making power from local residents. Common constraints include: difficulties faced in efforts to integrate the community in the design and management of the project; and the reduction of participation to sweat equity instead of an active role in decision-making (Choguill, 1996; Lizarralde and Massyn, 2008).

In reality, sustainable housing solutions in developing countries must account for two levels of strategic planning: responses to regular housing deficits (including not only quantitative but also qualitative deficits – and thus vulnerability reduction); and planning for, and anticipation of, post-disaster reconstruction. Considering the non-availability of information regarding the needs and expectations of individual users (Lizarralde and Bouraoui, 2010) decision-makers acting at both levels of strategic planning are required to learn from the adaptive solutions provided by the informal sector (Lizarralde et al., 2009). Therefore, a comprehensive understanding of the potential contribution and limitations of all project stakeholders is needed.

Stakeholder participation

The relationship between stakeholder participation within the TMO and project outcomes has been studied in the literature (de Blois and De Coninck, 2008; Friedman and Miles, 2006; Newcombe, 2003; Walker et al., 2008). Table 10.1 presents a summary of some of the relationships that have been identified by previous authors between participants' involvement and project processes and outcomes. It shows some of the most important findings with regard to dependent variables associated with the role of stakeholders in general project management, in construction management, in international development and in housing and post-disaster reconstruction.

Table 10. 1 shows that there is a general consensus regarding the positive effects of active and effective implication of stakeholders of the TMO: increased users' satisfaction, project performance; conflict prevention; construction industry development; and vulnerability reduction. However,

Table 10.1 Summary of the variables commonly discussed in the literature on stakeholder participation (source: after Lizarralde and Bouraoui, 2010)

Author	*Independent Variable (IV)*	*Dependent Variable (DV)*	*Results (IV-DV)*
Project management			
(McKeen et al., 1994)	Users' participation	Users' satisfaction	Positive, but the strength is different depending on contingency factors
(PMI, 2008)	Stakeholder involvement in project definition	Project performance	IV is necessary for DV
(Newcombe, 2003)	Stakeholders' different roles, interests and power	Stakeholder management by the project manager	IV justifies DV

Author	*Independent Variable (IV)*	*Dependent Variable (DV)*	*Results (IV-DV)*
Construction management			
(Thomson, 2010)	Facilitating stakeholder participation	Construction industry performance	Positive relationship
(Ross, 2009)	Stakeholder participation in partnering	Conflict prevention	Positive relationship subject to appropriate partnering methods
(Leung & Olomolaiye, 2010)	Stakeholder involvement	Project risk reduction	Positive relationship
(Al-Khafaji et al., 2010)	Stakeholder communication	Project success	Positive relationship
(Karlsen et al., 2008)	Stakeholder relationships	Increase of trust	Positive relationship
International development			
(Choguill, 1996)	Community participation	Project success	Depends on the efficient (or not) practice of community participation and involvement
(Lizarralde et al., 2008)	Organisational design including all stakeholders	Project performance in international construction	Positive relationship subject to appropriate integration and differentiation
(UN-Habitat, 2009)	Citizen control over decision making	Transformative and empowering interventions	Positive relationship subject to appropriate distribution of power
Housing and reconstruction			
(Davis, 1978, 1981)	Community participation	Users' satisfaction	Positive relationship
(Corsellis and Vitale, 2008; UNDRO, 1982)	Local community participation	Success of reconstruction projects	Positive relationship
(Maskrey, 1989)	Community participation components	Successful implementation	Positive relationship
(Oliver-Smith, 2007)	Effective community participation	Community satisfaction	Significant positive relationship

Author	*Independent Variable (IV)*	*Dependent Variable (DV)*	*Results (IV-DV)*
(Blaikie et al., 1994)	"Active measures" and approaches for the most vulnerable	Reduction of vulnerability and vigorous mitigation	Positive relationship
	Top-down approach	Increased vulnerability	Positive relationship
(Jigyasu, 2000)	Appropriate technologies; local skills' participation	Vulnerability reduction	Positive relationship
(El-Masri and Kellett, 2001)	Top-down approach	Users' self-reliance and participation	Negative relationship
	Bottom-up approach	Users' self-reliance, development	Positive relationship: IV helps build DV
(Alexander, 2004)	Consideration of users' physical, emotional and economic attachment	Project success	IV increases the chances of DV
(Lizarralde, 2004; Lizarralde et al., 2008; Lizarralde and Massyn, 2008)	Participation of users	Project performance	Weak positive relationship (project performance is more affected by strategic aspects)
	Organisational design including all stakeholders	Project performance	Positive relationship subject to appropriate organisational design
	Decentralisation of decision-making	Successful project management	Positive relationship
(Arslan and Unlu, 2006)	Community participation	Understanding of community needs	Positive relationship
(Monday, 2006)	Principles of sustainability applied to local actions and decisions	Holistic recovery from a disaster	Positive relationship
	Public involvement and participatory processes	Sustainable reconstruction and local sustainable recovery	Positive relationship
(Özden, 2006)	Community involvement	Success of reconstruction projects	Significant positive relationship

Author	*Independent Variable (IV)*	*Dependent Variable (DV)*	*Results (IV-DV)*
(Barenstein, 2006; Jha, 2009)	Participation of users	Users' satisfaction and project efficiency	Significant positive relationship
(Bosher, 2008)	Centralised approach	Social refusal	Positive relationship
Lyons (2009)	Active participation and involvement of users	Acceptability of users	Positive relationship

most of the literature has tended to focus on the role of stakeholders *during* the development of the traditional project (from initiation to occupancy). There is still insufficient knowledge on *how* the participation of stakeholders affects the project process including its complete lifecycle, particularly including the post-occupancy phase. This chapter analyses the role of stakeholders during the regular construction phase and their role after occupancy.

The research

This chapter attempts to answer the following question: "How do the different roles of project stakeholders influence both the project process and its outcome during the pre- and post-occupancy phases of housing projects in developing countries?" In order to answer this question, this chapter includes the results of three research projects conducted by the author between 2002 and 2010. The research projects were on:

1. Stakeholder participation in post-disaster housing conducted between 2002 and 2006. Additional information about this study, including a detailed description of the methods used, can be found in Davidson et al. (2007) and Johnson et al. (2005, 2006);
2. Formal and informal housing solutions in South Africa conducted between 2006 and 2008 (Lizarralde and Massyn, 2008; Lizarralde and Root, 2007, 2008);
3. Stakeholder participation in incremental housing conducted between 2007 and 2010 (Lizarralde, 2008).

Multiple research methods were applied in these three studies, including participatory research, observation, interviews and systematic analysis of organisational structures. However, the results presented in this chapter were particularly obtained from the qualitative analysis of three case studies of regular housing and two case studies of post-disaster housing. The cases and some of their characteristics are presented in Table 10.2.

Table 10.2 Description of the case studies analysed in this chapter

Project	*Location*	*Number of units*	*Start and end of the pre-occupancy construction phase*
Regular housing projects			
Case 10.1: Project "Juan Pablo II"	Facatativa, Colombia (urban project)	1262	2004–2008
Case 10.2: Project "Netreg"	Cape Town, South Africa (urban project)	192	1986–2006 (18 years for project initiation)
Case 10.3: Project "Mfuleni"	Cape Town, South Africa (urban project)	700	2004–2008
Post-disaster housing projects			
Case 10.4: Project "Nueva Choluteca"	Choluteca, Honduras (semi-urban project)	52	1999–2002
Case 10.5: Project "Fundasal"	La Paz, El Salvador (rural project)	4400	2001–2002

The research projects consisted of following and documenting the whole construction process through a period of time that varied from one to four years; and reconstructing the project processes that were already completed before the beginning of the study. These two activities were undertaken by preparing and validating diagrams of the organisational structure of the projects (at different times), diagrams of the sequence of activities conducted, and a quantitative and qualitative analysis of project outcomes. The sequence of activities and processes was then reported in detailed case studies covering the different aspects of project management suggested by the *Project Management Body of Knowledge* of the Project Management Institute (PMI, 2008). Information was gathered and analysed until it was possible to reconstruct the project process and create the diagrams, thus achieving saturation of required data. The analysis was based on different combinations of the following sources of information:

- Semi-directed interviews with project participants;
- Open interviews and discussions with beneficiaries;
- Observation of the advancement of projects;
- Participation in project meetings;
- Plans, contracts, budgets, timelines and other construction documents;
- Bills and municipal or regional codes;
- Construction manuals and guides offered to beneficiaries;
- Press releases and magazine articles;

- Websites of the project participants;
- Project reports;
- Minutes of the project meetings;
- Photographs of the housing units at different moments in the project lifecycle.

Stakeholders

The case studies enabled the researcher to identify six groups of stakeholders who play a fundamental role in the housing project TMO. These groups are:

1. Central and regional governments;
2. Municipalities;
3. Participants of the private sector, including companies that supply materials, equipment or services such as urban, architectural or landscape design, structural or mechanical engineering, management, logistics, accounting and construction;
4. Participants of the informal sector, including small non-registered construction companies (acting as contractors or subcontractors), construction workers not working under the contract of registered construction companies, and paid or voluntary workers who offer advice or services in design and construction;
5. Participants of the not-for-profit sector, including non-governmental organisations (NGOs) and not-for-profit organisations that participate in project financing, initiation, planning, management, design or construction;
6. Beneficiaries, more specifically selected or unselected households who benefit directly from housing subsidies, housing units or any other project outcome.

However, stakeholders played different roles in each of the project TMO, and their roles evolved in a different manner during the project process. Two phases were identified in each project. The first stage was the pre-occupancy phase that included the definition and initiation of the project, the design and the construction of the outputs that were offered to beneficiaries. The second stage was the post-occupancy phase, starting immediately after the delivery of project outcomes to beneficiaries.

It was found that stakeholders assumed one or a combination of the following activities in each of these phases: financing, particularly the delivery or transfer of funds from project sponsors; project procurement, including initiation, management and contracting; design, including urban design, housing design and design of renovations and changes; and construction, including assuring the supply of materials, labour or equipment.

Incremental construction

Incremental construction after project occupation occurred in all cases, as was expected according to the literature on the subject (Garcia-Huidobro et al., 2008). However, the scope and scale of interventions varied significantly. The most important interventions occurred in Case 10.1, which was the only one in which the 40m² housing unit was designed for being enlarged to an 80m² unit by building on the vacant space in the backyard. In this particular case, beneficiaries received a construction guide containing plans, specifications and construction details about the desirable additions to the units. However, interviews with residents and observations in-situ demonstrated that households hired informal workers for construction, who also contributed to the design of both changes to the existing layout of the units and new construction different from the design anticipated by the construction guide (see Figures 10.1 to 10.4). Interventions to the existing units, including upgrading and new construction, are summarised in Table 10.3.

In Cases 10.1 and 10.4 (in which important new construction was quickly developed by users), extensions to the original units often did not

Figure 10.1 New homeowner painting his house just after project delivery in Case 10.1 (photo: Lizarralde)

Figure 10.2 Transitional extension built on the available space in Case 10.1 (photo: Lizarralde)

Figure 10.3 Transformation to housing units to integrate productive activities at the household level. Left: example in Case 10.1. Right: example in Case 10.4 (photos: Lizarralde)

Figure 10.4 Layout of the units in Case 10.1. Left: Plans of first and second floor "as built" in the pre-occupancy phase (black walls), and as anticipated by the construction guide (white walls) for the post-occupancy phase. Right: example of a house transformed during the post-occupancy phase (images: Lizarralde)

respect disaster-resistant standards. In Case 10.1, they often did not respect the standards identified in the construction guide (see Fig. 10.4). Common structural mistakes which were found in all cases included: lack of structural joints between materials; reduced sections for columns and beams; insufficient bracing or reinforcement of the structure; and insufficient attachment of the envelope and partitions to the structure.

Upgrades and new construction included also the use of temporary solutions that followed the rhythm of available investment. Enclosures made of plastic, or corrugated sheets and a basic wooden structure, as well as fabric curtains, wood subdivisions and other improvised materials were commonly used in all cases for temporary separation of spaces or for building additional rooms in a temporary manner (see Fig. 10.2). These temporary solutions were mostly undertaken by users themselves, whereas a good part of new construction was done by paid workers, or completed with the help of unpaid labour (mostly friends and relatives of the beneficiaries).

An important and most common intervention was the transformation of the house into a productive family business. The following common productive activities were identified:

1 Convenient stores: Cases 10.1, 10.2, 10.3, 10.4;
2 Spaces for agricultural production or storage: Case 10.5;
3 Cafeterias: Cases 10.1, 10.2, 10.3 and 10.4;
4 Paper stores: Case 10.1;
5 Space for babysitting (informal kindergartens): Case 10.1;
6 Hair and beauty salons: Cases 10.2 and 10.3;
7 Internet cafés: Cases 10.1 and 10.3;
8 Commerce of miscellanea: Cases 10.1, 10.2, 10.3 and 10.4;
9 Bars and taverns: Cases 10.3 and 10.4.

Tables 10.4 to 10.8 present summaries of the role of each of the groups of stakeholders in each of these four activities (financing, project procurement, design and construction) during the two main phases of the project. The tables show that central governments play an important role in financing (through national programmes of subsidies) in Cases 10.1, 10.2 and 10.3. Municipalities have a strong influence on project initiation and procurement, even though their role can be overlooked by the not-for-profit sector (as in Cases 10.4 and 10.5). Actors of the not-for-profit sector assume in some cases important roles of project commissioning – particularly in the cases in which the municipality plays a less important role, such as in Cases 10.4 and 10.5. The private sector plays an important role in subcontracting, in construction and – in some cases – in design and administration (see Case 10.1). In all cases, the informal sector and beneficiaries undertake most of the work in the post-occupancy phase, when the rest of the participants are mostly absent.

Tables 10.4 to 10.8 also enable the roles that stakeholders *did not* assume during the post-occupancy phase to be identified. Some examples are:

1 Central and regional governments provided neither subsidies nor low-interest loans for housing upgrading;
2 Municipalities did not enforce supervision and respect of disaster-resistant standards for additions and upgrades;
3 Participants of the private sector did not participate in the design of upgrading or extensions. Except in Case 10.1, original designs did not anticipate incremental construction;
4 Participants of the informal sector were not involved in the pre-occupancy phase;
5 Participants of the not-for-profit sector did not follow up on projects or community empowerment "after" the construction project. They did not plan for a post-occupancy intervention;
6 Beneficiaries did not follow earthquake-resistant standards in building additions and extensions to the units.

It was found that there is an imbalance of responsibilities during the pre- and post-occupancy phases, and that participants of the formal sector neglect the importance of contributing to, and supervising, the post-occupancy interventions. Considering the lack of commitment of municipalities, NGOs, private companies and central governments in the post-occupancy

Table 10.3 Summary of patterns of transformation found in the case studies

	Upgrading										*New construction*		
Project	*Interior painting*	*Exterior painting*	*Tiles on floor*	*Tiles in WC*	*Tiles in kitchen*	*New internal partitions*	*Demolition of internal partitions*	*Fences*	*Security grids*	*Production and family business*	*Addition of new rooms*	*Addition of a veranda*	*Addition of temporary structures*
Case 10.1: Project "Juan Pablo II" 12 months after occupation	Y	Y	Y	Y	Y	Y	Y	Y	Y	Y	Y	N	Y
Case 10.2: Project "Netreg" 12 months after occupation	Y	Y	Y	Y	Y	Y	N	Y	Y	Y	Y	N	Y
Case 10.3: Project "Mfuleni" 6 months after occupation	Y	Y	Y	Y	N	N	N	Y	Y	Y	Y	N	Y
Case 10.4: Project "Nueva Choluteca" 45 months after occupation	Y	Y	Y	Na	Y	N	N	Y	Y	Y	Y	Y	Y
Case 10.5: Project "Fundasal", 6 months after occupation	N	N	N	N	N	N	N	Y	N	N	Y	Y	Y

Table 10.4 Summary of activities of project stakeholders, project "Juan Pablo II" – Case 10.1

Stakeholders	*Area of intervention*	*Pre-occupation*	*Post-occupation*
Central and regional government	Financing	Central government offers national subsidies	
Municipality	Financing	Offers subsidies	
	Procurement	Initiates and manages the project	
	Construction		Limited supervision and control over additions and modifications
Private sector	Financing	Private funds administer subsidies and offer matching funds	
	Design	Consultants design infrastructure and housing	
	Construction	Construction by contractor and subcontractors	Supply of materials and components
Informal sector	Financing		Offers informal financing
	Design		Design of additions and modifications
	Construction		Active engagement in construction
Not-for-profit sector	N/A		
Beneficiaries	Financing	Users' savings are required for eligibility to subsidies	Users use their own resources and loans
	Procurement		Users hire labour and purchase materials
	Design		Design of additions and modifications to existing units
	Construction		If required, contribution in labour for additions and modifications

Table 10.5 Summary of activities of project stakeholders, project "Netreg" – Case 10.2

Stakeholders	*Area of intervention*	*Pre occupation*	*Post occupation*
Central and regional government	Financing	Central gov. transfers subsidies to regional gov. and municipalities	
	Procurement	Regional gov. identifies priorities and plans	
Municipality	Financing	Offers subsidies	
	Procurement	Initiates the project	
	Construction		Limited supervision and control over additions and modifications
Private sector	Procurement	Contractors search for subcontractor	
	Design	Consultants design infrastructure	
	Construction	Construction by contractor and subcontractors	
Informal sector	Financing		Offers informal financing
	Design		Design of modifications and upgrades
	Construction		Construction and supply of materials
Not-for-profit sector	Financing	NGO provides additional funds	
	Procurement	NGO initiates the project, identifies private accountant and contractors	
	Design	NGO designs housing project	
Beneficiaries	Financing	Users are motivated to save	Users use their own resources and loans
	Procurement		Users hire labour and purchase materials
	Design		Design of modifications to existing units
	Construction		If required, contribution in labour for additions and modifications

Table 10.6 Summary of activities of project stakeholders, project "Mfuleni" – Case 10.3

Stakeholders	*Area of intervention*	*Pre-occupation*	*Post-occupation*
Central and regional government	Financing	Central gov. transfers subsidies to regional gov. and municipalities	
	Procurement	Regional gov. identifies priorities and plans	
Municipality	Financing	Offers subsidies	
	Procurement	Initiates the project	
	Construction		Limited supervision and control over additions and modifications
Private sector	Procurement	Contractors search for subcontractor	
	Design	Consultants design infrastructure	
	Construction	Construction by contractor and subcontractors	
Informal sector	Financing		Offers informal financing
	Design		Design of modifications and upgrades
	Construction		Construction and supply of materials
Not-for-profit sector	Financing	NGO provides additional funds	
	Procurement	NGO initiates the project, identifies private accountant and contractors	
	Design	NGO designs housing project	
Beneficiaries	Financing	Users are motivated to save	Users use their own resources and loans
	Procurement		Users hire labour and purchase materials
	Design		Design of modifications to existing units
	Construction		If required, contribution in labour for additions and modifications

Table 10.7 Summary of activities of project stakeholders, project "Nueva Choluteca" – Case 10.4

Stakeholders	*Area of intervention*	*Pre-occupation*	*Post-occupation*
Central and regional government	No activity		
Municipality	No activity		
Private sector	Construction	Construction by contractor and subcontractors	
Informal sector	Financing		Offers informal financing
	Construction		Construction and supply of materials
Not-for-profit sector	Financing	NGO provides funds	
	Procurement	NGO initiates and manages the project	
	Design	NGO designs housing project and infrastructure	
Beneficiaries	Financing		Users use their own resources and loans
	Procurement		Users hire labour and purchase materials
	Design		Design of modifications to existing units
	Construction	Contribution in labour	Contribution in labour for additions and modifications

phase, it is not surprising to find that upgrading and additions do not respect disaster-resistant standards and therefore that physical vulnerabilities are not fully reduced by the housing projects. Lack of planning for, and anticipation of, customisation, additions and upgrades were also found in five cases. Only the project in Case 10.1 includes the possibility of post-occupancy additions, and some measures were taken to influence the choice of beneficiaries towards safer construction.

Table 10.8 Summary of activities of project stakeholders, project "Fundasal" – Case 10.5

Stakeholders	*Area of intervention*	*Pre occupation*	*Post occupation*
Central and regional government	No activity		
Municipality	No activity		
Private sector	Construction	Construction by contractor and subcontractors	
Informal sector	Financing		Offers informal financing
	Construction		Construction and supply of materials
Not-for-profit sector	Financing	NGO provides funds	
	Procurement	NGO initiates and manages the project	
	Design	NGO designs housing project and infrastructure	
Beneficiaries	Financing		Users use their own resources and loans
	Procurement		Users hire labour and purchase materials
	Design		Design of modifications to existing units
	Construction	Contribution in labour	Contribution in labour for additions and modifications

Discussion and implications for construction industry development, policy development and application

Subsidised housing is one of the most important strategies of the formal sector for reducing the qualitative and quantitative deficits of affordable housing in developing countries. On the other hand, incremental construction is the most important strategy of the informal sector. The Colombian case study

(Case 10.1) shows that local governments can take advantage of both strategies simultaneously. Housing solutions can greatly benefit from three decisions which are now discussed:

1 Coordinating the roles of different public and private participants (including the end users) *during* and *after* the construction project. Formal and informal participants play a fundamental role in housing development. Even though it is largely accepted that stakeholder participation brings important benefits to both the construction projects and the building industry, little attention is often paid to the role of the stakeholders during the post-occupancy phase. However, case studies show that it is precisely during that phase that upgrading, customisation and additions are built. Therefore, it is of prime importance to integrate the roles of the stakeholders "after" the project, including the following responsibilities:

1. Central and regional governments can provide subsidies and support low-interest loans for the upgrading of housing units, an alternative that did not occur in the cases studied here;
2. Municipalities can provide supervision and assure the respect of disaster-resistant standards for additions and upgrades. This was hardly achieved in the case studies presented here;
3. Participants of the private sector supply materials, equipment or services (a common practice that is found in all cases);
4. Participants of the informal sector provide most of the labour during the post-occupancy phase (a common practice already found in all cases);
5. Participants of the not-for-profit sector can contribute to reinforce local capacities and community empowerment in order to reduce social and economic vulnerabilities and to support better practices of self-help and construction management and commissioning. This was targeted as an objective in Cases 10.2, 10.3, 10.4 and 10.5;
6. Beneficiaries have the capacity to make decisions regarding the upgrading, customisation and enlargement of the original units. However, this requires training and education on disaster-resistant standards and a clear understanding of the consequences of physical vulnerabilities associated with poor construction practices.

2 Including the participation of the informal sector. A great deal of housing construction, upgrading and addition is undertaken by the informal sector, notably through self-help, paid informal workers and unpaid labour (voluntary work by beneficiaries' friends and family members). Rather than being ignored, the informal sector needs to be considered as an important (also unavoidable) partner of housing development. Its participation in post-occupancy upgrading and housing completion can be planned in advance and managed in order to better contribute to the project.

3 Replicating the natural response of the informal sector to incremental housing development. Lower housing standards can be (and are) accepted by users if alternatives for incremental upgrading are available and properly planned in advance (such as in Case 10.1). However, these post-occupancy interventions need to be properly supervised and controlled in order to reduce physical vulnerabilities related to disaster risk.

Merging formal subsidised initiatives with incremental strategy requires careful planning of the roles of all project stakeholders. It requires particularly the re-valuation of three principles:

1. Contrary to common belief, informal housing construction is not the cause of disaster vulnerabilities. It is actually the consequence of the required adaptation of the building sector to hostile conditions and market failures;
2. The informal construction sector presents several advantages that can be used in the formulation and implementation of housing solutions. Some of them are: increased capacity of adaptation to hostile economic environments; increased capacity to deliver customised solutions increased capacity to deliver affordable solutions; and increased capacity to adapt work and objectives in the project lifecycle following the pace of available resources and partial commissioning;
3. The informal construction sector also presents several limitations that need to be addressed in building industry development. They include: limited access to financial solutions with competitive interest rates; limited capacity to respond to disaster-resistant standards of construction; and limited capacity to develop comprehensive solutions of infrastructure for public services.

Only by addressing these three conditions can housing projects contribute in a sustainable way to their two most important objectives: reduction of vulnerabilities and alleviation of qualitative and quantitative deficits of housing.

Conclusion

This chapter builds on the widely accepted premise that the participation of stakeholders greatly influences both the construction project process and its outcome. However, it also recognises that little evidence exists on *how* the different roles of project participants in developing countries influence low-cost housing projects, particularly during the post-occupancy phase. The chapter proposed the following research question: "How do the different roles of project stakeholders influence both the project process and outcome during the pre- and post-occupancy phase of housing projects in developing countries?" In order to answer this question, this chapter considered results of three research projects conducted by the author between 2002 and 2010. Even though multiple research methods were applied in the studies (including participatory research, observation and interviews), the results presented in this chapter were particularly obtained from the qualitative analysis of three case studies of regular housing and two case studies of post-disaster housing.

The chapter analysed the different roles of stakeholders in housing development projects and found that very little attention is often paid to planning the post-occupancy phase. This pattern becomes worrisome because most of the upgrading and extensions generated by users and informal labour force (usually without respecting disaster-resistant standards) occur precisely during that phase. More attention is required in the planning of activities that will occur after housing delivery. Architectural and urban designs need to anticipate and carefully consider the extensions and upgrading that occurs after the occupation of units. If reduction in vulnerability to natural disasters is to be achieved, architects, urban planners and project managers need to plan and organise the roles of project stakeholders during the post-occupancy phase. Finally, the chapter also highlighted the importance of incremental construction, a strategy that facilitates: the participation of end-users in the housing process, users' customisation of units, users adaptive response to changes and particular needs, users adaptive response to income availability, and the participation of informal local companies in housing construction.

References

Al-Khafaji, A.W., Oberhelman, D., Baum, W. and Koch, B. (2010) Communication in stakeholder management. In E. Chinyio and P. Olomolaiye (eds), *Construction Stakeholder Management*. Chichester: Blackwell Publishing; pp. 159–73.

Alexander, D. (2004) Planning for post-disaster reconstruction. *2004 i-Rec international Conference and Student Competition on post-disaster reconstruction "Planning for reconstruction"*. Retrieved from http://www.grif.umontreal.ca/pages/papers2004/Paper%20%20Enginoz%20E%20B.pdf/

Arslan, H. and Unlu, A. (2006) The evaluation of community participation in housing reconstruction projects after Duzce earthquake. *2006 i-Rec International Conference and Student Competition on post-disaster reconstruction "Meeting stakeholder interests"*. Retrieved from http://www.grif.umontreal.ca/pages/ARSLAN_%20Hakan.pdf/

Barenstein, J. (2006) Housing Reconstruction in Post-Earthquake Gujarat: A Comparative Analysis. *Humanitarian Practice Network Paper*, London: Overseas Development Institute.

Bhatt, V. and Rybczynski, W. (2003) How the other half builds. In D. Watson, A. Plattus and R. Shibley (eds), *Time-saver Standards for Urban Design*. New York: McGraw-Hill; 1.3.1–1.3.11.

Blaikie, P., Cannoon, T., Davis, I. and Wisner, B. (1994) *At Risk: natural hazards, people's vulnerability, and disaster.* New York: Routledge.

Bosher, L. (2008) *Hazards and the built environment: attaining built-in resilience.* New York: Taylor and Francis Group.

Chinyio, E. and Olomolaiye, P. (2010) Introducing Stakeholder Management. *Construction Stakeholder Management*. Oxford: Blackwell Publishing.

Choguill, C.L. (2007) The search for policies to support sustainable housing. *Habitat International*, 31(1), 143–9.

Choguill, M.B.G. (1996) A ladder of community participation for underdeveloped countries. *Habitat International*, 20(3), 431–44.

Corsellis, T. and Vitale, A. (Lead Eds.) (2008) *Transitional Settlement and Reconstruction after Disaster: Field Edition*. New York: United Nations.

Datta, K. and Jones, G.A. (2001) Housing and finance in developing countries: invisible issues on research and policy agendas. *Habitat International*, 25(3), 333–57.

Davidson, C.H., Johnson, C., Lizarralde, G., Dikmen, N. and Sliwinski, A. (2007) Truths and myths about community participation in post-disaster housing projects. *Habitat International*, 31(1), 100–15.

Davis, I. (1978) *Shelter after Disaster*. Oxford: Oxford Polytechnic.

Davis, I. (1981) *Disasters and the Small Dwelling*. Oxford: Pergamon Press.

de Blois, M. and De Coninck, P. (2008) The dynamics of actors and stakeholders participation: An approach of management by design. *Architectural Engineering and Design Management*, 4, 176–88.

Economist (2010) Slumdog millions. *The Economist*, 24 March.

El-Masri, S. and Kellett, P. (2001) Post-war reconstruction. Participatory approaches to rebuilding the damaged villages of Lebanon: a case study of al-Burjain. *Habitat International*, 25(4), 535–57.

Ferguson, B. and Navarrete, J. (2003) New approaches to progressive housing in Latin America: A key to habitat programs and policy. *Habitat International*, 27(2), 309–23.

Ferguson, B. and Navarrete, J. (2008). A financial framework for reducing slums. In S. Sernau (Ed.), *Contemporary Readings in Globalization*, (pp.183–96). London: Sage.

Ferguson, B. and Smets, P. (2009) Finance for incremental housing; current status and prospects for expansion. *Habitat International*, 34(3), 288–98.

Friedman, A. and Miles, S. (2006) *Stakeholders: theory and practice*. Oxford: Oxford University Press.

Garcia-Huidobro, F., Torres, D. and Tugas, N. (2008) *Time Builds! The Experimental Housing Project (PREVI) Lima: Genesis and outcome*. Barcelona: Gustavo Gili.

Gilbert, A. (2004) Helping the poor through housing subsidies: lessons from Chile, Colombia and South Africa. *Habitat International*, 28(1), 13–40.

Gough, K. V. and Kellett, P. (2001) Housing consolidation and home-based income generation: Evidence from self-help settlements in two colombian cities. *Cities*, 18(4), 235–47.

Harpham, T. and Boateng, K. (1997) Urban governance in relation to the operation of urban services in developing countries. *Habitat International*, 21(1), 65–77.

Jha, A. Barenstein, J.D., Phelps, P.M., Pittet, D. and Sena, S. (2009) *Handbook for Post-Disaster Housing and Community Reconstruction*. Washington, D.C.: Word Bank.

Jigyasu, R. (2000) From "natural" to "cultural" disaster: consequences of post-earthquake rehabilitation process on cultural heritage in Marathwada region,

India. *International Conference on Seismic Performance of Traditional Buildings*. Retrieved from http://www.icomos.org/iiwc/seismic/Jigyasu.pdf/

Johnson, C., Lizarralde, G. and Davidson, C. (2005). Reconstruction in developing countries – A case for meta-procurement. *Proceedings CIB conference on Construction Procurement* (pp. 87–97). Las Vegas: Arizona State University Press.

Johnson, C., Lizarralde, G. and Davidson, C. (2006) A systems view of temporary housing projects in post-disaster reconstruction. *Construction Management and Economics*, 24(4), 367–78.

Karlsen, J., Graee, K. and Massaoud, M. (2008) The role of trust in project-stakeholder relationships: a study of a construction project. *International Journal of Project Organisation and Management*, 1(1), 105–18.

Keivani, R. and Werna, E. (2001a) Modes of housing provision in developing countries. *Progress in Planning, 55*(2), 65–118.

Keivani, R. and Werna, E. (2001b) Refocusing the housing debate in developing countries from a pluralist perspective. *Habitat International*, 25(2), 191–208.

Kellett, P. and Tipple, A. (2000) The home as workplace: a study of income-generating activities within the domestic setting. *Environment and Urbanization*, 12(1), 203.

Kombe, W. and Kreibich, V. (2000) Reconciling informal and formal land management: an agenda for improving tenure security and urban governance in poor countries. *Habitat International*, 24(2), 231–40.

Leung, M.-Y. and Olomolaiye, P. (2010) Risk and construction stakeholder management. In E. Chinyio and P. Olomolaiye (eds), *Construction Stakeholder Management* (pp. 75-98). Chichester: Blackwell Publishing.

Lizarralde, G. (2004) Organisational System and Performance of Post-disaster Reconstruction Projects. Unpublished Ph.D. thesis, Université de Montréal, Montréal.

Lizarralde, G. (2008) The challenge of low-cost housing for disaster prevention in small municipalities. Paper presented at the 4th International i-Rec Conference 2008. Building resilience: achieving effective post-disaster reconstruction. Retrieved from i-Rec website : http://www.grif.umontreal.ca/i-Rec.htm/

Lizarralde, G. and Boucher, M.-F. (2004) Learning from post-disaster reconstruction for pre-disaster planning. Paper presented at the 2nd International i-Rec Conference 2004. Planning for Reconstruction.

Lizarralde, G. and Bouraoui, D. (2010) Users' participation and satisfaction in post-disaster reconstruction. Paper presented at the i-REC international conference 2010, Participatory Design and Appropriate Technology for Post-Disaster Reconstruction.

Lizarralde, G. and Davidson, C. (2006) Learning from the poor. Paper presented at the 3rd International i-Rec Conference 2006. Meeting Stakeholder Interests.

Lizarralde, G. and Massyn, M. (2008) Unexpected negative outcomes of community participation in low-cost housing projects in South Africa. *Habitat International*, 32(1), 1–14.

Lizarralde, G. and Root, D. (2007) Ready-made shacks: Learning from the informal sector to meet housing needs in South Africa. Paper presented at the CIB World Building Congress "Construction for Development"Cape Town, South Africa.

Lizarralde, G. and Root, D. (2008) The informal construction sector and the inefficiency of low cost housing markets. *Construction Management and Economics*, 26(2), 103–13.

Lizarralde, G., Davidson, C., De Blois, M. and Pukteris, A. (eds) (2008) *Building Abroad: Procurement of construction and reconstruction projects in the international context*. Montréal: Groupe de Recherche IF, grif, Faculté de l'aménagement, Université de Montréal.

Lizarralde, G., Davidson, C. and Johnson, C. (2009) *Rebuilding after Disasters: From emergency to sustainability*. London: Taylor and Francis.

Maskrey, A. (1989) Disaster Mitigation: A community based approach. Oxford: Oxfam.

McKeen, J., Guimaraes, T. and Wetherbe, J. (1994) The relationship between user participation and user satisfaction: an investigation of four contingency factors. *MIS quarterly*, 18(4), 427–51.

Monday, J. (2006) After Disaster – building a sustainable community. *Journal of Green Building*, 1(2), 86–97.

Morado Nascimento, D. (2009) Auto production housing process in Brazil: The informational practice approach. In H. Santosa, W. Astuti and D. Widi Astuti (eds), *Sustainable slum upgrading in urban area*. Surakarta: CIB; pp. 51–62.

Newcombe, R. (2003) From client to project stakeholders: a stakeholder mapping approach. *Construction Management and Economics*, 21(8), 841–8.

Oliver-Smith, A. (2007) Successes and failures in post-disaster resettlement. *Disasters*, 15(1), 12–23.

Özden, A.T. (2006) Developing a model for community involvement in post-disaster housing programs. *2006 i-Rec International Conference and Student Competition on post-disaster reconstruction "Meeting stakeholder interests"* Retrieved from http://www.grif.umontreal.ca/pages/OZDEN_Ali%20Tolga.pdf/

Project Management Institute (PMI) (2008) *A Guide to the Project Management Body of Knowledge* (4th edn). Upper Darby, PA: Project Management Institute.

Ross, D. (2009) The use of partnering as a conflict prevention method in large-scale urban projects in Canada. *International Journal of Managing Projects in Business*, 2(3), 401–18.

Strassmann, P. (1984) The timing of urban infrastructure and housing improvements by owner occupants. *World Development*, 12(7), 743–53.

Thomson, M. (2010) Uptake, applications and best practices in stakeholder management. In E. Chinyio and P. Olomolaiye (eds), *Construction Stakeholder Management*. Chichester: Blackwell Publishing; pp. 56–64.

United Nations Human Settlements Programme (UN-Habitat) (2009) *Planning Sustainable Cities – Global Report on Human Settlements 2009*. London: Earthscan.

UNDRO (1982) *Shelter after Disaster: Guidelines for assistance*. New York: United Nations.

Walker, D.H.T., Bourne, L. and Rowlinson, S. (2008) Stakeholders and the supply chain. In Walker, D.H.T. and Rowlinson, S. (Eds.) *Procurement Systems': A cross industry project management perspective*. Abingdon: Taylor and Francis; pp. 70–100.

11 Post-war reconstruction and civil society

Sarah Dix
National Research Institute, Papua New Guinea

Introduction

The past two decades have witnessed large-scale violent conflict in over 50 developing countries and territories, destroying societies and social fabric, as well as infrastructure and institutions. Most visibly, buildings are reduced to rubble, roads and bridges become impassable, and power grids are blown up. Recent examples of countries and territories which have witnessed large-scale violent conflict include Algeria, Angola, Burundi, Côte d'Ivoire, Democratic Republic of Congo, Haiti, Liberia, Nepal, Palestine, Sierra Leone, Sri Lanka, Sudan and Uganda, as well as Cassamance in Senegal and Chechnya in Russia. Although violence continues in Iraq, Afghanistan and Colombia, these are also considered "post-war". This is in contrast with Kashmir, Somalia and Chad, where massive reconstruction efforts have yet to be launched.

The construction industry has much to contribute to the rebuilding of areas that have experienced anti-colonial liberation, civil or international wars or genocide. In creating new infrastructure, or rebuilding or rehabilitating the damaged ones, construction stimulates the local economy, and provides employment to former combatants and members of their communities. This facilitates re-integration after war ends. However, at the same time, experience has shown that, in post-war reconstruction, mismanagement, waste and large-scale corruption is rife. The flood of foreign financial aid to be spent in short timeframes in the post-war context begs for contracting and procurement rules to be bent, and provides opportunities for significant leakage and profiteering (Galtung, 2004). The first section in this chapter considers why this is so, and discusses the ways in which the post-war environment is different from that which prevails after natural disasters or in other developing countries in "normal" circumstances.

The next sections examine challenges of post-war construction including the accountability deficit, and the potential for non-governmental organisations (NGOs) and civil society to expose and denounce waste and corruption in this context. The construction companies selected to deliver physical infrastructure after war ends may benefit from non-competitive and potentially lucrative contracts, but also operate in an unenviable environment that provides incentives to cut corners or make unethical choices. These companies face challenges in terms of procurement and project administration (involving informal systems that operate outside the law), poor local materials and manpower, intense political and time pressures, and physical insecurity (see Mashat et al., 2008).

As the experiences of civil society watchdogs and community-driven reconstruction in Afghanistan, Iraq and Timor Leste show, accountability from below has the potential in post-war contexts to complement and activate mechanisms from above, and generate a positive response from construction firms as well as government and donors. However, at the same time, national and international political actors may be more or less receptive to the public's claims. Change occurs to the extent that the interests of those in power coincide with the development of new state–society relations inherent in the post-war state-building process.

Post-war reconstruction

Nature of post-war reconstruction

Although this chapter focuses on physical reconstruction, "post-war reconstruction" includes not only rebuilding physical infrastructure but also the state's capacity to govern and to deliver basic services, for example security, public health and education. Arguably, without a functioning government and security, physical construction would end in failure (Cross, 2010). For example, it is necessary for administrative mechanisms to consider and approve the planning and design proposals for buildings and infrastructure, and to inspect the works at key stages during construction. These actions are required to ensure that quality, safety and health considerations are given due attention. Underlying the reconstruction of the state is the establishment of democratic institutions and legal framework, including the restructuring of the economy and accompanying societal changes (Pouligny, 2005).

Conceptually, reconstruction fits in between – and overlaps with – immediate humanitarian relief work, and longer-term development. In virtually all post-war cases in recent decades, the reconstruction process is supported, if not driven, by the international community. As such, reconstruction suffers from lack of alignment to national government priorities. Although this could also be said of development aid in general, the lack of alignment is more acute after war ends because there is initially a weak or no indigenous government. Even after elections and the development of a national plan, post-war governments rely heavily on donor funds and therefore their policies, programmes and operations are subject to donor priorities. In some instances, donors impose their preferences over those of the government; in other cases, donors simply do not consult because they do not have to.

Although post-war reconstruction is similar in some respects to reconstruction after natural disasters, or construction in other developing countries, the post-war context poses its own challenges and opportunities. For example, human resources for reconstruction are scarce after war ends, in part because young combatants are inducted into war and do not have other skills to fall back on after war ends; schools are closed during conflict and young people go unschooled; and skilled workers, managers and others who flee from the conflict may not want to return of their own accord, and may prove difficult to recruit and re-integrate into their societies when hostilities cease.

Physically, violent conflict destroys roads, bridges, power grids, schools and health clinics. While this also happens in natural disasters, war often has an impact on a greater part of a country, and services are absent for a longer period of time. Moreover, standing infrastructure may be abandoned and left to decay during wars, to the point of needing extensive rehabilitation or replacement on a massive scale. There is also the danger posed by ordinance. Unlike in disasters, landmines may litter the country, posing danger to reconstruction workers. Socially, the perpetrators of violence may be shunned by communities or even by their own families; their reintegration requires more than a roof and a means to earn income. As will be discussed further below, community-led reconstruction including small-scale local infrastructure projects may facilitate the acceptance of ex-combatants. In this sense, the construction industry has the potential to play a role in rebuilding society after a war.

During protracted violent conflict, the economy is transformed as war may be financed by one or more sides through drug trafficking, arms dealing, smuggling, and exploitation of timber, gas, oil, precious metals and diamonds. The levying of local "taxes" or "levies" by the various groups is also common. Commercial networks are damaged, market institutions are weakened, and manufacturing and industry die out, or at best operate below their installed capacity. Agricultural production shrinks as the rural population is conscripted, landmines threaten cultivation and harvest, and farmers are cut off from transport networks. Local materials for construction may be scarce due to war. In addition, materials, equipment and expertise that are needed for economic infrastructure may be diverted to the building of war structures and protective facilities. The wartime economy continues in the initial post-war period, often discouraging investment, limiting access to capital, and posing a constraint to growth (Rose-Ackerman, 2008).

Foreign aid in the "rentier" state

Foreign private capital is the first to leave the country once hostilities break out. Therefore, these funds are not readily available to help in the post-conflict reconstruction effort. As the productive sectors of the economy are lagging, and there is typically widespread unemployment, there is a thin tax base from which to collect revenues domestically. Furthermore, a post-war government's capacity for taxation is low because revenue-generating systems are not up and functioning, the legislation to back the systems has yet to be created, or enforcement is either weak or non-existent. To the extent that corporations and individuals pay taxes, all or much of these may be informal and go into non-state pockets.

Multilateral and bilateral donors are usually more than happy to fill the void, pumping in vast amounts to gain a stake in the post-war effort. Donor interests in reconstruction include economic, geopolitical and security concerns (see Yanguas, 2010; Hussman and Tisné, 2009). With large infusions of foreign aid, donor interests tend to take over local concerns (Maren, 1997). As a result, the reconstruction is strongly shaped by international interests. Arguably, this occurs to a much greater extent in these circumstances than in a "normal" development setting where it is more likely that there is a government capable of articulating a national vision and policies, and a civil society capable of holding government and even donors to account.

Overall, experience across the world has shown that greater reliance on foreign aid or other resource wealth leads to less watchfulness by citizens, and gives international actors more power (Moore and Unsworth, 2007). The "resource curse" (which refers to the paradox that countries and regions with an abundance of natural resources tend to have less economic growth and worse development outcomes than countries with fewer natural resources) is more pronounced after violent conflict ends. Resource wealth creates "rentier states" in which rents extracted from sources such as foreign aid or natural resources provide a significant share of the government's revenues (Dunning, 2005). A rentier state is one that receives external rent that liberates it from the need to extract income from the domestic economy (Beblawi and Luciani, 1987). In this context, foreign aid influences the public spending and revenue-generation of political regimes. It affects spending because it is allocated or distributed by political authorities, with little or no oversight from government or civil society.

The foreign aid influx negatively affects revenue-generation because it reduces pressures on the state to raise domestic funds through activities such as taxation of businesses and individuals (Chaudhry, 1997). This is problematic because, when people are not taxed, they are less likely to demand representation, and to receive enough basic public services including the construction of roads, health facilities and schools (Bräutigam, 2008). Taxation systems shape accountability relationships and strengthen state capacities. Securing larger tax bases, better tax compliance and comprehensive tax reform helps improve the state's responsiveness and accountability (OECD, 2008). It also builds incentives for citizens to pay taxes, so they see it as being in their interest to pay them (Das Gupta et al., 2004).

The flood of aid after war ends comes with a donor-driven "spending imperative" to use funds within one-year or shorter time frames that conclude with the donor's fiscal year, and show results to foreign constituents and achieve political goals. To get things done quickly and reliably in the construction sector, there is a tendency to use foreign contractors, from the donor's country when possible. This is particularly the case with US Government-funded construction projects. In sectors such as health and education, international, and in some instances, national NGOs are contracted to deliver services after war ends. These contractors are, in many cases, not selected through competitive bidding and it has been suggested that they tend to use marketing techniques to convince donors to fund relatively expensive projects. In the challenging environments in which they must operate, they may adopt practices that are not appropriate. For example, to get the job done on time, they may make payoffs to warlords to obtain right of passage for their materials or equipment, or protection for their workers and the works. This unwittingly serves to extend the wartime economy, and to perpetuate the warlords' power rather than consolidate a democratic state.

After war ends, public financial management systems may be weak or non-existent. Although in some cases donors seek to strengthen these systems, the donors tend to be more concerned about accounting for aid money than expenditures in the country as a whole. On paper at least, donors are able to successfully control their financial risks by creating parallel procurement, payment, accounting and auditing systems using consultants, fund managers and others managed by the donors.

In practice, the effectiveness depends on the extent to which the external actors get into the field to monitor expenditures despite security constraints. This has the potential to reduce misuse of funds and guarantee a reasonable degree of accountability to the donor. However, in the best of circumstances, parallel systems do not answer to domestic actors, and any national capacity they build is outside the government; so they are not sustainable after a project ends. Ultimately, they undermine the development of government systems.

In the area of construction in particular, infrastructure projects are often managed by donors directly through private companies, rather than by letting government manage the funds and contracting. For example, private contractors such as Halliburton and Bechtel Corporation have received billions of dollars in multilateral and bilateral aid money for infrastructure reconstruction in Afghanistan, Iraq and other countries. In some cases these are no-bid or closed-bid contracts, justified on the basis of security clearance or reputation.

In many recent post-war situations, the project management systems applied are different from any model which practitioners in both developed and developing countries would recognise. For example, on some occasions, corporations and international NGOs are hired to execute "quick impact" projects, including local road construction using local NGOs as "implementation partners".

It is reported by the US-based NGO, Center for Public Integrity (2003), that more than 70 US companies and individuals won up to US$8 billion in contracts for work in post-war Iraq and Afghanistan between 1990 and 2002, and those companies were donors to national political campaigns, employed former officials or had close ties to high-ranking officials. The Center's investigation also found that many US contractors had close working relationships with one another, and, in some cases, subcontracted each other.

Although the United States Agency for International Development (USAID), the World Bank and other donors ostensibly give preference to local contractors in the process of awarding contracts, the local firms in most post-war settings do not meet the minimum capacity requirements. There is a need for post-war reconstruction projects to factor in more ways and means of developing the ability of local construction enterprises and practitioners to participate in reconstruction (see Goovaerts et al., 2005). To some extent, they are subcontracted by foreign companies, to carry out discrete parts of the project. In some cases, sub-contracts are sold several times over, making it difficult to identify and hold the final sub-contractor accountable. Counter-productively, the subcontractor's profit margins may be so low that there is an incentive to reduce quality.

Attempting to ensure quality, donors might specify in the contracts the materials, techniques and even labour from abroad which they consider necessary, so that the schools, health clinics or roads will not fall apart or need rehabilitation a year or two after they are completed. However, these items might have to be imported. As a result, local materials, resources and technologies are disregarded in favour of the external approaches.

This has a number of negative consequences. First, projects calling for higher-quality foreign materials are more costly to finance, and few if any local construction firms are qualified to bid for them. This undermines the potential growth of this economic sector. In many developing countries, even under normal conditions, the supply systems for building, and for providing the necessary inputs to, large and complex construction projects are dominated by foreign players. Second, these projects fail to stimulate growth in the local production and supply of construction materials, components and equipment. Finally, when international assistance shifts to another country or crisis, local funds, materials and knowledge are insufficient for infrastructure maintenance to be sustained. Foreign aid is often seen as causing or aggravating corruption, although it may also be used to combat it. However, even if rules and procedures are simple and well understood throughout government – and this is most often *not* the case in post-war settings – they are of no use without enforcement. Without enforcement, informal rules and practices prevail. However, as the next section examines and seeks to establish, where enforcement and formal accountability mechanisms are weak, social accountability mechanisms have the potential to strengthen oversight and enforcement.

Addressing the accountability deficit

Post-war physical reconstruction is notorious for its lack of accountability, which gives rise to waste and corruption. There are a number of reasons for this. First is the nature of the government or ruling authority. If elections do not take place immediately after war ends, the transitional authority may be a foreign force, a United Nations Mission, national actors who are appointed, or some combination of these. The transitional authority may impose decisions unilaterally, without answering to, or being punishable by, either voters or an oversight body. It lacks accountability in so far as it is not constrained by actors charged by law to demand accounting or impose sanctions on public officials.

Moreover, even where there may be freedom of information laws, the transitional authority is likely to be exempt from their provisions and does not have to release documents it does not wish to share. It is also difficult to take legal action against corrupt actors, given the lack of security and functioning judicial systems. There is the threat of thugs and militias who informally make rules and enforce them outside the legal system. In this environment, before and after contracts are awarded, construction companies may be forced to make various payoffs, and potential whistle-blowers cannot count on being protected. Moreover, contracts are not enforceable where there is no effective legal recourse.

From project planning stages through to implementation, civil society may be bypassed if it is weak *vis à vis* the government. This means that communities have little input into social and environmental assessments, for example, if any are carried out. Even if civil society was strong during the conflict, its leaders may have taken up positions in the government, may be hired by foreign governments or international organisations, or may not be positioned substantively or ideologically to work on new agendas with new funding sources.

Even more concerning, government itself may be bypassed, as projects and contracts are negotiated between foreign donors and contractors. In the case of NGOs contracted as service providers, the government may then perceive them as competitors; this damages the future relationship between the state and civil society.

As a requirement of the donor, monitoring of post-war construction projects is often carried out by foreign experts who tend not to venture far into the field, for reasons of security as well as convenience. Local NGOs also face security threats and do not have the option to exit the country when things become difficult. In this context, it is not surprising that oversight of construction projects is weak. Yet, as discussed below, reconstruction also offers an opportunity to engage and work through community leaders and social groups to strengthen accountability and state–society relations.

Social accountability is an approach to enhancing government responsiveness, answerability and transparency through citizen actions, as well as actions on the part of government, media and other actors that promote or facilitate these efforts (McNeil and Malena, 2010). Social accountability strategies and tools such as participatory budgeting, public expenditure tracking, social audits and community report cards, help empower people to exercise their rights and to hold government accountable for its use of public funds and how it exercises authority.

Social accountability works by activating conventional mechanisms of accountability such as political checks and balances, accounting and auditing systems, administrative rules and legal procedures. Even where the state offers few formal accountability systems, citizens can use informal mechanisms to demand "rude" accountability, for example through threats of shaming or use of violence (Hossain, 2009). However, the results may not stick, so, in the long run, formal non-violent mechanisms need to be developed.

Strategies include legal action (taking cases to court), mobilisation (organising mass protests, marches, strikes and boycotts), and going to the media to get a message or complaint out, or have corruption investigated and exposed. Exposure can serve to control a specific issue, such as building a school that was funded but not constructed, or draw public attention to a broader number of issues, such as poor performance of public school teachers, lack of toilets, malnutrition or child labour, as well as corruption in a school construction project. Signalling a problem can also extend the scope of conflicts and issues. For example, denouncing specific cases of school roofs that collapse due to substandard materials can effectively raise the broader issue of security. Although citizens cannot formally impose punishments, they can impose symbolic sanctions (such as reputational costs). Social accountability works as a fire alarm to trigger or activate horizontal mechanisms – although these may not exist immediately after war ends.

After war ends, the lack of a functioning legal system means the media can make statements relatively freely without being sued for libel or slander. At the same time, however, there are extra-legal threats and repercussions that are paralysing. The next sections examine citizen efforts to promote accountability at the community level.

Community-driven accountability initiatives

Accountability is one of the most effective ways to combat corruption and promote good governance. It is based on information, justification and sanction (Ackerman, 2005). In post-war contexts, many accountability methods focusing on government institutions have limited success. These include "vertical" accountability through elections, and "horizontal" accountability through anti-corruption agencies, ombudsmen and law enforcement agencies. However, as discussed below, there has been some success in strengthening the "voice", or capacity, of ordinary citizens to directly participate in policy-making processes, and hold government to account.

One example of civil society's contribution to reconstruction has been that of community-driven accountability initiatives. The terms participatory, demand-driven, community-driven and community-led are used interchangeably in the international community, to refer to development approaches that give control over decisions and resources to local stakeholders (Goovaerts et al., 2005). In war-torn areas, community-driven accountability has the potential to restore trust in government and among neighbours. Combined with targeted local-government reforms, such a collaborative process can also moderate the expectations of citizens with the state's limited ability to deliver services (Galtung and Tisné, 2009).

While making a positive contribution at the local level, as the cases below show, "community-driven" accountability may be a bit of a misnomer. While being community-based, and carried out by local monitors, the initiatives are generally prompted by an external mobiliser or local NGO, the elite in the capital city are behind them (Mansuri and Rao, 2004). Moreover, communities are unlikely to participate in or influence the national-level decision-making process that determines how much funding goes where. To the extent that they participate, it is to monitor the execution of projects at the local level.

Also, as seen in the three countries studied here, there is the question of who is being held accountable, and by whom. In the post-war setting, private contractors are doing much of the physical reconstruction work, and may not answer to local or national government. The community-driven accountability approach starts with the local leaders and community members, and works its way up through different levels of government, and in some cases puts the case to the donor, to pressure the government. In some cases, a contractor only responds to the foreign donor. While this occurs to some degree in other developing countries (in normal circumstances), the greater donor presence and weight after war ends makes it more characteristic of post-war situations.

Case studies

Timor Leste

Timor Leste is Asia's youngest independent nation, following decades of occupation by the Indonesia. It is suggested that, when Indonesian military and militias withdrew in 1999, they destroyed up to 70 percent of the infrastructure. Since then, limited progress has been made on the country's twin objectives of attaining economic growth and poverty reduction. Timor Leste has a low life expectancy, a high rate of illiteracy and widespread food insecurity. The human development index (HDI) of Timor-Leste is 0.502, which gives the country a rank of 120 out of 169 countries with comparable data (UNDP, 2010a). About half of the population in Timor Leste lives below the basic-needs poverty line of $0.88 per person per day, and approximately half of all children were underweight in 2007 (UNDP, 2009).

A violent crisis in 2006 highlighted the state's low level of legitimacy, and the extent to which public institutions are captured by political actors at all levels. Elections in 2007 resulted in a new government, not unlike the previous one, in terms of the lack of capacity in transparent and accountable budget execution and monitoring. For example, the supplementary budget in 2008 was only submitted to parliament a few days before being passed, without public consultation. The situation had slightly improved by 2010, when the supplementary budget was debated by the legislature for ten days. Once funds are appropriated and executed, there is no control mechanism in place for the government to oversee the projects. Moreover, according to local NGOs, the Timorese public in general – and rural communities in particular – lack sufficient information, knowledge and skills to assure the transparent and accountable implementation of reconstruction budgets and projects.

The country has an abundant supply of oil and gas revenues that, along with foreign aid, allows the government to be less responsive to citizens. In 2008, oil revenues were high, so the government decided to increase the budget substantially beyond the 3 percent limit set by the Petroleum Fund Act (2005). The Act mandates that oil and gas revenue be deposited in a Petroleum Fund, managed through the Ministry of Finance (a statutory independent body, the Petroleum Fund Consultative Council – KKPF, is responsible for the Fund's oversight and monitoring). As seen in other resource-rich countries, high levels of spending and corresponding public works projects are not sustainable after resource levels fall. However, the Timorese Parliament appears to be able to violate its own laws with relative freedom from public scrutiny.

Luta Hamutuk is a Timorese national NGO that is working with communities in 13 districts to understand the national budget process and participate in district-level budget and reconstruction monitoring and policy-making processes (Timor Institute of Development Studies, 2007). As of November 2010, Luta Hamutuk had monitored development projects valued at US$7.4 million. As a result of the monitoring, the budgets and reconstruction process have begun to serve the developmental needs of the participating local communities. The work of Luta Hamutuk has enabled communities and citizens to recognise and exercise their civic rights within the rule of law. This is vital in Timor Leste, where peaceful dissent is not well integrated into civic life, and interactions between the government and citizens can be acrimonious. When the sub-district or district administrator will not listen to the focal points, Luta Hamutuk calls and facilitates a meeting between them.

With the national NGO's training and support, community focal points facilitate community meetings to identify which projects the community prioritises for monitoring, organise monitoring committees and discussion forums, and mobilise action. Focal points may be teachers, village chiefs, youth coordinators, women's group representatives, farmers or others. Their selection is endorsed by the elected village chief, whose cooperation in future monitoring and advocacy activities is essential. The focal points work with existing accountability systems, strengthening or creating linkages between local leaders and the community, and between the community and national policy-makers. They encourage their communities to claim their rights, put forward demands, and ensure that the state is responsive to their development needs.

Each chief is elected by the village to represent the people of the village. Each of them has an interest in the contribution that reconstruction projects make to the village, as the chief is then perceived by the village to be a good chief. Additionally, if projects are not successful the village people will complain to the village chief, so it becomes his or her "headache". In some cases, the village chiefs are also the contractors of reconstruction projects. However, when the village chief is the contractor, the chief is under a lot of scrutiny, and can feel the social pressure to a greater extent than external contractors do.

Generally, the focus of community monitoring committees in Timor Leste is on clinic and school buildings, electricity supply, irrigation systems, canalisation, water and sanitation projects, and roads. For example, the inhabitants of Homboe village monitored a water project valued at US$70,000, and inhabitants of Estadu village monitored a primary school rehabilitation project valued at US$26,503. Communities in Liquiçá district have monitored rural roads valued at close to US$5 million and serving a population of about 60,000. The monitors have investigated whether roads were in fact completed, or whether or not substantial development took place.

Over time, focal points have become local organisers, and their work extends to other issues beyond the monitoring of roads, health clinics and schools. For example, they have organised communities to monitor the construction of a police station, and the construction of a power station. The focal points also pass on the training on community mobilisation, development and monitoring that they have received from Luta Hamutuk to other community members who become active in furthering the needs of the village.

At the district level, Luta Hamutuk has established District Infrastructure Monitoring Committees in Liquica, Bazartete and Aileu/Dili districts. Monitoring committees are comprised of local government authorities and other community members. For example, the committee in Bazartete comprises Luta Hamutuk representatives, focal points, youth representatives, four village chiefs, the health department chief at the sub-district level, the police chief, and the sub-district administrator of Bazartete, who has assumed the role of Counsellor for the committee. In monthly meetings, monitoring findings and project updates are presented by focal points. The meetings are not open to the public, and the committee determines what is made public and when.

The committees are also used as complaints mechanisms, with community members bringing forward any concerns for the village chief to take up with other authorities. For example, in Lospalos a contractor was not paying local salaries because the government had not paid the contracted company. The committee members were able to discuss the issue, and it was finally resolved. In another case, the Bazartete committee heard a complaint from the Metago village chief, who raised a community concern that water had been piped to the health clinic, but not to others. The health department complained that villagers were cutting holes in the water pipes leading to the health clinic. The community then worked together to bring its water needs up at the district level, to be considered for inclusion in future development planning.

In Liquiçá district, focal points have been monitoring road and canalisation projects, which have been implemented as part of the Ministry of Infrastructure's Referendum Package, valued at approximately US$75 million. The Referendum Package is managed through an association of companies, whose head decides which companies are to be awarded contracts under this scheme at both the national level and throughout all 13 districts. Community focal points found that both the canalisation and the road construction projects had been poorly implemented, as the materials of the canalisation tubes were of bad quality and the standard of the roads was poor. Furthermore, the communities were never consulted; the project budget was never made available to the public and the labour force was generally poorly paid. As the Referendum Package does not include a grievance mechanism, people were never able to complain about any of the projects.

The focal points documented the faults in the implementation of the road project, and shared the report with Luta Hamutuk, which, after verifying the data, then submitted the findings as an official monitoring report to the state secretary of public works and the national director of road construction within the Ministry of Infrastructure. In order to ensure a response from the state officials, Luta Hamutuk simultaneously submitted the report to the press. At the same time, some of the focal points informally contacted a war veteran-turned-businessman, who is responsible for the same road construction and canalisation projects within Liquiçá district. They were able to approach him since they are related to him, and could therefore raise the community's concern about the low standards of the road construction and canalisation projects.

The businessman did respond to the community concerns raised by the focal points, because they are related. Using a recovery fund, the canalisation tubes were consequently replaced, and this improved the overall standard of the project. However, nothing has changed in regard to the road project, because the contractor was not from the area concerned.

Also, 24-hour surveillance by local focal points puts constant pressure on contractors to improve their performance on their projects. In general, if the owner of the contracting firm is from the village where the projects are located, he or she is likely to be more responsive and deliver better quality projects. Persuasion and advocacy through personal networks and relations remain effective ways to achieve results in Timor Leste.

Apart from Liquiçá district, Luta Hamutuk has monitored infrastructure projects under the Ministry of Public Works. They included: the Fuiloro–Tutuala road (US$1,598,000), the Lospalos–Illiomar road (US$91,574) and the Tutuala and Valu road (US$205,750) in Lautém district. It has also monitored the construction of a primary school in Lautém district. The job was only 20 percent completed by the contractor, so the children could not go to school. In addition, Luta Hamutuk monitored the construction of a hospital building (US$18,381) in Lospalos district; part of the building was damaged, and was never replaced with new material.

Luta Hamutuk monitored the building of housing for war veterans in Dili, Viqueque, Manatuto and Liquiçá districts. There were various problems; often the houses were built too small and in other cases there were just not enough houses built. As an example, in the sub-district Lospalos, in Lautem district, 19 houses (US$180,463) were built for war veterans. All the 19 houses were completed, but since many persons other than those for whom the houses were earmarked also claimed to be veterans, the issue of who was entitled to the houses remained a problem, given that it is difficult to define who exactly is a war veteran, and thus has an entitlement to a house. Meeting with Luta Hamutuk, the Secretary of State for Veterans and Former Combatants (SEAVAC) revealed that it is a real headache to deal with the issue of veteran housing, as the problem of classification of what constitutes a veteran remains.

As a result, Luta Hamutuk produced press releases on veteran housing issues and called upon the SEAVAC to give attention to the problems. They also offered recommendations to the Office of the Prime Minister, Parliament, SEAVAC, the Ministry of Infrastructure, and the High Court, which were shared with the media. As a result of their monitoring and advocacy, Luta Hamutuk was able to secure an increase in the budget allocation to veteran housing. However, this budget has not been executed yet.

Based on its experience, Luta Hamutuk reports that there is high demand for participatory development and accountability. Communities are very interested in hearing about whether there is any waste in construction, as they often have complaints about projects. So they want to understand why the waste occurs. Is it waste, is it corruption, is it mismanagement by the government officials, or is it the government's lack of capacity? As a result, many people are volunteering to become focal points. Local officials are also keen to participate, interacting in all aspects of the monitoring process. Through budget seminars at the national and regional levels, in which officials and community members participate, Luta Hamutuk's work also strengthens links between national and district-level government officials.

Many parliamentarians seek knowledge from Luta Hamutuk, and want to know what is really going on in the districts. The ministers spend most of their time in Dili, and do not travel much to the districts. The findings interest them as they reveal if there are problems with any of the project implementers, and if people are in fact satisfied with the projects. Luta Hamutuk finds that ministers often accept the validity of the findings, and the group has built good relations with some. However, this does not always guarantee action. So Luta Hamutuk gives the findings to the media at the same time, forcing a response. In the case of the Bazartete school, the Minister of Infrastructure was receptive to the presentation by focal points and Luta Hamutuk, and delivered what was needed.

Luta Hamutuk has filed some complaints about corruption with the Anti-Corruption Commission and the ombudsman, but there is seldom any follow-up. The Anti-Corruption Commission is not very active. The response to monitors' findings from others depends on the context. If there is collusion between the government and a contractor, there is no response. When the owner of the contracting firm is from the area concerned or feels strong pressure from the community, the contractor does respond. One strategy that makes it more likely that citizens will be heard is when a memorandum of understanding (MOU) is established between the community and the body in question. The community-based committee becomes an institution, and in this way participating bodies are held accountable.

Bringing the concerns of communities to those making decisions about the country's income and expenditure, and building citizens' skills and understanding of national decisions, are essential activities and arguably a pre-requisite for equitable and democratic development in Timor Leste. The monitoring of the reconstruction projects itself promotes a culture in which nepotism and corruption become more difficult and less profitable. Contractors are concerned that anything might be exposed on their projects. The monitors' findings are revealed in press conferences, and they are published in the national newspapers. With officials being aware that their work is under closer supervision from the local community with the support of a national NGO, the culture of corruption becomes a less easy option for those who would be attracted to it.

Afghanistan

Afghanistan is plagued by poor-quality buildings and roads that have been constructed under projects funded by foreign donors. The projects are often sub-contracted several times to a final implementer whose profits are increased if it uses cheap labour and sub-standard materials. Even with the best intentions, armed conflict continues across a large extent of Afghanistan's territory, making construction and quality control a challenging and risky endeavour for the construction companies and the government alike.

Following civil war, and then war between the foreign-backed Taliban and former Mujahidin groups, later known as the Northern Alliance, the United States invaded the country in October 2001. Two months later, the Bonn Accord was signed on behalf of Afghanistan by the Northern Alliance, among others. Although the Taliban had previously ruled the country and was the major party to the conflict, it was not represented in Bonn. A transitional Afghan Government was established in June 2002, and, as agreed in Bonn, the transitional administration used pre-existing, Soviet- and Mujahidin-era procedures, structures and laws that were often confusing. After the new Constitution was adopted and a new government established in 2003, President Hamid Karzai brought former provincial power holders including former warlords into his Cabinet.

Today, Afghanistan is often described as a "shadow state" controlled by many factions and drug interests. In this context, corruption has flourished. Although the country receives massive amounts of foreign aid each year, it is estimated that only as little as a quarter of aid money stays in the country, and of the billions of dollars spent on reconstruction, about a third is spent on bribes and protection money (Owen, 2010). The massive influx of aid, with extreme pressure to deliver results in short timeframes, compounded by non-functioning institutions, poorly paid public servants and limited government capacity have together seriously compromised the detection, prosecution and punishment of corrupt practices. The judiciary is reportedly perceived to be the most dysfunctional and corrupt institution in Afghanistan (Center for Policy and Human Development, 2007).

By the end of 2005, donors started to voice concerns about corruption. In 2006, President Karzai directed the Attorney General to "crack down, arrest, and prosecute the perpetrators of corruption in government offices … at all levels, even if its tentacles reach high levels of the government" (*Afghanistan News*, 2006). The Attorney General reportedly responded enthusiastically to his mandate, and was subsequently removed from office.

In 2008, an anti-corruption law was decreed, and the Afghan National Development Strategy (ANDS) with a complex Anti-Corruption Strategy was presented in Paris. Despite a number of new laws recently enacted that are favourable to governance, control over resources continues to be monopolised by local commanders, strongmen and networks of corrupt officials. Donor agencies also shield their programmes from Afghan government scrutiny, and to the extent that they are accountable, it is to their constituents.

Although corruption is well entrenched in Afghanistan, the public does not condone it. While the giving and taking of *baksheesh* (small gifts) is generally considered to be acceptable, corruption is morally rejected as being against the basic principles of Islam. Moreover, although security threats make it difficult for individuals to report specific cases, corruption is an issue that is discussed widely on radio and television by political figures, intellectuals and opinion leaders.

Integrity Watch Afghanistan (IWA) is a national NGO that, since 2007, has been supporting the development of community-based monitoring projects in Parwan, Balkh, Nangarhar and Herat provinces (Torabi, 2007). As a member of the Network on Integrity in Reconstruction supported by Tiri, IWA visited Lao Hamutuk in Timor Leste. IWA reports that, initially, Afghan people were sceptical of their ability to bring about social change, but now they are hopeful, and more communities are asking for the opportunity to participate. Monitors have detected serious deficiencies in about 30 percent of the projects they track. Half of these faults have been made good as a result of the evidence uncovered by the monitors. One example is the construction of a girls' high school North of Kabul in Jabal Seraj, Parwan province. The community succeeded in getting the building contractor to replace approximately 10,000 sub-standard bricks and double the thickness of the metal sheeting on the roof.

Perhaps more so than in Timor Leste, the Afghan NGO trains monitors not only to work with the community and government, but also on how to approach the donor when needed. As a result, the North Atlantic Treaty Organization Training Mission Afghanistan (NTMA) is working with the IWA to expand its monitoring of projects to NATO-financed police station construction projects. This is not only to ensure that the quality of construction is good, but also to build relations between the police and local communities.

IWA also carried out a study on corruption in the construction of two high-profile roads, the Bibi Mahro road and the Kabul–Gardez road, which were funded by USAID and contracted to the Louis Berger Group (IWA, 2007). The cost of upgrading the 2.83km-long Bibi Mahro road from Kabul city to Kabul airport topped US$2.4 million per kilometre. With a total project cost of over US$7 million, this road was approximately ten times more expensive than other road construction projects in Afghanistan. According to the report, the road cover, which was of a sub-standard grade, did not survive the first winter, when under normal circumstances it should have held up for at least a decade.

Sierra Leone

After ten years of violent conflict and four successive attempts at peace agreements, civil war in Sierra Leone was declared to have ended in 2001. In May 2002, President Ahmad Tejan Kabbah was re-elected to a five-year term in a landslide victory. The main former rebel group, the Revolutionary United Front (RUF) political wing, failed to win a single seat in Parliament. The cost of reconstruction in Sierra Leone, with a population 5.7 million, was estimated in 2004 at US$16.4 billion a year for the United Nations and £100 million per year for the British government (Baker and May, 2004). The US disbursed $45 million for the reintegration of former fighters, and better management of the diamond sector (Human Rights Watch, 2004). Following the return of over 300,000 internally displaced persons and the disarmament of over 70,000 ex-combatants, efforts focused on physical reconstruction (UNDDR, 2006).

Half of Sierra Leone's national budget is financed by aid (Eurodad, 2008). The UK's Department for International Development (DfID), the European Commission, the World Bank and the African Development Bank give little more than a quarter of their aid through budget support (Eurodad, 2008). Some international agencies have been commended for their coordination efforts. However, in the road sector for example, the main challenge is that most donors have no sector staff or no representation at all in Sierra Leone (EuropeAid, 2010). In reconstruction overall, untimely disbursement of aid, donor-driven programming, and the establishment of parallel governance structures have undermined the post-war state (Eurodad, 2008).

Two years after the war ended, approximately 70 percent of the population lived below the national poverty line and 26 percent of the population lived in extreme poverty (UNSTATS, 2004). Eight years after the war, Sierra Leone was experiencing economic growth, and was promoted as a tourist destination in West Africa. The UN Security Council lifted the last sanctions against Sierra Leone in September 2010, reporting that the government had fully restored control over its territory, and former combatants had been disarmed and demobilised (United Nations Security Council, 2010).

Post-war, Sierra Leone has made efforts to attract foreign investment and improve conditions for business. The country is one of Africa's most active reformers of legislation and business regulation, with improvements in the areas of exports, business start-up, construction permits, registration of property, and trading across borders (World Bank, 2010). Despite these advances, reconstruction at the local level has been hampered by a lack of transparency and accountability. There are reports that some contractors have abandoned their projects before completion, and that development funds have been siphoned off into private pockets.

For example, citizen reports of the Roads Authority, while recognising resource constraints, indicate poor quality and lack of responsiveness of service providers (Campaign for Good Governance, 2006). Others such as the National Accountability Group have highlighted problems of corruption. The Anti-Corruption Commission conducted comprehensive reviews of a number of ministries and agencies, highlighting conflicting roles in the collection of revenue, and weaknesses in budget execution, internal and external audit, stores management, procurement management and project implementation. In general, the citizens of Sierra Leone lack information and a mechanism to channel their concerns in the face of limited accountability.

Civil society organisations in the Network Movement for Justice and Development in Sierra Leone are engaging community monitors to address the gaps in transparency and accountability in reconstruction programmes and services (see Bu-Buakei Jabbi, 2007). They mobilise community members, map their priorities and concerns, and promote community learning, engagement, monitoring and advocacy as means to achieve desired development goals. Based on their experience, it has been crucial in this process to engage councillors, chiefs, organisational heads and administrators who have a direct influence on project outcomes, as well as the media to reach the public through, for example, interactive radio programmes.

Community monitors have noted a variety of concerns while conducting their assessments of reconstruction programmes and services, including abandonment of projects, delays in project completion, misappropriation of materials, and poor quality of implementation. In conducting their monitoring and advocacy, they work with responsible bodies to improve service delivery and administrative procedures. Monitors share their findings with implementing agencies, local communities and governments, as well as with actors such as the Anti-Corruption Commission and international NGOs.

In Sierra Leone, community monitoring and calls for greater integrity have led to contractors returning to abandoned sites, and projects being completed to the satisfaction of communities. For example, in Fiama chiefdom, sanitation facilities for the primary school at Bandasuma were found to be inappropriate. Through engagement of the district council, satisfactory toilets were constructed for the pupils.

As a result of monitoring and the identification of gaps and remedies, dormant management structures including School Management Committees have been revitalised and are taking an interest in the management of school resources. Greater participation of students and parents has also been observed with the monitoring of the Free Primary Education policy enshrined in the Primary Education Act 2004. In addition, transparent practices of headmasters have improved relations between headmasters, teachers and parents.

A constructive relationship has developed between the monitors, members of the district councils and the district administration and budget oversight committees, and this has led to the formation of a Civil Society Organisation Council Forum. This forum evolved as a platform for civil society organisations to discuss community concerns with the district administration. Such forums help build trust and confidence and ensure gains in public service delivery.

Civil society organisations that support the independent monitoring teams are linking up at the national level to attempt to influence policy and practice. Through national coalitions, they are building a movement to press for greater access to information and accountability of donors, government and NGOs. Connecting the voices and concerns of local communities with committed civil society organisations and other advocates at the national and international levels can play an important role in building the post-war state in Sierra Leone, ensuring that reconstruction funds and programmes reach the intended beneficiaries.

Conclusion

The key actors in reconstruction today are what Bello (2006) calls the "new establishment": the US military–political command, the World Bank, corporate contractors, international NGOs and local civil society. The national government or transitional authority and the UN participate in post-war reconstruction, but are arguably less influential in relation to the others. This configuration of power means that decisions are made, to a large extent, by foreign actors whose interests differ from, and may be in conflict with, the national interests of the post-war countries. Although civil society is a weaker partner in the "establishment", its inclusion allows for the possibility of a watchdog role.

The cases of Timor Leste, Afghanistan and Sierra Leone highlight the benefits of instituting social as well as other accountability mechanisms, and building local competencies to monitor construction projects. As seen in these countries, despite all the challenges, at the local level citizens are capable and keen to track where the money that is supposed to improve their wellbeing is going. Moreover, in many instances, the faults detected by monitors are corrected as a result of the evidence made public.

They also suggest that community-driven accountability has its limits. While it is able to get results in many cases at the local level, there was less evidence that aspirations to influence national policy were realised. Even the national NGOs themselves faced challenges in doing so, and in shaping, not just reacting to, donor preferences and *modus operandi*. All three countries lacked watchdog groups in numbers that would allow for protection, as well as multiply the effects of the messages and activate horizontal accountability mechanisms. A "thicker" as well as stronger civil society, with many layers, would be better able to exercise its voice in an insecure and risky post-war environment.

Another point that emerged from the cases is that, rather than continuing or defending practices that allow corruption to flourish, donors need to be more transparent in their own programmes. This is agreed by donors in principle, but the pressure to do so by local actors is often outweighed by the scandals that may ensue if their home constituencies delve into the balance sheets, or by the political backlash from supporters whose contracts are no longer secure when the details of bidding, procurement, contract amounts and quality are made public.

While the international community recognises the importance of civil society in reconstruction, it is unclear on how to foster its (re)emergence. It sees a role for citizens in state-building, but finds that the concept of citizenship is still under development in emerging nations (OECD and DAC, 2010). There is also a concern about "uncivil" society, groups that threaten stability and undermine democracy (Belloni, 2008). Development of social capital and civil society is not an end in itself but part of a process that has the potential to strengthen governance, increase state legitimacy and stability and inhibit a return to violence (UNDP, 2010b).

As Chabal (1998) points out, the legitimacy of rulers has been sustained in ways more complex than "multiparty rituals" such as elections. It is more important for people to believe that rulers are accountable in ways they believe to be legitimate, than to focus on the trappings of democracy. Supporting public participation in governance is likely to be more effective than social engineering (Pugh and Potter, 2003).

In working with civil society, the international community needs to consider to what extent its interventions strengthen or inadvertently undermine state-building. Direct funding to NGOs or to INGOs which support local groups is attractive because it results in immediate service delivery, yet these organisations may also lack integrity and are no panacea. In any case, the greater problem is in supporting parallel structures that compete with the state. Although services are delivered, the state is not providing them or managing the contracting out.

While public participation in governance and creating new space for citizens to engage with the state is necessary, it is not sufficient to (re) construct a nation. The development discourse may still be exclusionary, framed in donor terms, and inequalities may remain or be reinforced. And civil society growth without engagement with the post-war state can promote elite capture, clientelism or patronage. Where the state is non-existent or weak, non-state actors such as national NGOs may compete for power and gain people's support outside the formal system. This undermines the state-building process.

Along with attention to civil society, it is important for governments and donors to consider ways of building the capacity of the domestic private sector to lead the reconstruction effort. There is already some effort in this regard. For example, groups such as Peace Dividend Trust track local suppliers and encourage local procurement. Furthermore, incentives are needed to promote private-sector investment in infrastructure (the challenges of attracting private investment in infrastructure are discussed by Hoeffler, 1999).

Beyond participatory development and monitoring – engaging local populations formally and extensively in decision-making for needs assessment, project design, evaluation and monitoring – the three cases studied here suggest a role for monitoring agents (and civil society, more generally) in activating the social change that is necessary after war ends. Watchdogs in Timor Leste, Afghanistan and Sierra Leone have raised the costs of corruption for contractors and government and, to some extent, have activated other accountability mechanisms. What is more important is that they show that citizen monitoring initiatives can be successful in creating a space for the social contract with the state to be mediated and re-negotiated to fit the post-war landscape.

Acknowledgment

Claire Schouten and Katrine Jorgensen provided case material and introductions to partners from the International Network on Integrity in Reconstruction of the INGO, Making Integrity Work (Tiri), as well as helpful comments on this chapter. Many thanks to Mericio Akara and Emanuel Bria of Luta Hamutuk in Timor, and Yama Torabi of Integrity Watch Afghanistan, for their responses to questions, and to Edward Ssenyange for field reports from Sierra Leone.

References

Ackerman, J. (2005) *Social Accountability for the Public Sector: A Conceptual Discussion*, The World Bank, Social Development Paper No. 82. Washington, DC.

Afghanistan News (2006) 30 August 2006. http://www.afghanistannewscenter.com/news/2006/august/aug302006.html/

Baker, B. and May, R. (2004) Reconstructing Sierra Leone. *Commonwealth and Comparative Politics*, 42(1) (March), 35–60.

Beblawi, H. and Luciani, G. (eds) (1987) *The Rentier State, Volume 2, Nation, State and Integration in the Arab World*. London: Croom Helm.

Bello, W. (2006) Rise of the relief and reconstruction complex. *Journal of International Affairs*, Spring/Summer, Vol. 59, No. 2, pp. 281–96.

Belloni, R. (2008) Civil society in war-to-democracy transitions. In A. Jarstad and T.D. Sisk (eds) *From War to Democracy: Dilemmas of Peacebuilding*. Cambridge: Cambridge University Press.

Bräutigam, D., Odd-Helge, F. and Moore, M. (eds) (2008) *Taxation and State-Building in Developing Countries: Capacity and Consent*. Cambridge: Cambridge University Press.

Bräutigam, D. (2008) Taxation and state-building in developing countries. In Bräutigam, D., Fjeldstad, O-H. and Moore, M. (eds) *Taxation and State-Building in Developing Countries: Capacity and Consent*. Cambridge: Cambridge University Press.

Bu-Buakei Jabbi, S.-M. (2007) *The SABABU Education Project: A negative study of post-war reconstruction, Sierra Leone 2007*. Freetown: NAG; London: Tiri.

Campaign for Good Governance (2006) *Citizens Report Card Survey: An Assessment of the Sierra Leone Roads Authority (SLRA) and the National Power Authority (NPA)/BO Kenema Power Service (BKPS) in Sierra Leone*, 2006. http://www.slcgg.org/resources.html/

Center for Policy and Human Development (2007) *Afghanistan Human Development*

Report: Bridging modernity and tradition – rule of law and the search for justice. Islamabad: Army Press.

Center for Public Integrity (2003) *US contractors reap windfalls of post-war reconstruction*, 30 October. http://www.commondreams.org/headlines03/1030-10.htm/

Chabal, P. (1998) A few considerations on democracy in Africa. *International Affairs*, 74(2), April, 302–3.

Chaudhry, K.A. (1997) *The Price of Wealth: Economies and institutions in the Middle East.* Ithaca: Cornell University Press.

Cross, J.W. (2010) *Criteria for Post-War Infrastructure Reconstruction Efforts.* http://www.stormingmedia.us/01/0197/A019715.pdf/

Das Gupta, A., Ghosh, S. and Mookherjee, D. (2004) Tax administration reform and taxpayer compliance in India. *International Tax and Public Finance* 11, 575–600.

Dunning, T. (2005) *Crude Democracy: Natural Resource Wealth and Political Regimes.* Cambridge: Cambridge University Press; pp. 6 and 52.

Eurodad with the Campaign for Good Governance (2008) *Old Habits Die Hard: Aid and Accountability in Sierra Leone.* http://www.liberationafrique.org/IMG/pdf/Old_20habits_20die_20hard._20Aid_20and_20accountability_20in_20Sierra_20Leone.pdf/

EuropeAid (2010) Report. http://ec.europa.eu/europeaid/documents/aap/2010/af_aap-spe_2010_sle.pdf/

Galtung, F. (2004) *Corruption in Post-Conflict Reconstruction: Confronting the Vicious Cycle.* http://www.tiri.org/, p. 4.

Galtung, F. and Tisné, M. (2009) A new approach to postwar reconstruction, *Journal of Democracy*, 20(4), 93–107.

Goovaerts, P., Gasser, M. and Inbal, A.B. (2005) Conflict Prevention and Reconstruction. Community-Driven Development. World Bank Social Development Paper No. 29. Washington, DC.

Hoeffler, A. (1999) Challenges of Infrastructure Rehabilitation and Reconstruction in War-Affected Economies. African Development Bank Economic Research Papers No. 48. http://www.irisproject.umd.edu/ppc_ideas/FS_Strategy/Resources/articles_pdf/infrastructure_hoeffler.pdf

Hossain, N. (2009) Rude Accountability in the Unreformed State: Informal Pressures on Frontline Bureaucrats in Bangladesh, IDS Working Paper 319. Brighton: Institute of Development Studies.

Human Rights Watch (2004) Sierra Leone. http://hrw.org/english/docs/2004/01/21/sierra6989.htm/

Hussman, K. and Tisné, M. (2009) *Integrity in Statebuilding: Anti-Corruption with a Statebuilding Lens.* OECD Development Action Committee. http://www.oecd.org/dataoecd/26/31/45019804.pdf/

IWA (Integrity Watch Afghanistan) (2007) *Afghan Roads Reconstruction: Deconstruction of lucrative assistance*, Afghanistan. Kabul: IWA; London: Tiri.

McNeil, M. and Malena, C. (2010) *Demanding Good Governance: Lessons from Social Accountability Initiatives in Africa.* Washington, DC.: World Bank.

Mansuri, G. and Rao, V. (2004) Community-based and -driven development: A critical review. *World Bank Res Obs*, 19(1), 1–39.

Maren, M. (1997) *The Road to Hell: The Ravaging Effects of Foreign Aid and International Charity.* New York: Free Press.

Mashat, M., Long, D. and Crum, J. (2008) A Conflict-Sensitive Approach to

Infrastructure Development. United States Institute of Peace Special Report 197. http://www.usip.org/

Moore, M. and Unsworth, S. (2007) *How Does Taxation Affect the Quality of Governance?* Brighton: Institute of Development Studies.

OECD and DAC (Organisation for Economic Cooperation and Development Action Committee) (2008) *Governance, Taxation and Accountability: Issues and Practices*, Paris: OECD DAC Guidelines and Reference Series.

OECD and DAC (Organisation for Economic Cooperation and Development Action Committee) (2010) *The State's Legitimacy in Fragile Situations*. Paris. http://www.oecd.org/dataoecd/45/6/44794487.pdf/

Owen, J. (2010) Army launches investigation: Corrupt Afghans stealing millions of aid funds. *The Independent*, 7 March, http://www.independent.co.uk/news/world/asia/army-launches-investigation-corrupt-aghans-stealing-millions-from-aid-funds-1917436.html

Pouligny, B. (2005) Programmes aimed at building "new" societies. *Security Dialogue*, 36, pp 495–510.

Pugh, J. and Potter, R. (2003) *Participatory Planning in the Caribbean*. Aldershot: Ashgate Publishing.

Rose-Ackerman, S. (2008) Corruption and Post-Conflict Peace-Building. *Faculty Scholarship Series*, Paper 593. http://digitalcommons.law.yale.edu/fss_papers/593/

Timor Institute of Development Studies (2007) *Case Studies in Post Conflict Reconstruction, Timor Leste*. Dili: TIDS; London: Tiri.

Torabi, Y. (2007) *Assessing the National Solidarity Program: The Role of Accountability in Reconstruction, Afghanistan*. Kabul: IWA; London: Tiri.

UNDP (United Nations Development Programme) (2009) *MDG Report for Timor Leste*. New York: United Nations. http://hdr.undp.org/en/statistics/

UNDP (United Nations Development Programme) (2010b). *Fighting Corruption in Post-Conflict and Recovery Situations: Learning from the Past*. UNDP, New York.

UNDP (United Nations Development Programme) (2010a) *International Human Development Indicators*. Available at: http://hdr.und.p.org/en/statistics/

UNDDR (United Nations Disarmament, Demobilization and Reintegration Resource Centre) (2006) Sierra Leone Country Report. http://www.unddr.org/countryprogrammes.php?c=60/

United Nations Office for the Coordination of Humanitarian Affairs (2005) In-depth: Guinea – Living on the edge. 10 January http://www.irinnews.org/IndepthMain.aspx?reportid=62546&indepthid=17/

United Nations Security Council (2010) Security Council agrees to lift Sanctions Against Sierra Leone, extend Mandate for Integrated Peacebuilding Office in Country until 15 September 2011, Security Council SC/10044, 29 September. http://www.un.org/News/Press/docs/2010/sc10044.doc.htm

UNSTATS (United Nations Statistics Division) (2004) *Poverty Measurement in a Post-Conflict Scenario: Evidence From the Sierra Leone Integrated Household Survey 2003/2004*. http://unstats.un.org/unsd/methods/poverty/AbujaWS-SierraLeone.pdf/

World Bank (2011) *Doing Business 2011: Sierra Leone – Making a difference for entrepreneurs*. World Bank, Washington, D.C. available at http://www.doing-business.org/~/media/fpdkm/doing%20business/documents/profiles/country/db11/sle.pdf

Yanguas, P. (2010) *Aid and State Reconstruction in Post-Conflict Sierra Leone*. http://www.polisci.ufl.edu/rfap/site%20files/uf_yanguas_2010.pdf/

Part V

Project management

12 Construction project performance in developing countries

P.D. Rwelamila
University of South Africa

Introduction

The construction industry and its activities play an important role in socio-economic development and improvement in the quality of life. According to Nguyen et al. (2004), construction is considered to be one of the most important industries in the economy. It interacts with nearly all fields of human endeavour. In developing countries, this role is significant especially when one reflects on the characteristics of these countries, as outlined below, with respect to the economic development equation.

The developing countries have different climatic, cultural and economic conditions, yet they have many common characteristics. Some of the main characteristics are now outlined.

- The majority of the people have low incomes; they are unable to gain access to capital, and they have poor economic prospects;
- Developing countries are characterised by a lack of sufficient infrastructure and basic services, as well as the capacity and resources to improve and maintain the existing infrastructure; they also have to deal with the demands of rapid urbanisation;
- While the developing countries consume far less resources, and release far less green-house gases than the developed nations, the environmental degradation experienced has a more direct and visible impact on, and presents a more immediate threat to, the physical survival of the poor;
- There are high levels of inequity within developing countries; many of the countries have developed a dual economy with a wealthy elite that has developed consumption patterns similar to those in developed countries, and the majority of the rest of the population living in poverty (see Box 12.1);
- The main sources of foreign income for most developing countries remain agricultural products and raw materials, and with fluctuating

prices of these commodities, these countries find it increasingly difficult to gain access to the financing necessary to move towards industrialisation and a knowledge economy;

- There is strong grassroots ability for innovation. For example, in the field of construction, this is evident in the use of building materials, settlement development and institutional structuring. These can be regarded as being among the most important resources of developing countries;
- Developing countries have strong traditions of a co-operative society and have developed sophisticated methods of conflict resolution.

Levels of inequalities in developing countries

- Eighty percent of the world's gross domestic product belongs to the 1 billion people living in the developed world and *the remaining 20 percent is shared by the 5 billion people living in developing countries*;
- While China and India have seen considerable economic growth, the gap between rich and poor in those countries remains wide;
- In Zambia, the richest 10 percent of the population earns a total income that is 42 times larger than that of the poorest 10 percent;
- In Cameroun, a child born into the poorest 20 percent of families is more than twice as likely to die before the age of five as one born into the top 20 percent;
- In Ghana, the incidence of poverty in Accra, the capital, is only 2 percent, but in the dry rural savannah regions in the north, it is 70 percent. In Accra, poverty has been declining, but in the savannah regions the levels have not changed;
- South Africa has one of the most unequal distributions of income in the world, with approximately 60 percent of the population earning less than R24,000 per annum (about US$7000), whereas 2.2 percent of the population have an income exceeding R360,000 per annum (about US$50,000).

Source: United Nations Development Programme (2005)

The need for the construction industry to contribute towards addressing the challenges relating to the first six characteristics highlighted above is fundamental. Hence, it is necessary for construction projects to be successful. It is through successful projects that the construction industry will contribute to the three objectives of development in developing countries, which are posited by Todaro (2000: 18) as:

- *to increase the availability, and widen the distribution, of basic life-sustaining goods* such as food, shelter, health, and protection;

- *to raise levels of living*, including, in addition to higher incomes, the provision of more jobs, better education, and greater attention to cultural and humanistic values, all of which will serve not only to enhance material wellbeing but also to generate greater individual and national self-esteem;
- *to expand the range of economic and social choices* available to individuals and nations by freeing them from servitude and dependence, not only in relation to other people and nation-states, but also to forces of ignorance and human misery.

This chapter considers construction project performance and identifies some accepted project Key Performance Indicators (KPIs). It reflects on KPIs identified in terms of their relevance to developing countries and their effect on the United Nations (UN) Millennium Development Goals (MDGs). Based on key research reports on the performance of the construction industries in developing countries, three countries are assessed on their status with regard to project performance using the identified KPIs. Common trends of project performance are identified and used as a basis for a discussion across the developing countries. Finally, some recommendations are made towards addressing challenges relating to construction project performance in developing countries.

Project performance – determining success or failure of a project

According to Nguyen et al. (2004), a construction project is commonly considered to be successful when it is completed on time, within budget, and in accordance with specifications, and to the stakeholders' satisfaction. Toor and Ogunlana (2010, p. 228) note that on projects in Thailand, considerations now go beyond the "iron triangle" to cover factors such as: safety, efficient use of resources, effectiveness, satisfaction of stakeholders, and reduced conflicts and disputes. Criteria such as functionality, profitability to contractors, absence of claims and court proceedings, and "fitness for purpose" for occupiers or users of the facility have also been used as measures of project success (Takim and Akintoye, 2002). Success on a project, according to Sanvido et al. (1992), means that certain expectations of a given participant are met, whether this is the owner, planner, architect, engineer, contractor or operator.

It is clear that project success, as a concept, means different things to different people, and hence performance assessment remains ambiguously defined among construction researchers and practitioners. Freeman and Beale (1992) argue that this topic is still attended to in an intuitive and ad hoc manner by many project managers as they attempt to manage and allocate resources across various areas of a project. No agreement has been achieved on this concept. Hence, it is necessary to review the theory and practice on measuring construction project performance.

What are the criteria of project performance (success or failure)?

A project is considered as the achievement of a specified objective, which involves a series of activities and tasks that consume resources (Munns and Bjeirmi, 1996). Lim and Mohamed (1999) consider a criterion as a principle or standard by which anything is, or can be, judged. Based on these arguments, Chan and Chan (2004) define the criteria of project success as:

> ... the set of principles or standards by which favourable outcomes can be completed within a set of specifications.

Success is considered as an intangible perceptive feeling, which varies with different management expectations, among persons, and with the phases of projects (Parfitt and Sanvido, 1993). Parfitt and Sanvido (1993) further argue that consultants, owners, contractors and subcontractors have their own project objectives and criteria for measuring success. They are convinced that definitions of project success are dependent on factors including: project type, size and sophistication, project participants, and experience of owners. Chua et al. (1999) propose a hierarchical model for construction project success. They contend that the objectives of budget, schedule and quality are key measures that contribute to the goal of "*construction project success*" – the top of the hierarchy. Similarly, Long et al. (2004) argue that the four main project aspects – *project characteristics*, *contractual arrangements*, *project participants* and *interactive process* – measure the success of each of the three distinct objectives.

This section reviews different measures of project performance and identifies the appropriate KPI for measuring construction project performance in developing countries. The review is based on research work by Chan (1996, 1997), Chan et al. (2002), Chan and Chan (2004) and others.

Chan (1996, 1997) reviewed the measurement of project success around the world in the late 1980s and early 1990s. He found that project success was considered to be tied to performance measures that, in turn, were tied to project objectives. The work of Navarre and Schaan (1990), which focused on the measurement of success, supported his finding. They found project duration and project costs to be the key indicators of success. A number of researchers, including Atkinson (1999), Walker (1995, 1996), Belassi and Tukel (1996) and Hatush and Skitmore (1997), have identified time, cost and quality as the basic criteria of project success. These three parameters are referred to by Atkinson (1999) as the "iron triangle". While there are other definitions of success in project management, the components of Atkinson's (1999) "iron triangle" are always included among the definitions.

Psycho-social project outcomes, which refer to the inter-personal relations between project team members, have been advocated in addition to these basic criteria in the work of Pinto and Pinto (1991). The inclusion of satisfaction as a success measure was suggested by Rwelamila and Khumalo (2002) and Wuellner (1990). Pocock et al. (1995) suggested that the absence of legal claims should be considered an indicator of success. This then, according to Tembo and Rwelamila (2008) and Chan and Chan (2004), calls for the inclusion of *safety* as an indicator of project success as well. They argue that it is reasonable to expect that, if accidents occur, both contractors and clients may be subject to legal action, as well as financial loss and delays to progress on the construction project.

Using a more comprehensive approach to assess project success, Kometa et al. (1995) extended the criteria domain. Their criteria included: safety; economy (construction cost); running/maintenance cost; and time and flexibility to users. Rwelamila and Savile (1994) identified four parameters for measuring project success: quality; cost; schedule; and utility, as well as overarching requirements such as stakeholder management and meeting health, safety and environmental requirements. Songer and Molenaar (1997) consider a project as successful if it is completed on budget and on schedule, conforms to users' expectations, meets specifications, attains quality workmanship, and minimises construction aggravation. In Kumaraswamy and Thorpe's (1996) study, a variety of criteria of project valuation were considered. These included meeting budget, schedule, quality of workmanship, client's and project manager's satisfaction, transfer of technology, friendliness to the environment, health and safety.

According to Shenhar et al. (1997), a project process is divided into four dimensions, as shown in Figure 12.1. These dimensions are time-dependent. The first dimension is the period during project execution and right after project completion. The second dimension follows shortly afterwards, when the project has been delivered to the customer. The third dimension proposed by the authors is after a significant level of sales has been achieved (1–2 years). Finally, they propose that the fourth dimension should be some 3–5 years after the completion of the project.

A similar study was carried out by Atkinson (1999), who divided the project process into three stages, as shown in Figure 12.2. The first stage is *the delivery stage*: the process – "doing it right"; the second stage is the *post-delivery stage*: the system – "getting it right"; and the third stage is *the post-delivery stage*: the benefits – "getting them right".

Lim and Mohamed's (1999) work was based on the belief that project success should be viewed from the different perspectives of the individual owner, developer, contractor, user, the general public, and so on. Thus, they proposed that project success should be evaluated from both the macro- and micro-level perspectives, as shown in Figure 12.3.

Sadeh et al. (2000) considered project success as having four dimensions. These success dimensions and their respective measures are shown in Table 12.1. The first dimension was considered to include *meeting design goals*, which applies to the contract that is signed by the customer. The second dimension includes the *benefit to the end user*, which refers to the benefit

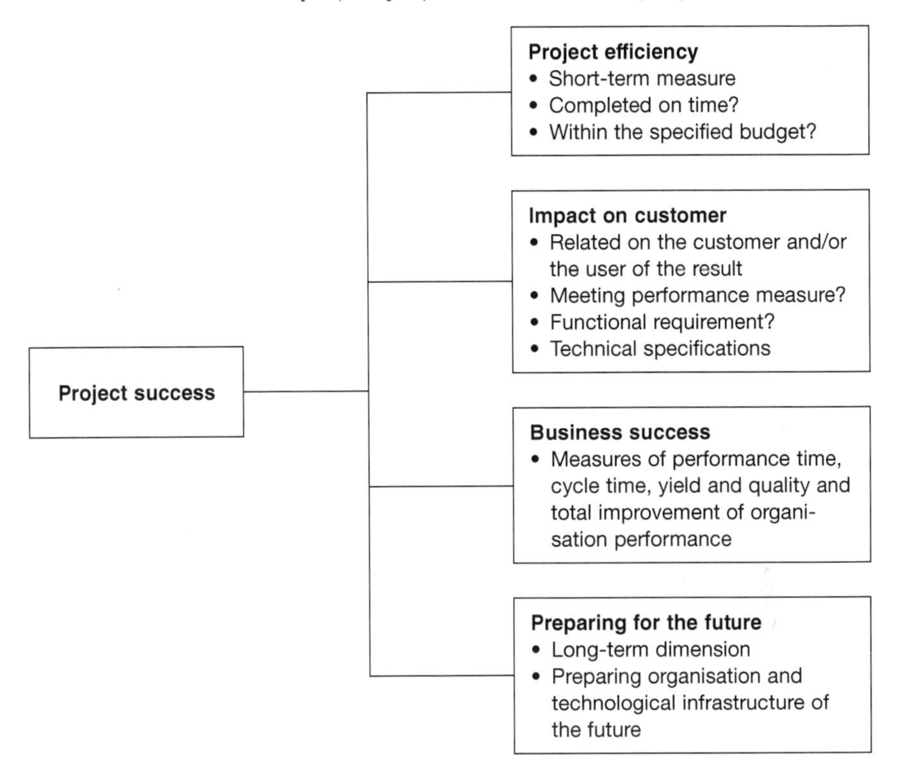

Figure 12.1 The four dimensions of project success (source: Shenhar, et al., 1997)

to the customers from the end product(s). The third dimension was found to include the *benefit to the developing organisation*, which refers to the benefit gained by the developing organisation as a result of the execution of the project. The fourth dimension was considered to include the *benefit to the technological infrastructure of the country and of firms involved in the development process.*

Kagioglou et al. (2001) proposed a conceptual framework for performance management. They described this framework through a performance management and measurement process – through an output model, as shown in Figure 12.4. Kagioglou et al. (2001) used the definition of a performance measurement system proposed by Bititci et al. (1997) that linked performance management and measurement. According to Bititci et al. (1997), a performance measurement system:

> ... is the information system which is at the heart of the performance management process and it is of critical importance to the effective and efficient functioning of the performance management system.

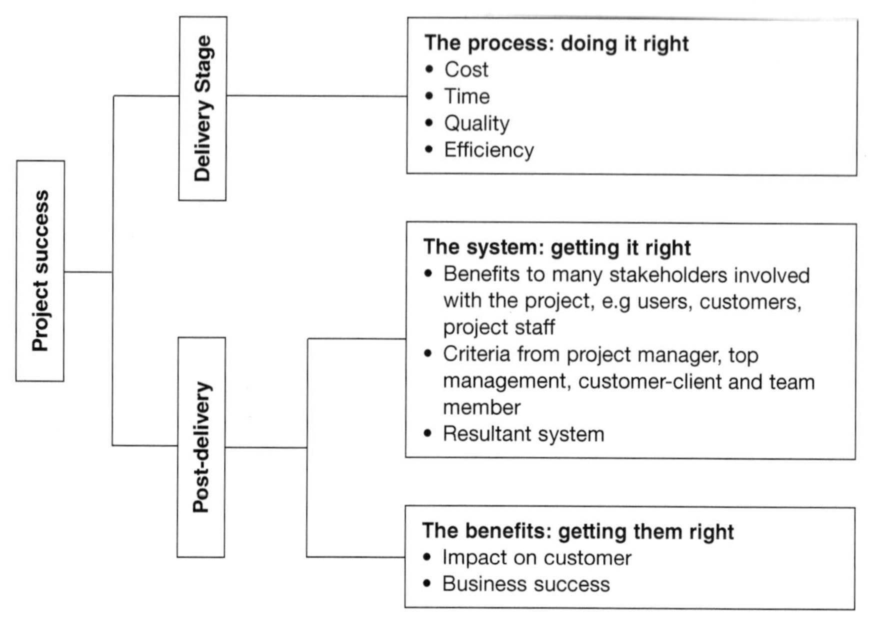

Figure 12.2 Atkinson's model for measuring project success (source: Atkinson, 1999)

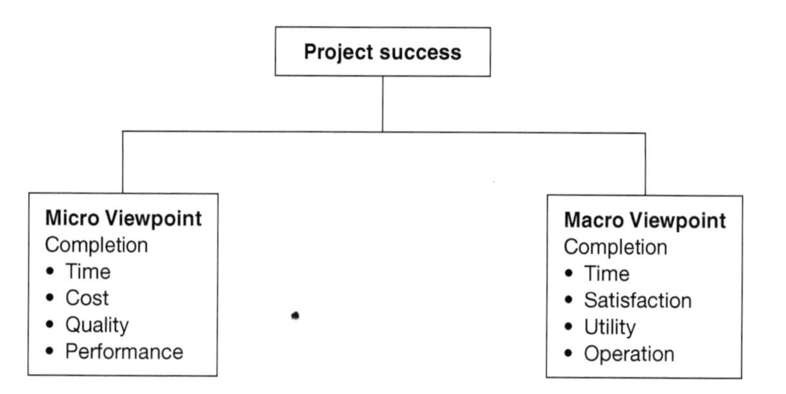

Figure 12.3 Micro and macro viewpoints of project success (source: Lim and Mohammed, 1999)

Evangelidis (1992) defined performance measurement as:

> ...the process of determining how successful organisations or individuals have been in attaining their objectives (and strategies).

Although Bititci et al. (1997) and Evangelidis (1992) defined a performance measurement system and performance measurement respectively from the point of view of an organisation, their definitions and arguments can be used when analysing project performance. The same applies to the work of

Hronec (1993) and Euske (1984) on company performance measurement and management control respectively. In order to measure project performance, the output of strategic and operational project processes (from project conception and development to implementation) are measured in a quantifiable form, to monitor what Hronec (1993) and Euske (1984) refer to as the "vital signs" of the project. The relationship between performance management and measurement can be seen in a wider context from a process (input-process-output) view, as shown in Figure 12.4.

A review of the different criteria for measuring project success that various authors have proposed reveals different ways of considering project performance measurement. There is no single work that can be considered to cover all the criteria for measuring project success. Hence, it is necessary to have a combination of all the proposed dimensions in order to formulate the overall assessment of project success. Chan and Chan (2004) provided a consolidated framework for measuring the success of construction projects as presented in Figure 12.5. It is arguably the most comprehensive representation of the criteria for measuring construction project performance.

To provide the framework for the discussion of the measurement of project performance in developing countries, the key performance indices (KPI) developed for the UK Ministry of Construction by the KPI Working Group (2000), as shown in Figure 12.6, are used. Details on how these KPI were developed are provided by Collin (2002).

Table 12.1 Success dimensions and measures
Source: Sadeh et al., 2000

Success dimensions	*Success measures*
Meeting design goals	Functional specifications Technical specifications Schedule goals Budget goals
Benefit to the end user	Meeting acquisition goals Answering the operational need Product entered service Reached the end user on time Product has a substantial time for use Meaningful improvement of user operational level User is satisfied with product
Benefit to the developing organisation	Had relatively high profit Opened a new market Created a new product line Developed a new technological capability Increased positive reputation
Benefit to the defence and national infrastructure	Contributed to critical subjects Maintained a flow of updated generations Decreased dependence on outside sources Contributed to other projects
Overall success	A combined measure for project success

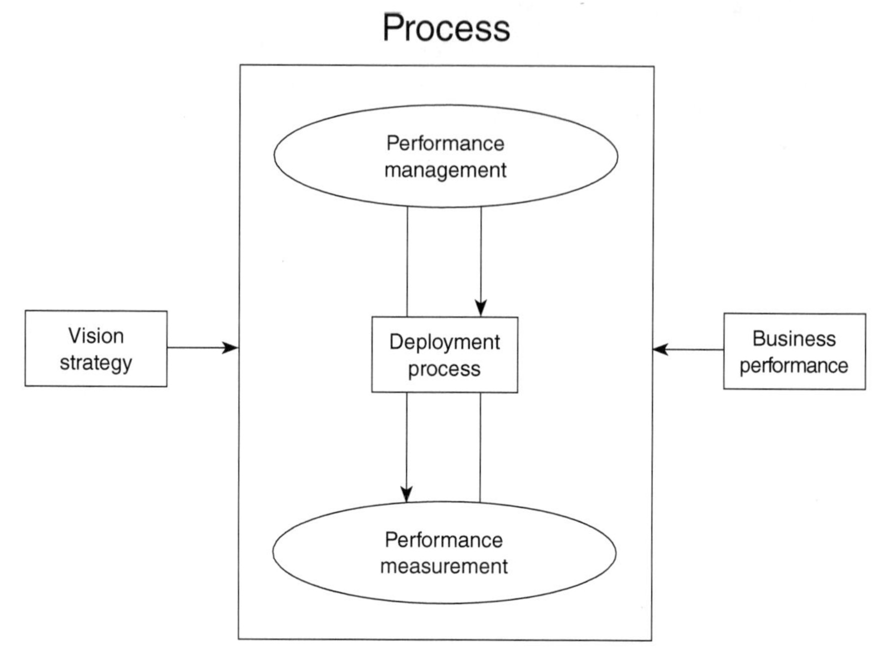

Figure 12.4 Performance management/measurement process (source: Kagioglou et al., 2001)

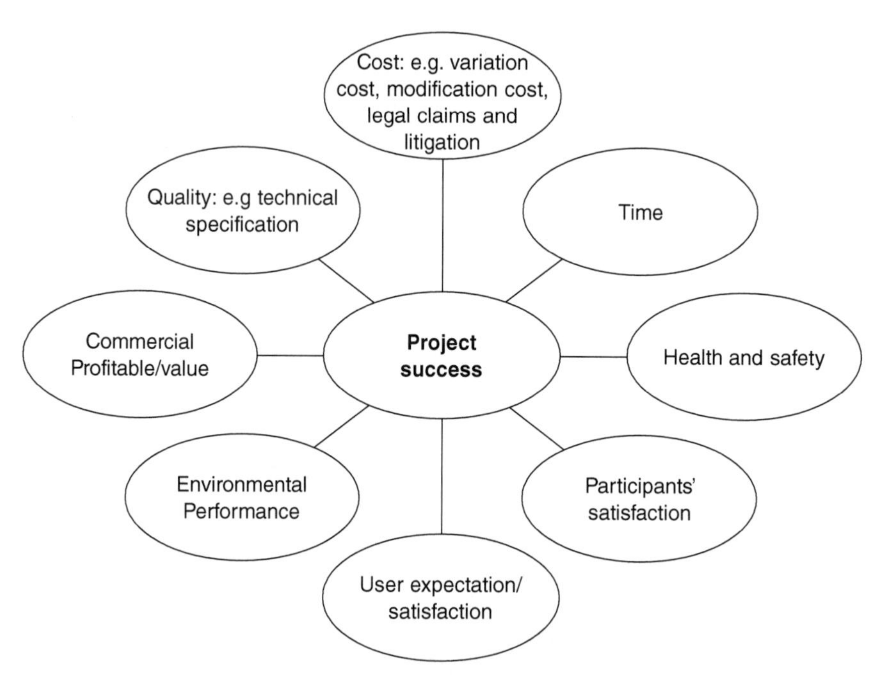

Figure 12.5 Consolidated framework for measuring project success (source: Chan and Chan, 2004)

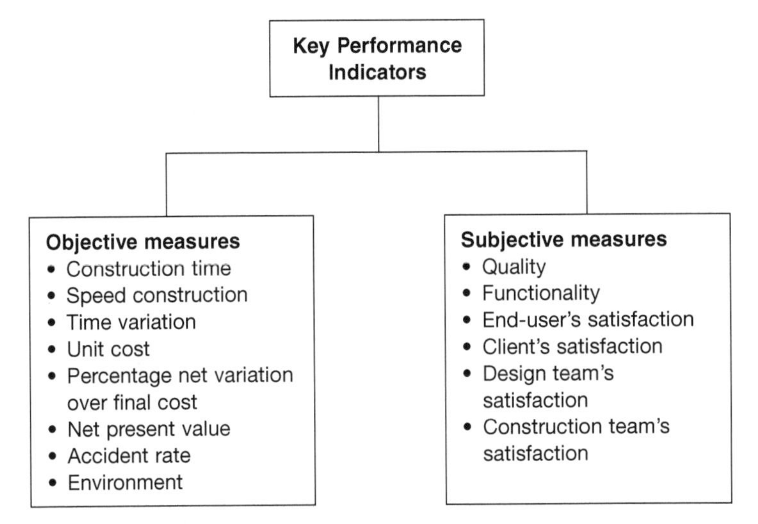

Figure 12.6 KPIs for project success (source: Chan and Chan, 2004)

Following on the work of Naoum (1994) and Chan and Chan (2004), each KPI shown in Figure 12.6 is discussed in detail and practical approaches that can be adopted to measure these KPIs are discussed. The methods are divided into two groups. The first group uses mathematical formulae to calculate the respective values. No formulae are presented in this chapter, but detailed explanations of each KPI are given. Calculations for these KPIs can be found elsewhere (see, for example, Chan and Chan, 2004). The second group uses the subjective opinions and personal judgement of the project stakeholders. Again, detailed explanations of each KPI are given.

Project construction time

Time refers to the period for completing the project. According to Hatush and Skitmore (1997), it is scheduled to enable the constructed asset to be used by a date determined by the client's plans. According to Chan and Chan (2004), project construction time is the *absolute time* – the number of days, weeks or months – from the start of the project on site to its practical completion.

Speed of construction

Speed is the relative time, which is defined by unit of work done by the construction time that has elapsed (for example, square metres of floor of building or kilometres of road per day, per week or per month).

Time variation

Time variation is the percentage increase or decrease in the estimated duration of the project in days, weeks or months, discounting the effect of any extension of time granted by the client.

Project cost

Bubshait and Almohawis (1994) suggest that project cost is the degree to which the general conditions promote the completion of a project within the estimated budget. According to Rwelamila and Savile (1994) and Chan and Chan (2004), cost is the overall cost incurred on the project from inception to completion, including any costs arising from variations or modification(s) during the construction period, and that arising from contractual claims, including that of any related arbitration or litigation.

Value and profit/fulfilling the mandate

Alarcon and Ashley (1996) proposed that to measure project value is to evaluate the satisfaction of the client's needs in a global sense. Their argument is in line with the theme of this chapter. They argue that evaluating the satisfaction of the client's needs includes realisation by the client of the quantity of work produced, operational and maintenance costs, and flexibility. In the private sector, it can be considered as the "business benefit" derived from the completed project. In the public sector, it can be considered as "fulfilling the mandate" derived from the completed project. For public-sector projects, "fulfilling the mandate" means the government department or statutory agency completing the project as promised to respective groups or individuals.

Health and safety

According to Bubshait and Almohawis (1994), the health and safety performance on a project is the degree to which the general conditions promote the completion of a project without major accidents or injuries.

Environmental performance

Construction is regarded as a major contributor to the negative impacts of human activity on the environment (Chan and Chan, 2004). The UNIDO (United Nations Industrial Development Organisation) (1985) sectoral study report reported that about 62–86 percent of domestic production of non-metallic minerals such as glass, cement, clay and lime in developing countries are consumed by the construction industry. The ISO14000 series provides a benchmark of proper environmental practice. Environmental Impact Assessment (EIA) ordinances are now widely accepted statutory frameworks for predicting and assessing potentially adverse environmental impacts from development projects around the world. Several developing countries such as Indonesia, Malaysia and Vietnam have mandatory requirements for an EIA to be undertaken for each development project.

As a rule of thumb, the application of ISO14000, the EIA score (depending on the extent and nature of the project) and the total number of complaints received during the construction project can be used as an indicator to reflect the environmental performance of a given project.

Quality

Parfitt and Sanvido's (1993) definition of quality is a good starting point here. They define *quality* as the totality of features required by the product or service to satisfy a given need: fitness for purpose. Wateridge (1995) considers this requirement as *meeting specifications*. According to Chan and Chan (2004), the measure of technical specifications is the extent to which the technical requirements specified are achieved. Rwelamila (1996) argues that the specifications should be satisfied the first time (avoiding re-work). Hence, meeting the requirements of the specifications and doing it right the first time should be sufficient in determining project quality.

Functionality

Functionality is related to the expectations of the project's internal stakeholders and can best be measured by the degree of conformance to all technical performance specifications (Chan et al., 2002). Quality, technical performance and functionality are closely related.

User expectation and satisfaction

It is important that the completed asset meets the expectations and satisfaction of those who work in, or use, the constructed item. The suggestion by Torbica and Stroh (2001) that, if the users of the asset are satisfied, the project can be considered as having been successfully completed in the long run, is a sufficient measure for this attribute.

Client, design team leader and construction team leader satisfaction

The satisfaction of the key participants in the project is considered as an important measure of project success by a number of researchers (see, for example, Rwelamila and Savile, 1994; Task Force on Road Construction

Costs, 2010; Parfitt and Sanvido, 1993; Sanvido et al., 1992). The level of satisfaction of the participants is considered here as an indicator of project success.

How relevant are the KPIs for developing countries?

Taking a leaf from Legum's (2002) "*Washington Consensus*" argument, it is important to reflect on the KPIs identified above (see Figure 12.6) and assess their applicability to developing countries. According to Legum (2002), the "*Washington Consensus*" declared that the whole world is *necessarily* one global market, in which *all countries and regions and all producers of all goods and services must compete.* Because the "consensus" was seen as a solution to the problems of the developed countries in finding markets, it was assumed that it would benefit everyone (including developing countries). The above KPIs are accepted as measures of construction project performance in developed countries. In accordance with the "*Washington Consensus*", each KPI and its suitability to developing countries should be scrutinised.

Each of the KPIs has an objective measure. Construction schedule; speed of construction; time variations; unit cost; variations over final project cost; value of investment; accident rate; and environmental impacts are not drawn from a narrow perspective of narrow interests such as the "*Washington Consensus*", as for example, they are not tied with issues of capital generation without considering the impact on the environment and how people will be affected. Each of the KPIs' objective measures will be useful in assessing project performance in both developed and developing countries. The uniqueness of every project will influence the interpretation of the impact of each KPI on the project. For example, in most developing countries, a significant number of projects are funded by donor agencies, and excessive cost overruns could lead to a situation where a project could be terminated by the donor(s). This argument also applies when considering subjective measures (Chan and Chan, 2004): quality; functionality; end-user's satisfaction; client's satisfaction; design team's satisfaction; and construction team's satisfaction.

From the above discussion, it can be argued that the KPIs shown in Figure 12.6 are sound and applicable to construction projects in developing countries. These KPIs are used to assess construction project performance in developing countries below.

KPIs and their relationship with MDGs

Project success (positive measures on the KPIs) should be a positive contribution to national social and economic development, whereas a failed project (negative measures on KPIs) reduces the volume from one side of the development equation. It is necessary to link each KPI with the respective MDGs in order to reflect on how project performance in developing countries will affect the achievement of the targets of the MDGs. Table 12.2 maps the MDGs to the construction project KPIs.

As indicated in Table 12.2, the attainment of each of the MDGs is affected by construction project performance. The attainment of most of the targets under the MDGs, such as those relating to education and healthcare improvement, requires the provision of infrastructure (schools, healthcare facilities, roads, and water and sanitation facilities). If infrastructure projects are constructed within established levels of the KPIs, the possibility of reaching the MDGs by 2015 would be enhanced.

Assessing project performance in developing countries

Which countries make the developing countries group?

While Todaro's (2000) characteristics of developing countries provides a good start in understanding the dynamics of these countries, a universally acceptable definition of a developing country remains elusive. Such a term would appear to imply that there is homogeneity among the countries in

Table 12.2 Relationship between MDGs and KPIs

Millennium Development Goals (MDGs)	*Project Key Performance Indicators (KPIs)*
1 Eradicate extreme hunger and poverty	Labour-based construction – will be affected by all KPIs (OM1–OM8 and SM1–SM6)
2 Achieve universal primary education	Construction of infrastructure: school buildings and offices – will be affected by all KPIs (OM1–OM8 and SM1–SM6)
3 Promote gender equality and empower women	N/A
4 Reduce child mortality	Construction of healthcare facilities – will be affected by all KPIs (OM1–OM8 and SM1–SM6)
5 Improve maternal health	Construction of healthcare facilities – will be affected by all KPIs (OM1–OM8 and SM1–SM6)
6 Combat HIV/AIDS, Malaria and other deseases	Construction of healthcare facilities, water and sanitation facilities – will be affected by all KPIs (OM1–OM8 and SM1–SM6)
7 Ensure environmental sustainability	The need to make sure that Environmental Impact Assessments (EIA) are undertaken for every construction project – affected by OM8
8 Develop a global partnership for development	

Objective measures (OM) and Subjective measures (SM) – from Figure 12.6, each bullet indicator is represented by a number from '1'. OM = OM1–OM8; SM = SM1–SM6 the group. However, the levels of development vary among these countries (Arthur and Sheffrin, 2003). It is also important to note that the term "developing" implies mobility and does not acknowledge that development may be static, or in decline in some countries, such as those Southern African countries that are worst affected by HIV/AIDS.

The World Bank (2003) classification where Gross National Income (GNI) comparisons are used to rank countries was applied to identify developing countries for this study. Detailed discussions on the various ways in which this classification is done (exchange rate method and purchasing power parity method), and the merits and weaknesses of each method, can be found elsewhere (see, for example, Todaro, 2000).

After identifying countries falling within the developing countries band, the World Bank (2003) groups these countries into regional aggregates that include:

- Arab States (20 countries – including the Saharan African countries);
- Asia and Pacific – divided into two areas: (i) East Asia and the Pacific (28 countries); and (ii) South Asia (9 countries);
- Latin America and the Caribbean (33 countries);
- Southern Europe (2 countries);
- Sub-Saharan Africa (45 countries).

For the purpose of this study, the World Bank's five regional aggregates were further divided into six domains in order to give each aggregate an equal chance of being selected in the study. The six research domains were: (i) Arab States; (ii) Asia Pacific (East Asia and the Pacific); (iii) Asia Pacific (South Asia); (iv) Latin America and the Caribbean; (v) Southern Europe; and (vi) Sub-Saharan Africa.

Methodology

In order to interpret, verify, evaluate and establish the nature of project management performance in developing countries, a content analysis technique was used. Insch et al. (1997) noted that content analysis is the prevalent analytical method used in interpreting documents to find underlying themes or messages. Bryman and Bell (2003: 195) referred to the approach as that where "documents and text are analysed to quantify content in terms of predetermined categories and in a systematic and replicable manner". Neuman (2006) noted that measurement in content analysis uses four attributes: frequency, direction, intensity, and space. In this study, two of these attributes were employed: frequency and intensity. Frequency simply means how often something occurs in the content, while intensity is the strength or depth of the theme in the content (Neuman, 2006).

Two research reports on construction industry practices and project performance in developing countries were identified. These are Ssegawa and Ngowi's (2008) report on the causes of poor performance of public construction projects in Botswana, and the Engineers Against Poverty and

Institution of Civil Engineers (2006) report on the utilisation of infrastructure procurement to enhance social development.

Prior to the actual analysis, two cycles of random selection were carried out. In the first cycle, two countries were selected from each of the six research domains described above. A total of twelve countries were selected, as indicated in Table 12.3.

In the second cycle, one country from the two countries in the each of the research domains was selected. Hence, six countries were randomly selected, as presented in Table 12.4. From these four countries, three were randomly selected – Indonesia, Botswana and Nigeria.

While it could be argued that these three countries represent only two of the six research domains – Asia Pacific and Sub-Saharan Africa – and thus are not representative of all the area covered by the six domains, the fact that most developing countries share fundamental characteristics as indicated above should dilute this criticism. Furthermore, these three countries could be described as strategic regional leaders with the capacity to promote the implementation of the recommendations emerging from this review and to influence other countries in the region. These three countries are analysed in the succeeding section.

Table 12.3 First cycle randomly selected countries

Research area domain	*Country selected*
Arab States	United Arab Emirates Morocco
Asia Pacific (East Asia and the Pacific)	Indonesia Malaysia
Asia Pacific (South Asia)	India Sri Lanka
Latin America and the Caribbean	Brazil Chile
Southern Europe	Cyprus Turkey
Sub-Saharan Africa	Botswana Nigeria

Table 12.4 Second cycle randomly selected countries

Research Area Domain	*Country Selected*
Arab States	Morocco
Asia Pacific (East Asia and the Pacific);	Indonesia
Asia Pacific (South Asia)	India
Latin America and the Caribbean	Chile
Southern Europe	Turkey
Sub-Saharan Africa	Botswana

Project management performance in selected countries

A detailed scrutiny of the reports was conducted to establish the extent of project performance in the six countries selected (Table 12.3). The situation in each country is discussed below.

Botswana

According to Ssegawa and Ngowi (2008), the construction industry in Botswana has been at the forefront of the process of national development in building the national infrastructure since the country attained independence in 1966. The contribution of the industry has been considerable as it produced the productive facilities for the local mining and manufacturing industries that have been the mainstays of the economy; and the housing units, schools and hospitals which have improved the quality of life. The industry has also provided significant levels of employment. Ssegawa and Ngowi (2008) further argue that the growth of Botswana's construction industry in the past three decades has been phenomenal. However, the industry has also experienced a number of challenges. There have been reports of poor project performance and there have even been situations where projects have been abandoned (Ssegawa, 2004; Ssegawa, 2002; Sentongo, 2005).

In what is considered to be one of the authoritative studies on the construction industry in Botswana, Ssegawa and Ngowi (2008) selected a sample of 323 public-sector construction projects from the Department of Building and Engineering Services (DBES), Roads Department, Ministry of Local Government and various local authorities. The sample took cognisance of the nature of the various clients. The key performance indicators indicated in Figure 12.6 under the broader parameters of cost and schedule (time) were applied to examine how projects performed, with respect to the terms agreed upon at the tender award stage.

The overall performance of the 323 public-sector construction projects at the construction stage showed the following results (Ssegawa and Ngowi, 2008):

- On 92 percent of projects (297 out of 323 projects) the agreed cost and time were exceeded;
- On 65 percent of projects (210 out of 323 projects) the agreed cost was exceeded;
- On 85 percent of projects (275 out of 323 projects) the agreed time was exceeded;
- 13 percent of projects (42 out of 323 projects) were abandoned by the contractor.

From the above results, Ssegawa and Ngowi (2008) estimated that the probability of a public construction project in Botswana being finished within agreed schedule and cost was 0.08 (or 8 percent).

The above results have implications for all the objective and subjective KPIs of the projects. Failure to complete a project within the agreed duration means that speed of construction is below the agreed levels; time variations are common features of most of the projects; end users' and clients' satisfaction levels are likely to be low when these basic parameters are not positively achieved; and conflicts resulting from poor schedule performance will have a negative effect on the satisfaction levels of both the design and construction teams. Similarly, failure to meet project cost requirements has a strong chance to lead to a high percentage of net variation over final costs, raising the capital outlay on the project.

The poor construction project performance in Botswana appears to have been caused by a number of factors. The salient factors that were highlighted by Ssegawa and Ngowi (2008) are now discussed.

Challenges from the project demand side

The major findings of the study indicated that, in the public-sector client organisations in Botswana, there was a lack of the following: adequate human capacity; adequate initial project details; proper project budgeting; project management approach; a prompt payment system for suppliers; coordination regime for the major parties to a project; effective monitoring during construction; appropriate contract; and an effective tender evaluation and adjudication system. These issues are now considered.

1 *Lack of human capacity* – there is a lack of adequate professional capacity to plan for projects at both the procuring and project management entity levels. This has led to the existing professionals being unable to cope with their workload;
2 *Lack of adequate initial project details* – in both the procurement and project management entities, there is lack of sufficient details at the pre-design stage to develop comprehensive design briefs for the projects. For some projects, neither the procuring entity nor the project management entity had the necessary expertise to provide a comprehensive brief;
3 *Lack of project budgeting* – the basis on which funds are allocated to projects appears to be flawed owing to the inaccuracies which existed initially in the "project memorandum", and later in the bills of quantities (BoQs). The inaccuracies arose from a number of factors including inflation and an inability to identify and document all the requirements of the projects;
4 *Lack of project management approach* – a project management approach emphasises at least two issues. The first of these is that a project should be treated as a succession of continuous phases, in its metamorphosis. The second issue is that there is the need for a single point of responsibility to coordinate the contributions of all the project stakeholders, and provide centralised communication and effective

decision-making. The fundamental principles do not appear to be understood by the public clients in Botswana and their advisers;

5 *Lack of a prompt project payment system for project services and goods suppliers* – unlike the situation in many other developing countries, in Botswana government officials did not report any lack of money for public projects. However, suppliers of goods and services continue to receive their payments beyond the period of time indicated in the relevant contracts. The causes of the late payments appear to be inefficiency and excessive bureaucracy in the system for processing the payments on the client's side;

6 *Lack of a coordination regime for the major parties to a project* – due to the lack of application of a project management approach, there is a lack of coordination of parties related to a project. On most projects, there is no effective coordination and teamwork. This can have an adverse impact on many aspects of the project owing to the inter-dependence of the activities of the participants in construction projects;

7 *Lack of effective monitoring during construction* – the client organisations rely on the clerk of works and consultants to monitor and control progress, costs and workmanship on the project. However, most contractor representatives (site agents) rarely listen to the clerk of works. The main reason is that there is usually a discrepancy in the level of authority and expertise between them. Thus, the instructions of the clerk of works are ignored. This sometimes leads to conformity with project parameters being compromised. Therefore, the effectiveness of the project monitoring and control process is reduced and, consequently, the project fails to meet the requirements of the client;

8 *Lack of appropriate contract* – the choice of the contract form used on the project is based on convenience to the client organisation, and does not take into consideration the benefits to all participants, the relationship with the procurement system selected, or the overall performance on the project. This has an adverse impact on the relationships, levels of cooperation and tendency to dispute;

9 *Inefficient and ineffective tender evaluation and adjudication process* – two main issues are relevant here:

- ***discrepancy between tender evaluation and award***: there are often discrepancies between the tender award decisions and the recommendations made in the evaluation reports (this is most common in the local authorities). There are instances where the construction firm awarded the contract is different from that recommended in the evaluation, without any plausible explanation being given;
- ***inability to award tender within reasonable time***: the evaluation and adjudication processes seem to take such a long time that the prices that were used to prepare the bid change. Although bidders are advised to revise their bids in such circumstances, most of them do not do so for fear of raising their prices and having bids that are not competitive.

Challenges from the project supply side

The report also highlighted deficiencies on the part of the construction industry in Botswana, including consultants, contractors, materials suppliers and utility providers.

1 *The role of project consultants*: None of the professions in the construction industry in Botswana has a statutory body to regulate the professional conduct of practitioners. Hence, the qualifications and experience of construction professionals are not thoroughly checked to determine their appropriateness to offer the services for which they claim to be competent; in particular, there appears to be a number of incompetent construction consultants in the country;
2 *The role of contractors*: Many contractors in Botswana (both main and specialist sub-contractors) lack skills in key project processes: estimating and price analysis, project planning, and organisational management including site management, contract administration and financial management. Moreover, most of the contractors lack discipline when dealing with company and project-related funds. Furthermore, they do not have adequate numbers of personnel with the relevant skills; and the resources of the companies are often over-stretched, as they take on many jobs at a time, or seek to undertake projects that are too distant from their bases and beyond their logistical capacity;
3 *The role of material suppliers*: The construction companies in Botswana do not have a good supporting industry. The suppliers often fail to deliver the quantities of certain materials at the time they are required, although there had been a prior undertaking to supply them. Other common problems include: failure to supply materials that meet the specifications, particularly in properties such as strength, which cannot be checked or verified by visual inspection. The prices of materials are also high, and this, in turn, raises contractors' bids and renders the project bidding platform uneven;
4 *The role of utility companies*: The utility companies (electricity, water and telephone suppliers) are slow in providing services related to projects, even where they are paid in full and informed in advance of the commissioning time, and in relocating their services when there is a development which affects their service lines;
5 *The role of the industry's environment*: The construction industry in Botswana lacks an umbrella organisation to cater for the interests of the stakeholders and coordinate the industry's activities towards a culture of good practice. This situation has led to a number of undesirable attributes which include: poor assessment and monitoring of service suppliers; inadequate assessment of the needs of the construction industry, leading to lack of information on various issues (such as the supply and demand of resources, and price and cost indices); poor relations between the industry and the wider community; and poor

public perception of the construction industry owing to a lack of understanding of the way the industry operates;

6 *Perceived corruption in the construction industry*: There is an increasing tendency towards corruption in the construction industry in Botswana. The Directorate of Corruption and Economic Crime report (DCEC, 2006) disclosed that 874 cases were reported between 2000 and 2006. While 60 percent were considered to be not "concrete" cases, the remaining cases are still under investigation. Most of these cases are related to the tendering and the construction phases of projects.

Indonesia

The report by Engineers Against Poverty and Institution of Civil Engineers (2006) on the Indonesian construction industry focused on identifying opportunities to improve the delivery of social development objectives by modifying the way in which infrastructure projects are procured. When it is compared with that of Ssegawa and Ngowi (2008) on the construction industry in Botswana, it is evident that the two works show two different approaches to dealing with the same issue – project performance. While the latter report was directly focusing on construction project performance in Botswana, the former focused on identifying opportunities to improve the delivery of social development objectives on infrastructure projects.

The report concentrates directly on two main areas of social development opportunity: the impact/performance of the asset and the service it delivers (the product); and the opportunities during the project's construction and operation (the process). It is possible to assess project performance KPIs in Indonesia as described in Figure 12.6 by analysing the findings presented in the report.

According to Engineers Against Poverty and Institution of Civil Engineers (2006), the project value creation regime in the Indonesian construction industry was, for some time, prone to misuse and abuse. Thus, prior to 2000, the system failed to result in the best value for money for public clients. Public procurement was regulated through a mixture of presidential decrees, ministerial directives, letters of information and other decrees and instructions by governors, mayors and *bupatis* (heads of district government), thus creating a confusing system of overlapping regulations. During that period, construction projects were characterised by: considerable delays; budget overruns; poor quality; complaints from clients and users; and dissatisfied internal project stakeholders.

Following a World Bank (2001) report and the introduction of new procurement guidelines in a presidential decree, *Keppres* 80/2003, the Indonesian project procurement regime has been improved in terms of efficiency and transparency. However, the construction industry in Indonesia still faces a shortage of project procurement expertise. Moreover, there are a number of factors negatively affecting project performance in

Indonesia (Engineers Against Poverty and Institution of Civil Engineers, 2006). They found the following factors:

- Lack of public consultation, national plans or other clear criteria for project identification – this has led to poor satisfaction of end users;
- Failure to plan and budget for operation and maintenance – this factor has affected clients' and end users' satisfaction;
- Inflexible procurement strategies and adversarial contract forms – this has had negative effects on construction time, speed of construction, time variations, and construction team dissatisfaction (especially that of main contractors);
- Failure to address corruption which is a major inhibitor at every stage of the project procurement cycle – this has an effect on the quality of products and services, end-users' satisfaction, accident rates and speed of construction.

Indonesia, like Botswana, has been facing a number of challenges that urgently need to be addressed to ensure an effective and sustainable project procurement regime.

The salient challenges are (Engineers Against Poverty and Institution of Civil Engineers, 2006):

- Lack of human capacity (in terms of sufficient numbers) to plan and monitor on-going projects. For example, in 2000, some 4000 project managers were required to carry out public-sector procurement at the national level. This number increased to 30,000 when considering the number of persons required to serve on the project procurement committees;
- Lack of a project management approach – the need to treat projects as successions of continuous phases; and the need for a single point of responsibility to act as a leader and the focal point of centralised communication and effective decision-making;
- Lack of effective monitoring during project implementation on site – this often resulted in delays in decision-making, poor workmanship and continuous project challenges and problems;
- Lack of appropriate contract forms – these are often based on the convenience of client organisations instead of balancing the interests of all project participants.

Nigeria

Obasanjo's (2004) statement summarises the features of the project procurement regime in the Nigerian construction industry:

> Sadly, principles of transparency, fair play, accountability and open competition were replaced with a broken down competition system

> that entrenched opaqueness, inefficiency, influence peddling and inflated costs with the attendant incidences of corruption. Over the past two decades estimates suggest a considerable portion of our public treasury was lost due to the poor contracting system.

Engineers Against Poverty and Institution of Civil Engineers (2006) observed that the above statement indicates how far the Nigerian public procurement system had degenerated. The National Economic Empowerment and Development strategy supports this observation (Nigerian National Planning Commission, 2004). It describes a procurement system of inflated contract costs and processes that were at best closed, discretionary and had well-designed conduits for the abuse of public power.

The Nigerian procurement regime suggests that all project performance KPIs indicated in Figure 12.6 are negatively affected on construction projects in that country. The following factors are inhibiting project success in Nigeria (Engineers Against Poverty and Institution of Civil Engineers, 2006):

- Lack of good practice in project identification processes – the non-existence of public consultation on public projects and national plans has led to poor satisfaction of end users of products and services;
- Failure to consider operational and maintenance issues during project planning and budgeting – this factor has negatively affected clients' and end users' satisfaction;
- Inflexible project procurement strategies and adversarial contract forms – this has negative effects on construction time, speed of construction, time variations, and construction team dissatisfaction (especially main contractors);
- Failure to address corruption, which is a major inhibitor at every stage of the project procurement cycle – this has an effect on the quality of products and services, end-users' satisfaction, accident rates and speed of construction.

Various interventions focusing on procurement reform and aiming to improve governance have been proposed (Engineers Against Poverty and Institution of Civil Engineers, 2006). One of these is the introduction of a certifying mechanism referred to as "Due Process", managed by the Budget Monitoring and Price Implementation Unit. Due Process certifies each public procurement transaction at the federal government level to ensure that it complies with the stated criteria. The Nigerian federal government is seeking to establish a permanent legal framework through a Public Procurement Bill that is currently under review in the country's parliament. These interventions are intended to create a project procurement environment that is characterised by good practices. If the interventions are successful, most of the project KPIs would be affected in a positive manner. However, the depth of the challenges suggests that implementation will be a major task.

Common findings from the three developing countries

A number of common findings emerged from the study of the major reports:

1 In all three countries, there is a *lack of sufficient numbers of experts* to support the planning, design and implementation of the construction projects which the countries need for their long-term socio-economic development (Engineers Against Poverty and Institution of Civil Engineers, 2006; Ssegawa and Ngowi, 2008);
2 *Neglect of maintenance considerations* – in all three countries, there is a poor state of planning for the maintenance of the completed built asset (Engineers Against Poverty and Institution of Civil Engineers, 2006; Ssegawa and Ngowi, 2008);
3 There is *lack of a sound project management approach* – the coordination of the contributions of the parties related to the project is a challenge in these countries, and this has led to poor project performance (Engineers Against Poverty and Institution of Civil Engineers, 2006; Ssegawa and Ngowi, 2008);
4 *Corruption across project value chain* – all three countries appear to be faced with corruption, stemming from insufficient transparency in the selection of projects which allows corruption to appear in a variety of forms (for example, in Nigeria, media reports of the government identifying and funding projects that do not actually exist demonstrate one extreme form of corruption at the initial project identification stage); and with the introduction of some decentralisation policies, the ability of both national and local political leaders and government officials to influence project selection seems to have increased (Engineers Against Poverty and Institution of Civil Engineers, 2006; Ssegawa and Ngowi, 2008);
5 There is a *lack of a good practice culture* in two of the three countries (Nigeria and Botswana) – a significant part of this challenge is caused by the lack of an umbrella organisation to cater for the interests of the stakeholders of construction projects and coordinate the activities of the construction industry. This has led to undesirable attributes such as: failure to maintain robust registers of consultants, contractors and other service providers; inadequate assessment of the industry's needs, leading to lack of information on various aspects (such as price and cost indices; and information on the supply of, and demand for, resources) (Engineers Against Poverty and Institution of Civil Engineers, 2006; Ssegawa and Ngowi, 2008).

Conclusion

A review of the literature on and practice of construction project performance and two reports on performance in the construction industries in developing countries brought forward a number of salient issues, some of which are now discussed.

1 There are a number of inhibiting factors in construction project procurement procedures and the role of the various stakeholders that appear to be stumbling blocks to project success. These include:
   - inflexible procurement strategies and adversarial contracts;
   - government failure to maintain robust registers of contractors;
   - confusion over the roles of the client and donor which seems to lead to a lack of project leadership and poor implementation;
   - government failure to enforce various regulations relating to project delivery;
   - failure to plan and budget for costs-in-use (operations and maintenance) of infrastructure facilities; and
   - failure to address corruption, which is a major contributor to malpractice and mismanagement at all stages of project delivery.
2 There is a scarcity of knowledgeable project delivery experts to support the implementation of construction projects. For example, despite the heightened need for competent construction project managers, in most cases individuals are appointed as project managers only because they have qualifications in the same field as the project's core technical content – building or civil engineering;
3 There appears to be a lack of public consultation – national voluntary and statutory professional groupings seem to play a limited role in the creation of national development plans although they have significant knowledge regarding infrastructure planning, design, implementation and operation. These professional groupings can make positive contributions towards sustainable infrastructure development, and project(s) that would provide the maximum socio-economic benefits to the local population;
4 Many developing countries have construction industries which lack organisations to provide effective leadership and vision – apart from a handful of countries which have construction industry development organisations (such as Malawi, Malaysia, South Africa, Tanzania and Zambia), most countries do not adopt holistic approaches when addressing most of the challenges facing the industry. The challenges are tackled in an ad hoc fashion by various parties, leading to short-term solutions.

Some recommendations may be made with regard to appropriate actions to improve construction project performance in developing countries.

1 There is a need to have a sustainable construction industry embracing best practices across organisations and projects – the establishment of a statutory body as an umbrella organisation to oversee the activities of the construction industry and provide leadership for all its stakeholders should be considered. The main roles of such an organisation

will include provision of advice to government and other stakeholders on issues concerning the industry;

2. In order to contribute to a construction industry best practice culture, clients at the national, regional and local government levels, need to increase their capacity and efficiency when managing the planning and implementation phases of projects through a number of actions such as:
   - recruit, train and motivate adequate qualified and experienced personnel in various departments and agencies which deal with project planning and implementation;
   - re-design the project value creation processes (project planning and implementation systems) to make them more efficient and to reduce bureaucracy;
   - monitor and evaluate projects (project audits and close-out) to ensure the required throughput of work and efficiency are achieved; and
   - adopt project management fundamentals and view a project life-cycle as being made up of inter-linked and inter-dependent phases.
3. There is a need for a formal process of project identification, planning and design – project identification should be in line with national, local or sector plans and/or based on public consultation; the broad stakeholder requirements should be identified at the planning phase and incorporated into the design; and the whole lifecycle of the asset should be considered during planning and design, and an operation and maintenance strategy developed for each project;
4. Appropriate frameworks are needed to deal with project finance, procurement and contracts – there is the need to set aside funds in the budget of each project for the realisation of the project's social objectives. It is also necessary to consider alternative project procurement strategies to ensure the appropriate approach to meet specified stakeholder requirements; and there is the need for the key project participants to agree on contractual mechanisms to deliver the project according to stakeholders' requirements.

References

Alarcon, L.F. and Ashley, D.B. (1996) Modeling Project Performance for Decision Making, *Journal of Construction Engineering & Management*, 1222(3), 265–73.

Arthur, S. and Sheffrin, S.M. (2003) *Economics: Principles in Action*, Upper Saddle River, NJ: Pearson Prentice Hall.

Atkinson, R. (1999) Project Management: Cost, Time, Quality, Two Best Guesses and A Phenomenon, it is Time to Accept Other Success Criteria, *International Journal of Project Management*, 17(6): 337–42.

Belassi, W. and Tukel, O.I. (1996) A New Framework for Determining Critical Success/Failure Factors in Projects, *International Journal of Project Management*, 14(3): 141–51.

Bititci, U.M., Carrie, A.S. and McDevitt, L. (1997) Integrated Performance Measurement Systems: an audit development guide, *The TQM Magazine*, 9(1): 46–53.

Bryman, A. and Bell, E. (2003) *Business Research Methods*. Oxford: Oxford University Press.
Bubshait, A.A. and Almohawis, S.A. (1994) Evaluating the General Conditions of a Construction Contract, *International Journal of Project Management*, 12(3): 133–5.
Chan, A.P.C. (1996) Determinants of Project Success in the Construction Industry of Hong Kong, Unpublished PhD Thesis, University of South Australia, Adelaide.
Chan, A.P.C. (1997) Measuring Success for a Construction Project, *The Australian Institute of Quantity Surveyors*, 1(2): 55–9.
Chan, A.P.C. and Chan, A.P.L. (2004) Key Performance Indicators for Measuring Construction Success, *Benchmarking: An International Journal*, 11(2): 203–21.
Chan, A.P.C., Scott, D. and Lam, E.W.M. (2002) Framework of Success Criteria for Design/Build Projects, *Journal of Management in Engineering*, 18(3): 120–8.
Chua, D.K.H., Kog, Y.C. and Loh, P.K. (1999) Critical Success Factors for Different Project Objectives, *Journal of Construction Engineering & Management*, ASCE, 125(3): 142–50.
Collin, J. (2002) *Measuring the Success of Building Projects – improved project delivery initiatives*, July, Brisbane, Australia, Queensland Department of Public Works.
DCEC (Directorate of Corruption and Economic Crime) (2006) Report, Gaborone, Botswana, DCEC.
Engineers Against Poverty and Institution of Civil Engineers (2006) *Modifying Infrastructure Procurement to Enhance Social Development*, London: EAP and ICE.
Euske, K.J. (1984) *Management Control: Planning, Control, Measurement and Evaluation*, Reading, MA: Addison-Wesley.
Evangelidis, K. (1992) Performance Measured is Performance Gained, *The Treasurer*, February: 45–7.
Freeman, M. and Beale, P. (1992) Measuring Project Success, *Project Management Journal*, 23(1): 8–17.
Hatush, Z. and Skitmore, M. (1997) Evaluating Contractor Prequalification Data: selection criteria and project success factor, *Construction Management and Economics*, 15(2): 129–47
Hronec, S.M. (1993) *Vital Signs: Using Quality, Time and Cost Performance Measurement to Chart Your Company's Future*, New York: Amacon.
Insch, G.S., Moore, J.E. and Murphy, L.D. (1997) Content analysis in leadership research: Examples, procedures, and suggestions for future use. *Leadership Quarterly*, Vol. 8, No. 1, pp. 1–25.
Kagioglou, M., Cooper, R. and Aouad, G. (2001) Performance Management in Construction: a conceptual framework, *Construction Management & Economics*, 19: 85–95.
Kometa, S., Olomolaiye, P.O. and Harris, F.C. (1995) An Evaluation of Clients' Needs and Responsibility in the Construction Process, *Engineering, Construction and Architectural Management*, 2(1): 45–56.
Kumaraswamy, M.M. and Thorpe, A. (1996) Systematizing Construction Project Evaluations, *Journal of Management in Engineering*, 12(1): 34–9.
Legum, M. (2002) *It Doesn't Have to be Like This – a New Economy for South Africa and the World*, Cape Town: Ampersand Press.
Lim, C.S. and Mohamed, M.Z. (1999) Criteria of Project Success: an exploratory re-examination, *International Journal of Project Management*, 17(4): 243–8.
Long, N.D., Ogunlana, S.O. and Do Thi, L.X. (2004) A Study on Project Success

Factors in Large Construction Projects in Vietnam, *Engineering, Construction & Architectural Management*, 11(6): 404–13.

Ministry of Infrastructure Development (2010) Task Force on Road Construction Costs (2010) Study on Road Construction Costs Trend. Dar es Salaam: Ministry of Infrastructure Development.

Munns, A.K. and Bjeirmi, B.F. (1996) The Role of Project Management in Achieving Project Success, *International Journal of Project Management*, 14(2): 81–7.

Naoum, S.G. (1994) Critical Analysis of Time and Cost of Management and Traditional Contracts, *Journal of Construction Engineering & Management*, 120(3): 687–705.

National Economic Empowerment and Development (2004) Nigerian National Planning Commission (2004) *Meeting Everyone's Needs: National Economic Empowerment and Development Strategy*, Abuja: Nigerian National Planning Commission.

Navarre, C. and Schaan, J.L. (1990) Design of Project Management Systems from Top Management's Perspective. *Project Management Journal*, Vol. 21, No. 2, pp. 19–27.

Neuman, L.W. (2006) *Social Research Methods: Qualitative and quantitative approaches*. Pearson Education, Boston, Mass: Pearson Education.

Nguyen, L.D., Ogunlana, S. and Lan, D.T.X. (2004) A Study on Project Success Factors in Large Projects in Vietnam. *Engineering, Construction and Architectural Management*, 11(6), 404–413.

Parfitt, M.K. and Sanvido, V.E. (1993) Checklist of Critical Success Factors for Building Projects, *Journal of Management in Engineering*, 9(3): 243–9.

Pinto, M.B. and Pinto, J.K. (1991) Determinants of Cross-Functional Cooperation in the Project Implementation Process, *Project Management Journal*, 22(2): 13–20.

Pocock, J.B., Hyun, C.T., Liu, L.Y. and Kim, M.K. (1996) Relationship Between Project Interaction and Performance Indicator, *Journal of Construction Engineering and Management*, 122(2): 165–76.

Rwelamila, P.D. (1996) Quality Management in the Public Building Construction Process, Unpublished PhD Thesis, Department of Construction Economics & Management, University of Cape Town.

Rwelamila, P.D. (2010) Impact of Procurement on Stakeholder Management. In E. Chinyio and P. Olomolaiye (eds), *Construction Stakeholder Management*, Oxford: Wiley-Blackwell.

Rwelamila, P.D. and Khumalo, R. (2002) The Significance of "*ubuntu*" in Meeting Project Goals in Africa. In R. Fellows and D.E. Seymour (eds), *Perspectives on Culture in Construction*, Rotterdam: CIB Publications.

Rwelamila, P.D. and Savile, P.W. (1994) Hybrid Value Engineering: the Challenge of Construction Project Management in the 1990s. *International Journal of Project Management*, 12(3): 157–64.

Sadeh, A., Dvir, D. and Shenhar, A. (2000) The Role of Contract Type in the Success of R&D Defence Project Under Increasing Uncertainty, *Project Management Journal*, 31(3): 14–21.

Sanvido, V., Grobler, F., Parfitt, K., Guvenis, M. and Coyle, M. (1992) Critical Success Factors for Construction Projects, *Journal of Construction Engineering & Management*, ASCE, 118(1): 94–111.

Shenhar, A.J., Levy, O. and Dvir, D. (1997) Mapping the Dimensions of Project Success, *Project Management Journal*, 28(2): 5–13.

Songer, A.D. and Molenaar, K.R. (1997) Project Characteristics for Successful Public Sector Design-Build, *Journal of Construction Engineering & Management*, 123(1): 34–40.

Sentongo, J.L. (2005) Causes of Poor Performance and Sometimes-Outright Failure Among Citizen Contractors in Botswana, Unpublished MBA Dissertation, De Montfort University, Leicester.

Ssegawa, J.K. (2002) The Effect of the Unique Characteristics of the Construction Industry on Information Management in Construction Firms, Proceedings of the First International Conference on Construction in the 21st Century (CITC 2002), 25–26 April, Miami: 505–12.

Ssegawa, J.K. (2004) Strategic Planning Systems in Small and Medium Construction Firms, Proceedings of the International Conference of CIB W107 – Globalisation and Construction: Meeting the Challenges, Reaping the Benefits, Bangkok.

Ssegawa, J.K. and Ngowi, A.B. (2008) Causes of Poor Performance of Public Construction Projects in Botswana, Faculty of Engineering & Technology, Department of Civil Engineering, Gaborone, Botswana.

Takim, R. and Akintoye, A. (2002) A Conceptual Model for Successful Construction Project Performance, Paper presented at the Second International Postgraduate Research Conference in Built and Human Environment, 11–12 April, University of Salford.

Tembo E. and Rwelamila P.D. (2008) Project Management Maturity in Public Sector Organisations: the case of Botswana. In K. Carter, S. Ogunlana and A. Kaka (eds) *Transformation Through Construction*. Proceedings of the CIB W55/W65 Joint Symposium: 15–17 November, Dubai.

The KPI Working Group (2000) KPI Report for the Minister for Construction, Department of the Environment, Transport and the Regions, London.

Todaro, M.P. (2000) *Economic Development*, 7th edn, Harlow: Addison-Wesley Longman.

Toor, S.R. and Ogunlana, S. (2010) Beyond the 'Iron Triangle': Stakeholder perception of key performance indicators (KPI) for large-scale public sector projects. *International Journal of Project Management*, Vol. 28, Issue 3, 228–236.

Torbica, Z.M. and Stroh, R.C. (2001) Customer Satisfaction in Home Building, *Journal of Construction Engineering Management*, 127(1): 82–6.

United Nations Industrial Development Organisation (UNIDO) (1985) *The Building Materials Industry: The sector in figures*. Sectoral studies series, 16(2), UNIDO, UNIDO/IS.512/ADD.1.

United Nations (2007) *The Millennium Development Goals Report*, New York: United Nations.

United Nations Development Programme (UNDP) (2005) *Human Development Report 2005 – International Cooperation at a Crossroad: Aid, trade and security in an equal world*, New York: UNDP.

Walker, D.H.T. (1995) An Investigation Into Construction Time Performance, *Construction Management & Economics*, 13(3): 263–74.

Walker, D.H.T. (1996) The Contribution of the Construction Management Team to Good Construction Time Performance – an Australian Experience, *Journal of Construction Procurement*, 2(2): 4–18.

Wateridge, J. (1995) IT Projects: A Basis for Success, *International Journal of Project Management*, 13(3): 169–72.

World Bank (2001) *Country Procurement Report*: *Indonesia:* Washington, D.C.: World Bank.

World Bank (2003) *2004 Human Development Report*, New York: World Bank.

Wuellner, W.W. (1990) Project Performance Evaluation Checklist for Consulting Engineers, *Journal of Management in Engineering*, 6(3): 270–81.

13 Project risks faced in international construction projects in developing countries

Case study of Vietnam

Florence Yean Yng Ling
National University of Singapore
and
Vivian To Phuong Hoang
Rider Levett Bucknall, Singapore

Introduction

The construction industry is subject to more risk and uncertainty than many other industries (Flanagan and Norman, 1993). International construction involves the risks common to domestic construction projects as well as uncertainties associated with international transactions (Han et al., 2005), and operating in an unfamiliar overseas environment (He, 1995). International projects usually involve elaborate financial provisions, participants with different political ideologies, and multiple owners or stakeholders (Gunhan and Arditi, 2005). Hence, foreign architectural, engineering or construction (A/E/C) firms on international construction projects face more risks than they would face in domestic projects.

International construction projects in developing countries might involve even more risks. This is because developing countries face, among other things: frequent shortages of materials, plant and equipment, and spare parts; inappropriate contract documentation and procedures; scarcity of construction professionals and skilled workers; and weak regulatory and administrative frameworks (Ofori, 1991). There is also the issue of poor project management competencies of public-sector organisations responsible for infrastructure development (Rwelamila, 2007).

Risk is multi-faceted, and risks may be loosely categorised into those generated by the project (comprising financial, cultural, design and construction related risks), and non-project risks (comprising political, economic and legal risks). Ling and Hoang (2010) examined non-project risks in developing countries. This chapter investigates project risk management undertaken by foreign A/E/C firms in developing countries, using Vietnam as a case study. The specific objectives of this chapter are: to investigate project risks faced by foreign A/E/C firms in Vietnam and the risk response activities adopted; and to develop a risk management framework to help foreign firms manage project risks in developing countries.

Vietnam ranked third among Asian nations in attracting foreign direct investments (FDI) during the period 2007–9 (Asia Business Council, 2007). Vietnam attracted US$21.48 billion of FDI in 2009, with US$7.6 billion being invested in real estate (Foreign Investment Agency, 2010). Foreign A/E/C firms are expected to be involved in many of these construction projects. However, many of these firms are not familiar with the construction environment in Vietnam and the risks that they might face. Thus, it is pertinent to study these risks and consider the ways and means by which the foreign firms can address them.

Literature review

Construction projects are usually one-off endeavours with a long duration, complicated and interdependent processes, challenging operating environments, financial intensity, and a combination of organisational and technological complexity that generates substantial risks for all parties involved (Zou et. al, 2007). When the construction project is located outside the home countries of the A/E/C firms, the risks faced might be more severe. Unlike domestic construction, international construction involves managing both mobile factors of production (for example, the relocation of experts to host countries) and location-bound support industries (such as the local labour force) (Enderwick 1993).

International construction in developing countries has additional sets of difficulties because, as many researchers have found, the construction industries in developing countries face systemic problems stemming from the adoption of inappropriate systems and procedures (Aniekwu and Okpala, 1988). In developing countries, there is a wide gap between large and small firms, and a lack of large local firms; thus, foreign firms working temporarily on large and technologically complex projects are present (Ofori, 1991). The many problems faced in these construction industries culminate in poor performance on the projects undertaken in the countries (Ofori, 2000). Developing countries also have weak economies and poor economic prospects; under-developed legal and administrative systems; and challenging business environments (Ofori, 2007). Finally, these countries have a large and growing informal sector with both informal firms and informal labour, constituting an informal construction system (Wells, 2007), which foreign A/E/C firms have to deal with.

Risk management is important, and involves setting objectives, identifying and evaluating risks, designing a comprehensive risk management programme, implementing the programme and monitoring the results (Hampton, 1993). The Construction Industry Institute (2004) identifies the stages of risk management and assessment as: early identification of hazards and opportunities; communication of risk between project participants; identification and management of uncertainty; acknowledgment of risk issues and mitigation actions; and enhancement of risk-based decision-making.

Risk response approaches have been identified as risk elimination, risk transfer, risk retention and risk reduction (Flanagan and Norman, 1993). Risk reduction is found to be the risk response approach which is most commonly applied in the construction industry, followed by risk transfer, risk retention, and risk elimination (Baker et al., 1999). Risks that cannot be eliminated, transferred or retained should be reduced by good project management practices (Kartam and Kartam, 2001). Construction delay risks can be reduced to a certain extent by proper planning and scheduling with updated project information, close monitoring and coordination with sub-contractors, marshalling of sufficient personnel and equipment on site, and provision of close supervision to ensure that the programme is being maintained (Kartam and Kartam, 2001). Ensuring that orders for materials and deliveries are made in a timely manner and planning for alternative or back-up supplies might also reduce delay risks.

Risks may be transferred by obtaining insurance coverage. When contractors pay an insurance premium, insurers accept the risks that are beyond the control of the main participants or beyond their financial capacity to bear (Orman, 1991). Some common types of insurance utilised by the construction industry are fire insurance, public liability insurance, workmen's compensation insurance, and completion risk insurance covering project delays. Transferring risks to others was found not to be an effective way to reduce construction delay risk in Kuwait (Kartam and Kartam, 2001).

When risks are retained, a contingency sum should be set aside for this. The amount of the contingency sum varies among projects, operating environments, the decision-makers' risk attitudes and the nature of the risks retained. For contractors, the size of the contingency sum would affect the bid price and might decrease the probability of winning a job in a highly competitive construction market (Kartam and Kartam, 2001).

Risk elimination should be carried out at the early stages of a project (Thompson and Perry, 1992). For example, at the proposal stage, there is opportunity for adopting more flexible plans or construction methods, or designing out the risks.

Edwards and Bowen (1998) grouped risks into: natural risks caused by weather and geological systems; and human risks comprising social, political, economic, financial, legal, health, managerial, technical and cultural risks. Risk factors may also be categorised into three main groups: internal, external and project-specific (Bing et al., 1999). On international projects, internal risks stem from the nature of the foreign A/E/C firm's internal operations, whereas external risks are due to the competitive macro-environment the firm operates in. Project-specific risks refer to developments particular to the project which impact the cost, time and quality performance on the project. In the next sections, some of the important project risks such as design, construction, financial and cultural risks are discussed.

Design risk

Design risk may be due to defective design or design changes. Defective design and unavailable or inappropriate design details could be due to incomplete design scope, incomplete or erroneous geological and geotechnical exploration, and inadequate interaction of the design with methods of construction. This risk might be due to defective or incomplete design documents, and inconsistencies and flaws among the plans and specifications (Petrov, 2006). In developing countries, design risk might also be due to errors, omissions or a lack of technical competence (Mbachu and Nkado, 2007).

Other than defective design and deficiency in drawings, designers might not issue design documents on time (El-Sayegh, 2008). Problems of defective design are complex and deep rooted and some of the causes include insufficient time for preparation of drawings, low fees for designers, insufficient project budget, and changes in clients' needs after the design was prepared (Andi and Minato, 2003). Mbachu and Nkado (2007) found that consultants' acts of omission or commission are influential and most frequently occurring factors constraining successful project delivery in South Africa.

Late amendments to the design arise when clients' requirements change due to changing needs. Such changes result in abortive work and rework, requiring the contractor to re-schedule procurement and construction activities or change methods of working and execution plans, thus affecting productivity, and usually leading to the contractor seeking compensation (Petrov, 2006).

Construction-related risk

Construction-related risks include difficulties in site access, unforeseen ground conditions, construction accidents, poor construction management, poor work quality and delays in project implementation. Construction accidents; poor quality; low productivity; technical problems; incompetence, lack, or departure, of qualified staff are common construction risks (El-Sayegh, 2008).

In developing countries such as China, incompetence of sub-contractors and suppliers is a major construction risk for contractors (Bing et al., 1999). In China, poor management, out-of-date or low technology, and poor quality of materials are also significant risks (Fang et al., 2004). In Vietnam, the construction companies use obsolete technology and do not have adequate modern equipment (Long et al., 2004b). Unexpected delays in deliveries of materials, breach of contract by sub-contractors, disputes between main and sub-contractors contribute to construction risks. Therefore, it is important to select contractors and sub-contractors carefully by evaluating their construction and management abilities (Fang et al., 2004).

In many developing countries, the construction industry has a poor accident and safety record. In Vietnam, the problem is compounded by a lack of safety management systems and poor safety awareness. In India, safety equipment is not always provided, and when it is provided, it might not be used by the workers (Ling and Hoi, 2006).

Financial risk

Financial risk on international construction projects might include delayed payments or non-receipt of payment, and financial failures (Ling and Lim, 2007). Studies in China found that lack of credit facilities is a major constraint on the construction industry (Chen, 1998; Smith et al., 2004). In Vietnam, late payment by owners for completed works is a major threat (Luu et al., 2008). Delays in making payment, irregular payments or non-payment are common causes for disputes on construction projects.

Foreign firms might face the risk of financial failures of their own firms, or of their business partners or sub-contractors. As a construction project requires substantial amounts of financing, the threat of bankruptcy can have a serious impact on the progress of the project. Hence, careful selection of consultants and contractors is important. However, it is not easy to evaluate this risk because bankruptcy of one of the firms participating in the project might not necessarily be related to the project; it could be due to other business activities of the firm (Xenidis and Angelides, 2005).

Delays and failures of contractors or sub-contractors in construction projects generally lead to higher cost of financing, loss of earnings from delayed completion, and liquidated damages being imposed. For clients, it is important to select contractors and sub-contractors carefully, paying attention to their construction and management abilities, and then monitoring their performance closely during construction (Fang et al., 2004). For contractors, incompetent sub-contractors and suppliers is a major risk (Bing et al., 1999).

Cultural risk

Cultural differences among people of different nationalities include differences in attitude, belief, rituals, motivation, perception, morality, truth and superstition (Jain, 1996). Cultural differences might also be attributable to individual characteristics such as gender, age, job experience and race (Earley and Mosakowski, 2004). Hofstede and Hofstede (2005) categorised national cultures using the following dimensions: power distance; individualism versus collectivism; masculinity versus femininity; uncertainty avoidance; and long- versus short-term orientation. On an international construction project, according to the common definition, at least one of the major players (such as the lead consultant, client or contractors) would be working outside its home country. The cross-cultural encounter and cultural difference amongst the participants are expected to contribute to conflict and increased difficulties in the management of the project (Fellows and Hancock, 1994).

The success of international construction projects hinges on the practitioners' appreciation of the host country's culture (Low and Shi, 2002). Being sensitive to the differences between the cultures of their home and host countries, even if they do not know the similarities, will help to reduce conflict and resolve differences and disputes on construction projects (Low and Shi, 2002).

Educational background, beliefs, art, morals, customs and laws also affect culture (Evans et al., 1989). Dissimilarity in language, background, perceptions and mentalities are perceived as cultural differences (Swierczek, 1994). Wong et al. (2007) found that in Hong Kong, the local project managers in multinational construction firms based in that territory and their expatriate counterparts from Western countries differ in the power relationships with their subordinates. Loosemore and Chau (2002) studied racial discrimination towards Asian operatives in the Australian construction industry and found that cultural differences are not effectively managed and Asian operatives are expected to blend into the mainstream white society.

Differences between the cultures, language, religion and custom of foreigners and local people lead to problems in communication and in working together, often in challenging conditions, over extended periods or prolonged working hours (Swierczek, 1994). Cultural differences have also been found to inhibit innovation on construction projects (Ngowi, 1997).

In developing countries, where sizeable projects often involve both foreign and local companies and professionals working together, it is critical to recognise and deal with cultural issues in a sensitive manner (Ofori, 2007). The mode of entry for many foreign firms into Vietnam is through joint ventures with local firms to take advantage of the local partners' knowledge of the local market, laws and regulations (Lai and Truong, 2005). However, the possible cultural difference between the local and foreign joint venture partners might significantly affect their approaches to problem solving (Lai and Truong, 2005).

Gap in knowledge

Operating in foreign countries is generally considered more risky than working at home. The review above shows that there is a lack of studies regarding the risks foreign A/E/C firms face when operating in developing countries. Therefore, the extent to which any of the risk management activities discussed above is applicable to developing countries like Vietnam is not known.

It is pertinent to study risk management in Vietnam because knowledge and experiences obtained from other countries might not be sufficient when working in Vietnam. One of the reasons why this country is different is that the government is controlled by the Vietnamese Communist Party (VCP). The VCP plays a primary role in all state activities, and members of the party fill key positions in all government agencies; the most trusted party cadres control all mass organisations (Cima, 1987). The party plays a leading role in mapping strategies for economic development, setting growth targets, and launching reforms (Cima, 1987). Since 1986, the government of Vietnam has tried to embrace free trade. Vietnam joined the World Trade Organization (WTO) in 2007. Its open market environment has led to an increase in FDI, creating many opportunities for foreign firms to participate in constructing its buildings and items of infrastructure. The foreign firms' lack of knowledge of the relevant policies, regulations and procedures of a centrally controlled government makes it challenging for them to decide what risk management activities would be effective in Vietnam.

Long et al. (2004b) found that large construction projects in Vietnam face the following problems: incompetent designers and contractors; poor estimation; weak change management; social and technological challenges; site-related challenges; and application of improper techniques and tools. Long et al. (2004a) grouped the critical success factors in large construction projects in Vietnam into: comfort, competence, commitment and communication issues. Luu et al. (2008) proposed a framework for evaluating the strategic performance of large contractors in Vietnam by integrating the balanced scorecard and strengths, weaknesses, opportunities and threats (SWOT) matrix. None of the respondents in all the studies outlined in this paragraph were foreigners. Hence, foreign firms' views on risk management were not studied. A knowledge gap exists in the risks faced by foreign firms when operating in Vietnam's construction industry, and the fieldwork was undertaken to identify project risks faced by such firms in Vietnam.

Research method

The qualitative research method was adopted to study foreign firms' experience in the social and institutional context. The data collection instrument is a specially designed interview questionnaire comprising open-ended questions; this enabled the interviewees to share their experiences and opinions without constrained alternatives. This study adopted the face-to-face interview technique, as this allowed for extensive discussion and immediate clarification of the questions as well as responses. Respondents were asked to base their responses to the questions on one completed project of their choice on which they had worked in Vietnam.

The population for this research was comprised entirely of foreign main contractors, sub-contractors, consultants, developers and suppliers which have completed at least one construction project in Vietnam. The sampling frame was obtained from the Department of Planning and Investment in Ho Chi Minh City, the *Yellow Pages*, the European Chamber of Commerce in Vietnam, newspaper articles and the Internet. A total of 73 firms were identified from these sources, and requests for interviews were made by email messages and telephone calls to all these firms.

Characteristics of the sample

Out of the 73 invitations, 18 experts working in foreign firms agreed to be interviewed, giving a response rate of about 25 percent. The characteristics of the interviewees and their firms are shown in Tables 13.1 and 13.2. The majority of the interviewees were from middle or upper management and had been working in the construction industry for an average of 14 years, and working on projects in Vietnam for an average of 7.5 years.

The interviewees revealed that their companies ventured into Vietnam mainly to increase sales volume, gain competitive advantage and diversify their operations geographically. The types of projects undertaken by the interviewees' firms include residential buildings, commercial buildings, industrial parks and factories. The majority of the projects were in southern Vietnam. They were procured through open tenders, and based on the design-bid-build contractual arrangement. However, the interviewees reported that direct negotiation during the tender evaluation stage was common, indicating the importance of networking and maintaining good relationships to win construction contracts in Vietnam. At the same time, it also meant that there were possibilities of lack of transparency and potential for malpractice in the project procurement process.

Results and discussion

The data collected from the interviewees are presented in Table 13.3, and depicted in Figure 13.1. The project risks identified and the relevant risk response methods are discussed below.

Design risk (Dr)

Table 13.3 shows that design risk in Vietnam comprises defective designs and design changes.

Table 13.1 Profile of interviewees and their firms (note: WOFS = wholly owned foreign subsidiary; JV = Foreign–Vietnamese joint venture)

Code	*Designation*	*Work experience (years)*	*Work experience in Vietnam (years)*	*Service provided by firm in Vietnam*	*Firm's operation in Vietnam (years)*	*Home country*	*Entry mode*
C1	Regional director	9	3	Contractor	3	Singapore	WOFS
D1	Regional director	8	3	Developer	3	France	Project JV
D2	Project manager	10	5	Developer	12	France	Project JV
D3	Deputy director	6	6	Developer	15	US	WOFS
D4	Director	26	12	Developer	15	US	WOFS
D5	Director	20	12	Developer	12	Singapore	WOFS
D6	General manager	19	8	Developer	8	Singapore	WOFS
E1	Project manager	7	3	Engineering consultancy	14	France	Equity JV
E2	Chief engineer	13	13	Engineering consultancy	14	France	Equity JV
E3	Director	17	14	Engineering consultancy	14	France	Equity JV
E4	Project manager	9	2	Engineering consultancy	14	France	Equity JV
E5	Partner	16	1	Project management	12	Singapore	WOFS
R1	Account manager	23	23	Consultancy (risk advisory)	25	Hong Kong	WOFS
Q1	Project director	20	3	Quantity surveying	16	Singapore	WOFS
Q2	Project director	21	15	Quantity surveying	17	Malaysia	WOFS
Q3	Managing director	20	2	Quantity surveying	5	Malaysia	WOFS
A1	Architect	5	5	Architecture consultancy	10	US	Branch office
A2	Architect	5	5	Architecture consultancy	15	France	WOFS

Note: WOFS = wholly owned foreign subsidiary. JV = Foreign-Vietnamese joint venture.

Table 13.2 Profile of projects

Location	*Frequency*	*Percentage*
North	1	7.1
Central	2	14.3
South	11	78.6
Project type		
General building	8	57.1
Manufacturing plant	6	42.8
Floor area (m2)		
Up to 25,000	4	28.6
25,001–50,000	7	50.0
>50,000	3	21.4
Contract sum (USD)		
Up to $10M	5	35.7
$10.01–$30M	6	42.8
>$30M	3	21.4
Year completed		
2007	3	21.4
2008	9	64.3
2009	2	14.3
Ownership of project		
Private foreign	6	42.8
Private Vietnamese	2	14.3
Private foreign – private Vietnamese JV	6	42.8
Tendering method		
Open	9	64.3
Selective	5	35.7
Contract arrangement		
Design-bid-build	13	92.9
Turnkey	1	7.1

Note: Four interviewees (D5, D6, E3 and R1) did not provide information on their projects.

Defective design (Dr1)

The interviewees gave mixed responses regarding timeliness in producing and issuing design information as well as the quality of the design information produced by Vietnamese architects. Interviewees D1 and D2 were

Table 13.3 Project risks identified and risk response techniques

Risk code	*Project risks*	*Response code*	*Risk response techniques*
	Design risk		
Dr1	Defective design	PS	Supervise the consultants' work closely.
		CN	Conduct regular design coordination meetings involving the client, consultants and contractors.
		CT	Specify in the contract who bears the cost of the rectifications due to defective design.
Dr2	Late design changes	FX	Prepare a flexible design to cater for changing requirements.
		CT	Use design-build contractual arrangement.
	Construction related risk		
Cr1	Poor worksite safety	PS	Engage or require contractors to engage safety specialist to undertake safety management on site.
		SN	Verify contractor's ability to comply with safety requirements during tender evaluation.
		PS	Remind contractors to comply with safety requirements.
		CT	Transfer risk to contractors and subcontractors.
Cr2	Natural risk	PS	Build in adequate float in the schedule.
		PS	Conduct thorough studies in the planning phase.
Cr3	Poor management capability of local partners	SN	Select contractors based on technical ability, reputation and financial soundness.
		PS	Close monitor local contractor's work.
Cr4	Poor workmanship quality	TN	Conduct construction training courses for local workers to improve the level of skills and safety awareness.
		CT	Insert contractual requirement for contractors to remove workers with improper conduct, poor work attitude or poor safety awareness.
		SN	Evaluate the competencies of the main contractors and their subcontractors before award of the main contract.
		CT	Insert contractual requirement of "no sub-letting without approval".

Risk code	*Project risks*	*Response code*	*Risk response techniques*
Cr5	Poor materials and equipment quality	LO	Use local materials as far as possible.
		SN	Select suppliers who are reliable and honourable in complying with specified quality requirements.
		RP	Appoint locals with good relationship with port authorities and who are familiar with port clearance.
Cr6	Poor quality management	CT	Ensure compliance with minimum contractual requirements.
		CT	Ensure contractors and subcontractors are aware of the quality standard specified.
		CT	Require performance bonds from contractors and subcontractors.
Cr7	Poor time management	PS	Require contractors to pay close attention to key activities on critical path and pre-empt potential problems.
		PS	Require contractors to prepare, adhere to, control and revise project schedule.
		CT	Require contractors to accelerate work when it is delayed.
	Cultural risk		
Tr1	Cross-cultural issues	TN	Train Vietnamese to respond quickly to instructions.
		TN	Learn about Vietnamese way of life and working culture.
		RP	Establish relationships with the locals.
Tr2	Miscommunication	TN	Conduct English language training courses.
		CM	Prepare contracts and correspondence in both English and Vietnamese.
		CM	Insist on English being spoken in meetings.
		CM	Issue confirmation letters after verbal discussions.
Tr3	Difficulty in posting foreign staff to Vietnam	PA	Pay a premium to foreign staff posted to Vietnam.
		LO	Employ minimum expatriate staff.
		TN	Train local staff to do the work.

Risk code	*Project risks*	*Response code*	*Risk response techniques*
	Financial risk		
Fr1	No or delayed payment	SN	Work with reputable clients with strong financial backing from their parent companies.
		SN	Select privately funded projects instead of those funded by government.
		CT	Use "pay when paid" agreements between main contractors and subcontractors.
		PS	Follow-up reminders when payment is due or delayed.
		PA	Work out partial payments schemes with clients.
		LL	Take legal action for long overdue payments.
		RP	Negotiate for win-win solutions to avoid delays to projects due to unresolved payment disputes.
Fr2	Financial failures	SN	Pre-qualify business partners by assessing financial status and project commitments.
		PA	Make prompt or advance payments to ease other parties' financial difficulties.
		LL	Replace wayward contractor or subcontractor.
		CT	Ensure contract has provisions for dealing with financial failures of the other party.

not satisfied with the quality of architectural drawings produced by the Vietnamese architects engaged by their companies. D2 felt that they produced insufficient drawings; the drawings contained inconsistent information and were not coordinated. He reported that the designs produced by his firm's Vietnamese architect "were not buildable and alternative design solutions were not explored". D1 also complained about the lack of skill and experience of his company's Vietnamese architect.

To reduce the risk of defective design, D2 had to supervise the architect's work closely. A1's firm cross-checked the drawings among the different designers, and requested experienced designers to identify inconsistencies; rectifications were then carried out. Q1 raised many questions during the tender preparation stage of the project to reduce the problems during the post-contract stage. D1 and D2 had regular coordination meetings involving the client, consultants and contractors, so that concerns and problems could be resolved before the drawings were issued to contractors.

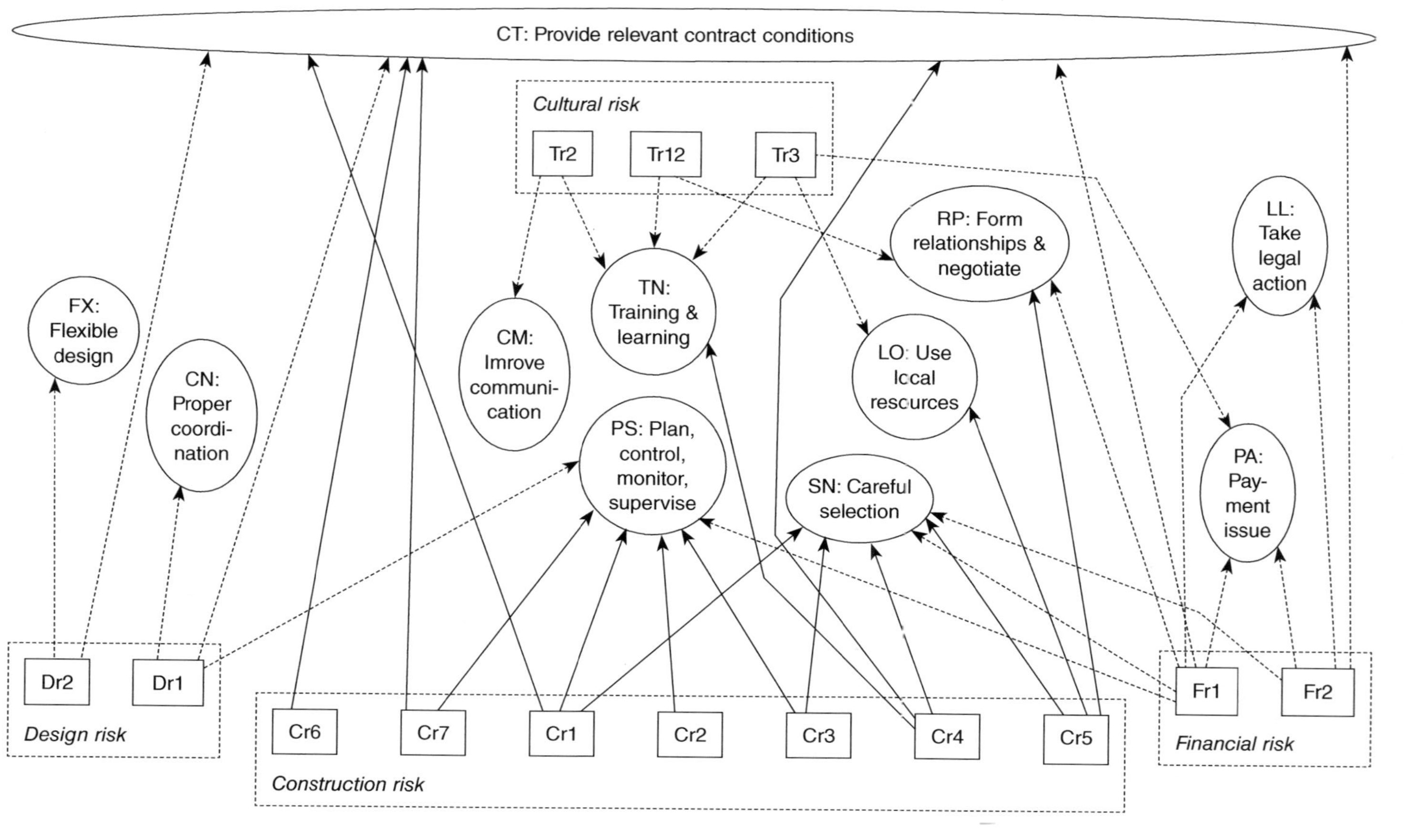

Figure 13.1 Framework to be used by foreign firms to manage project risks in Vietnam
Legend: □ Project risks; ○ risk response techniques. Note: see Table 13.3 for full descriptions.

E1 and E2 reported that there were inconsistencies and errors in the drawings produced by the consultants on their projects. E2 explained that defective design discovered before construction could be remedied by correcting and re-issuing the drawings, but if the construction work had already been carried out based on the defective design, there would be a need to decide who should pay for the rectification works. The interviewees suggested that clear contractual provisions on the party that should bear the cost of rectification works due to defective design should be provided. E2 felt that making consultants and the contractor share the cost of rectification works due to defective design might encourage both of them to be vigilant. Such a measure would also be likely to ensure that consultants' drawings are accurate and well coordinated, and that contractors would verify the drawings before commencing construction.

Late design changes (Dr2)

Interviewees shared that late design changes were prevalent on their firms' projects in Vietnam. D1, D2 and D3 admitted that they initiated design changes by changing materials and technology used on their companies' projects to reduce construction costs. However, E4's client was forced by circumstances to make a design change due to non-availability of the materials that had originally been selected. E1 and E2 reported that design changes were mainly due to their clients' vague or inaccurate requirements. D1 and E2 commented that, due to the time lag between the design and construction phases, some design changes might be necessary owing to technological changes in the equipment required for the client's operations. Other reasons for late design changes in Vietnam include the need to accommodate site conditions, fit the work to the budget and available equipment, or to meet the client's changing needs.

C1 did not encounter problems with design changes, as the company's project was a design-and-build waste treatment factory. Since the firm had full responsibility for both design and construction, it had better control over the coordination of the design and construction. In this contractual arrangement, the client was less involved in the detailed design phase and, therefore, had less opportunity to initiate design changes. Therefore, C1 recommended that the design-and-build contractual arrangement be used in order to minimise late design changes.

Construction related risk (Cr)

Table 13.3 shows that some foreign interviewees' projects were plagued by poor worksite safety, natural risk, weak management capability, poor workmanship, poor material and equipment quality, and poor time and quality management. These issues are discussed below.

Poor worksite safety (Cr1)

Many interviewees reported that poor site safety was the most severe problem they faced in Vietnam. Many local contractors do not comply with safety requirements, and safety awareness in the industry is generally inadequate. To combat this, E2's client engaged a safety specialist to ensure safe working conditions on the project site. D2 verified the contractor's ability to comply with safety requirements during the tender evaluation phase by making detailed background checks and requiring bidders to provide safety management plans. C1 issued constant reminders to the firm's workers to put on personal protection equipment such as safety boots and helmets. D3, E3 and E4 transferred the poor worksite safety risk to their contractors and sub-contractors by requiring them to engage safety specialists and implement safety measures under their contracts.

Natural risk (Cr2)

In Vietnam, adverse weather conditions and unforeseen ground conditions are natural risks that are commonly encountered in construction work. D2 and E3's projects encountered heavy rainfall for a number of days that could have resulted in delays. They avoided delays by providing adequate float in the schedules of the projects. D3 carried out careful and thorough site survey and weather studies in the planning phase in anticipation of flooding risks and heavy storms. D1 dealt with soft soil conditions by undertaking a detailed study and proper planning, and selecting the most suitable type of foundation design for this type of soil.

Poor management capability of local partners (Cr3)

D2 reported that the management capability of its local contractor was poor. The contractor was not willing to take responsibility for its tasks, and had a poor work attitude. D2 revealed that some local contractors would rather forego the retention sum, which is normally 5 percent of the contract amount in Vietnam, than rectify the defects in their completed projects during the maintenance period. The local contractor of D2 went so far as to bribe D2's project manager, which led to lax supervision, and hence to D2's interests being compromised.

Interviewees shared that the management capability of their local partners has an impact on the success of the project. They selected local contractors based on their technical ability, reputation and financial soundness. D1's main criteria for selection of local contractors were the quality of their previous projects, financial strength and current work commitments. D2 placed high importance on the contractor's ability to comply with safety requirements, followed by current work commitments, offer price and technical competence. In addition, C1 and E2 monitored the site closely so that problems could be resolved immediately.

Poor quality of workmanship (Cr4)

C1, D3 and E3 reported that, although labour is easily available in Vietnam, the workers lack job knowledge, job experience and technical ability. Several interviewees felt that many Vietnamese workers lack the necessary skills and professionalism. E4 was of the view that the workers' skills have generally improved but they are still unable to handle the more complex technologies adopted on the firm's projects.

To E1, many of the problems relating to poor workmanship are caused by contractors engaging unskilled workers, and to constant changes in personnel due to their redeployment on other ongoing projects. C1 found that low productivity and poor workmanship were due to lack of proper training. E3 suggested that there should be basic construction training courses for Vietnamese workers to improve their skill levels and safety awareness. D2 stipulated in its contract that contractors must remove workers who have improper conduct, poor work attitude or poor safety awareness.

E2 commented that many of the problems were caused by incompetent sub-contractors. E2's contractor submitted a list of skilled workers together with its bid, but after the contract was awarded, the contractor sub-let the work to sub-contractors with unqualified workers, whose quotations were low, in order to increase its profit margin.

Poor material and equipment quality (Cr5)

All the interviewees agreed that run-of-the-mill materials, plant and equipment are available in Vietnam at affordable prices. A1 and A2 shared that plant and equipment are in a poor state of maintenance. D2 commented that many of the local materials are of poor and inconsistent quality and tend to deteriorate within a period of around five years. The consequence of this is that the final constructed facility may have inconsistent quality.

D2 had difficulty finding high-quality materials in Vietnam, and, when available, such materials were expensive. Nevertheless, D2 and D3 specified local materials as much as possible to keep costs within budget. E4 reported that 80 percent of materials selected during the design phase of the project were not available in Vietnam during the construction phase. As a result, 70 percent were changed and 10 percent were imported from other Asian countries. E4 suggested that consultants should make a thorough study of locally available materials before preparing specifications and calling for tenders. To reduce the risk of using poor-quality materials, D2 and D3 selected suppliers which were reliable and had complied with specified quality requirements on previous projects.

In certain cases, special machinery and materials had to be imported. C1 reported that levies imposed on materials and equipment imported by foreign companies were the same as those imposed on those of local firms. This is in line with the WTO's "national treatment" principle of giving firms from WTO member countries the same treatment as those from the country concerned (WTO, 2005). However, Q1 reported that port clearance of goods takes a long time and there is a need to cut through the red tape. Q1's foreign contractor overcame this by appointing local companies that have good relationships with port authorities and are familiar with the procedures.

Poor quality management (Cr6)

Each of the interviewees confirmed that the quality of the constructed facility in Vietnam is not as high as that in his home country. According to A2, quality control is difficult to implement. D1 and E4 adopted the Vietnamese construction quality standard, whereas C1 specified its firm's in-house quality standard. In addition, D2 implemented good manufacturing practice standards required by the company's parent company. Q1 suggested that, if the home country's standard is expected, then international contractors should be engaged. E1, E2 and D3 did not insist that the quality of workmanship must match the standard in their home countries, although they did require the final product to comply with specifications.

C1 and D2 ensured that the contractors and sub-contractors were aware of the quality standard specified. This was done by going through every page of the specifications with the contractors and sub-contractors.

While it is not the usual practice in Vietnam, interviewees suggested that contractors be required to produce performance bonds issued by banks or insurance companies. These financial institutions would only provide performance bonds to contractors which have been assessed to be financially sound, with good track records, management capability and technical expertise for the particular project (Meng, 2002) and this is an additional layer of checks for clients. D3 required a 15 percent performance bond from its main contractor as a deterrent against delays and poor workmanship.

Poor time management (Cr7)

Delays in construction projects are costly; they increase financing costs and reduce the flow of income from the product. While liquidated damages are common provisions in construction contracts, several interviewees faced difficulties in imposing them as this was not common practice in Vietnam. C1's local partners completely ignored the implications of liquidated damages in the contract. Since it is difficult to impose liquidated damages in Vietnam, E1, E2 and C1 avoided delays by paying close attention to key activities on the critical path and addressing potential problems expeditiously. They required their contractors to compare actual work on site against the construction programmes and suggested alternative plans to catch up with the schedule or accelerate the work when necessary.

Cultural risk (Tr)

Table 13.3 shows three types of cultural risks encountered on projects. These are discussed below.

Cross-cultural issues (Tr1)

The interviewees reported that they faced cross-cultural issues because they were not familiar with Vietnamese culture. Cross-cultural differences identified include work style, values, custom, tradition, education and language. When asked to evaluate their working relationships with Vietnamese (on a scale ranging from 1 = "we quarrel almost daily", 4 = neutral, and 7 = "we got along like best friends"), the interviewees rated working relationships between foreign and local staff from 2 to 6 with a mean of 4.45. None of the interviewees chose "7", suggesting that foreigners and locals have not completely bridged the cultural gap. Several interviewees reported serious cross-cultural clashes between their foreign consultants and Vietnamese. For example, D3's German architects were not flexible in solving problems and accommodating changes even though Vietnam's economic environment was changing quickly.

Q1 felt that one of the major differences in working culture is that Vietnamese practitioners lack urgency in responding to foreigners' directions and instructions. Q3 considered this to be probably due to differences in expectations of speed of response. Q3's firm overcame this by being more tolerant and providing training to its Vietnamese personnel.

R1 suggested that foreign consultants should try to understand Vietnamese culture to ensure a smooth working relationship. Most of the interviewees overcame cultural risk by gaining a better understanding and appreciation of the Vietnamese way of life and working culture, and establishing relationships with the locals. Many interviewees suggested that there should be team-building activities to develop relationships, improve team performance and bridge the cultural gap.

E5 consulted local personnel on customary practices in Vietnam. A1 learnt about key aspects of Vietnamese culture from after-office social activities. E5 reported that weekly lunch meetings and informal discussions were effective ways to foster closer ties with locals. Nearly all interviewees reported that they had many meals with their project team members to engender team bonding.

Miscommunication (Tr2)

All the interviewees reported that English was the main language used on their projects in Vietnam. However, many Vietnamese practitioners do not have good command of English, and hence interpreters are engaged. A1 noted that differences in working language prolonged the communication process, since "We need to do everything in both English and Vietnamese." It was indicated that some nuances might be lost in translation.

D1, E1 and E5 experienced problems in their communication with their local joint-venture partners, consultants and architects. E1 highlighted that, during meetings, the locals might revert to speaking Vietnamese among themselves. The firm had to insist that English be used. After meetings and discussions, D1, D3 and E1 sent letters to relevant parties to confirm what was agreed and to specify the duties and responsibilities of each party, to ensure that the Vietnamese parties had full understanding of what was agreed.

In earlier projects, D3 experienced problems with conditions of contract that were drafted in Vietnamese only. Subsequently, D3 required contracts to be prepared in both English and Vietnamese. A1's firm provides all documents in both English and Vietnamese.

Difficulty in posting expatriate staff (Tr3)

More than half of the interviewees had difficulty convincing staff from their home countries to relocate to Vietnam to manage projects. C1 indicated that expatriate staff had to be paid a premium, as a hardship allowance, when they were posted to Vietnam. As this increases cost, E3's company maintains only a small number of expatriate staff in Vietnam, and employs and trains Vietnamese to carry out the work.

Financial risk (Fr)

The interviewees had anticipated financial risk to be severe when undertaking construction projects in Vietnam and put many mitigation measures in place to deal with the various types of financial risks. These are shown in Table 13.3 and discussed below.

No/delayed payment (Fr1)

The interviewees agreed that payments that are not received or are delayed would have a negative impact on a company's cash flow, and that a healthy cash flow is critical to all construction firms. Many interviewees noted that they did not face delays in getting payment because they had put in place strong measures to prevent this from happening.

E1, E2, E3, E4 and C1 had chosen to work only with reputable project owners from France, Singapore and the US. They carefully chose only to work with project owners with strong financial backing from their parent companies and that are known to make prompt payments upon presentation of invoices.

A2 observed that "Delayed payments are rampant – especially in government projects." E3 reported that the firm systematically chose projects that are privately funded and it did not encounter late payments on these projects. The interviewees suggested that one of the ways to reduce the risk of delayed, or no, payment is only to bid for, and undertake, projects that are initiated by private owners.

D1 and E5 encountered situations in which clients withheld payments from consultants or contractors when the output failed to meet stipulated requirements or when clients were dissatisfied with the quality and progress of the works. C1 prevented this from happening by complying with clients' requirements. C1 reduced the consequences of delayed, or no, payment risk with "pay when paid" agreements between its company and its sub-contractors.

While in developed countries firms may threaten legal action when payments are not received, E2 advised that litigation is not suitable in Vietnam as it causes alienation and affects long-term relationships. R1 also advised against using aggressive action to obtain payments from Vietnamese clients except for long-overdue payments. Several interviewees followed up on payment certificates and sent constant reminders to clients. They also worked out partial payment schemes with clients.

E3 used negotiations to resolve payment disputes and found win-win solutions to ensure that payment disputes did not lead to delays in project completion. It appeared that negotiation skills are crucial for success in managing projects in Vietnam.

Financial failures (Fr2)

While no interviewee's firm met serious financial difficulty in Vietnam, other parties involved in the projects of several interviewees failed financially (Fr2). R1 explained that the bankruptcy of a project owner would result in the project being abandoned. E4 encountered contractors which were over-committed to projects and subsequently faced cash-flow problems because they did not have sufficient working capital to run their projects.

All the interviewees adopted the risk mitigation measure of pre-qualifying their local partners to decrease the likelihood of choosing partners who might face financial trouble. Most of the interviewees suggested that pre-qualifying partners should involve examining their financial status, finding out about pre-existing financial difficulties or cash-flow problems, and assessing the financial ability of the business partners to undertake a new project over and above their current workloads. C1 pre-qualified sub-contractors by assessing their financial soundness, track records, type and scale of previous projects and quality of previous work. However, C1 still encountered the financial failure of a sub-contractor, which also incurred losses on other projects. Therefore, C1 cautioned that pre-qualification by itself is not a foolproof method.

Several sub-contractors working on the projects of C1 and E3 were on the brink of financial failure. C1 responded to this risk by making prompt and sometimes advance payments to ease their financial difficulties. In E3's case, the firm had to replace the sub-contractor because the financial assistance it provided failed to rescue the situation. E5 relied on the contract provisions to manage the financial failure of the firm's client.

Recommendations and policy implications

Specific types of risk were investigated and the risk response activities to help foreign firms manage these risks when undertaking projects in Vietnam were identified (see Table 13.3). Figure 13.1 shows the four main groups of project risks: design, construction, financial and cultural risks. The methods that were used by the respondents' firms to effectively manage the risks are also shown in Figure 13.1, and discussed below.

Figure 13.1 shows that design risk may be managed by having a flexible design. In contrast, Ling and Hoi (2006) found that the occurrence of frequent design changes in India is managed by "freezing" the design or not allowing further design changes after a certain point. Generally, in developing countries, "freezing" the design is not practical as the economy is in a state of flux and clients need to respond to the environment. Design should be based on local circumstances in the developing country, and it should also consider possible changes in the environment (Rwelamila, 2007). Therefore, the effects of the risk of design changes could be reduced by making provisions for changes in the first place. Flexibility should be built into the design so that the project scope may be expanded or reduced when market conditions change.

Poor construction safety is a perennial problem in developing countries. In India, if accidents do happen, these may be quietly settled by compensating the victim's family, and compensation amounts are usually low (Ling and Hoi, 2005). It is not good policy to leave construction safety management to market forces as contractors are likely to minimise their expenditure on safety provisions. The governments of developing countries need to do more through education. Laws could be passed to require all construction workers to attend training programmes to help inculcate safety awareness and safety culture. Governments could offer a combination of regulation and incentives. For example, if contractors manage to achieve zero accidents within a period of time, some rewards could be given, whereas contractors which flout safety rules could be fined. The fines could be used to reward the accident-free contractors. Foreign firms should pay close attention to site safety to ensure that the local firms that they employ comply with safety requirements.

The risks faced in Vietnam with respect to the level of skills and quality of workmanship, and the mitigation measures suggested by the interviewees, are similar to those proposed in other developing countries such as India (Ling and Hoi, 2006) and China (Ling et al., 2009), and those in Long et al.'s (2004b) earlier Vietnam-based study. When operating in developing countries, foreign firms should check the quality of labour and equipment, so that the finished product is up to expectations and meets clients' requirements. Governments of developing countries should equip their workmen with the necessary skills, and have testing, certification and registration or accreditation systems in place.

The problem of using incompetent sub-contractors in Vietnam is similar to that in China, where projects awarded to pre-qualified contractors are subsequently dismembered and parcelled out to sub-contractors without pre-qualification (Sha, 2004). In response to this risk, several interviewees assessed the competencies of main contractors as well as their sub-contractors during the tender evaluation phase. Foreign firms face the risk of multi-level sub-contracting when they operate in developing countries. The conditions of contract should specify that sub-letting to firms outside the approved list of sub-contractors without approval might lead to termination of the main contract. A code of practice could be introduced to require main contractors to provide the names of sub-contractors along with their bids. After contract award, the sub-contractors cannot be changed without the approval of the project owner. There should also be monitoring to ensure that further sub-contracting is not practised.

As regards the financial risk of late or no payment by clients in developing countries, foreign firms should overcome it by: working for reputable private clients; pre-qualifying local partners and contractors; requiring contractors to submit performance bonds; and making payments promptly (Xenedis and Angelides, 2005). Developing countries could also enact regulations to provide for property developers and contractors on projects above a stated value to enter into an escrow agreement. The proceeds from any pre-completion sale of units in the development should be deposited into a special escrow account that would be managed by a reputable bank. With this safeguard, foreign A/E/C firms would be less likely to face financial risks.

To minimise cultural risks, foreigners need to have a good understanding of Vietnam, prepare contracts in both English and Vietnamese, and follow up all verbal agreements with written confirmations. The firms could also organise English language courses for their Vietnamese employees and partners. It should be noted that, even when people of different cultures are speaking the same language, they think in their native language; hence, effective communication could be hampered by the need to translate from their native idioms and thought processes (Chua et al., 2003). Therefore, foreigners should be mindful of cross-cultural situations in order to work effectively with the locals (Thomas and Inkson, 2004; Ling et al., 2007). Keeping an open mind and being receptive to another person's culture is in agreement with Chevrier's (2003) study on effective cross-cultural teams in the construction industry. In managing cultural risk, the strategy for foreign A/E/C firms is not to get locals to adjust to foreigners, but the other way round. This is possible, as Wong et al. (2007) have found that expatriate managers from Western countries have undergone inter-cultural adjustment and adaptation when they operate in Hong Kong. Governments of developing countries should continue to welcome foreign A/E/C firms because local firms may benefit from doing business with foreign firms, learning from them, competing with them, collaborating with them and adopting them as role models and their performance on key criteria as benchmarks (Ofori, 2007). They should also equip local firms with the ability to adopt a strategic response to the presence of foreign firms in their midst (Ofori, 2007).

Figure 13.1 shows that several of the risk management activities are used frequently. The most often-used method is to provide relevant contract conditions. Long et al. (2004a) found that comprehensive contract documents are important in Vietnam. In addition, Bing et al. (1999) found that, in international joint ventures, contractual terms and conditions should be clearly stated, and authority and responsibilities should be clearly defined. While this is the most frequently cited risk management activity, the study also found that Vietnamese do not comply strictly with contract conditions, and it is not the norm to resort to arbitration or litigation. C1 reported that the working culture in Vietnam is based on relationships and not on contractual terms. This characteristic is also observed in other cultures in Asia. For example, Chinese regard long-term relationships, trust and friendship as very important (Ang and Ofori, 2001). Thus, it is essential to rely on good relationships and clear explanations to lessen disputes. C1 built relationships with all sub-contractors involved in the project to foster greater cooperation. This is similar to the practice in China, where long-term relationships are established through trust building (Jin and Ling, 2005).

The next two frequently cited risk management methods are careful selection of team members, and close planning, control, monitoring and supervising. Careful selection of team members ensures that competent designers, contractors, sub-contractors and suppliers are engaged. Bing et al. (1999) also found that international joint-venture partners should be selected based on credit-worthiness, financial soundness and strong connection to the host government. Luu et al. (2008) also found that it is important to carefully evaluate customers to avoid projects that do not have access to funding in Vietnam.

The next important risk management activity is the basic project management function of “plan, monitor and control”. Luu et al. (2008) recommended that the construction management process at sites should be improved. Planning should include setting clear objectives and scope (Long et al., 2004a) and having alternative methods as a stand-by (Kartam and Kartam, 2001). Monitoring should include close coordination with sub-contractors and close supervision of abortive works (Kartam and Kartam, 2001) and frequent progress meetings (Long et al., 2004a). The controls needed include cost control and quality control (Luu et al., 2008).

Limitations

The main limitation of this chapter is the relatively small sample size of only 18 experts. However, these 18 experts included a good distribution of different parties in a project team. They are also experienced and shared important insights into the risks faced on their projects. They also provided examples of practical risk mitigation measures to be taken in Vietnam. Another limitation of the study is that it focused on project risks. The other risks that might be encountered by foreign companies in developing countries such as political, economic and legal risks, are not discussed here. Furthermore, it is acknowledged that there is a strong interplay among these risks.

Conclusion

This chapter investigated project risks faced by foreign firms when undertaking construction projects in developing countries, focusing on Vietnam. Interviews were conducted with 18 foreign experts who are experienced in managing construction projects in Vietnam. Four categories of project risks – design, construction, financial and cultural risks – were investigated. Risk response activities were identified, and foreign firms may adopt them to lessen the impact of project risks that they may face in developing countries. Three main actions to take are: select project, clients, consultants, contractors and suppliers carefully; provide relevant and comprehensive contract conditions to address risks that may be faced; and undertake basic project management functions of planning, monitoring and controlling.

The importance of a well-drafted contract document is often taken for granted in developed countries. The study found that Vietnamese may not pay much attention to the written contract, and it is not the custom to take disputes to arbitration or litigation. Instead, the working culture is based on relationships. The implication of this is that the contract conditions should be carefully explained to the Vietnamese participants in the project before the contract is entered into, and steps should be taken to foster trust and cooperation.

In some developing countries, project management is still in its infancy, and hence this basic function is not always carried out in a competent manner. Foreign consultants and contractors should work closely with the locals to ensure that the basic project management functions are carried out. There is also merit in providing training in project management to local professionals in the developing countries.

Acknowledgement

Ms Jina Boo's and Ms Tan Bee Lian's assistance in some of the data collection and analysis is acknowledged with thanks.

References

Andi, A. and Minato, T. (2003) Design documents quality in the Japanese construction industry: factors influencing and impacts on construction process, *International Journal of Project Management*, 21(7), 537–46.

Ang, Y.-K. and Ofori, G. (2001) Chinese culture and successful implementation of partnering in Singapore's construction industry, *Construction Management and Economics*, 19, 619–32.

Aniekwu, A.N. and Okpala, D.C. (1988) The effect of systemic factors on contract services in Nigeria, *Construction Management and Economics*, 6, 171–82.

Asia Business Council (2007) *Chief Executive Perspectives: 2007–Results from the 2007 Annual Member Survey),* 12. Retrieved on 7 Oct 2008 from http://www.asiabusinesscouncil.org/docs/Council2007Survey.pdf/

Baker, S., Ponniah, D. and Smith, S. (1999) Risk response techniques employed currently for major projects, *Construction Management and Economics*, 17, 205–13.

Bing, L., Tiong, R.L.K., Wong, W.F. and Chew, D. (1999) Risk management in international construction joint ventures, *Journal of Construction Engineering and Management*, 125(4), 277–84.

Chen, J.J. (1998) The characteristics and current status of China's construction industry, *Construction Management and Economics*, 16, 711–19.

Chevrier, S. (2003) Cross-cultural management in multinational project groups, *Journal of World Business*, 38(2), 141–9.

Chua, D.K.H., Wang, Y. and Tan, W.T. (2003) Impacts of obstacles in East Asian cross border construction, *Journal of Construction Engineering and Management*, 129(2), 131–41.

Cima, R.J. (1987) *Vietnam: A Country Study*. Washington: GPO for the Library of Congress. Retrieved on 18 June 2009 from http://countrystudies.us/vietnam/

Construction Industry Institute (2004) *Risk assessment of international projects: a management approach*. Austin: University of Texas.

Earley, P.C. and Mosakowski, E. (2004) Toward cultural intelligence: turning cultural differences into a workplace advantage, *Academy of Management Executive*, 18(3), 151–7.

Edwards, P.J. and Bowen, P.A. (1998) Risk and risk management in construction: A review and future directions for research, *Engineering, Construction and Architectural Management*, 5(4), 339–49.

El-Sayegh, S.M. (2008) Risk assessment and allocation in the UAE construction industry, *International Journal of Project Management*, 26(4), 431–8.

Enderwick, P. (1993) Multinational contracting, in K.P. Suavant and P. Mallampally (eds), *Transnational Corporations in Services*. New York: Routledge; 186–203.

Evans, W.A., Hau, K.C. and Sculli, D. (1989) A cross-cultural comparison of managerial styles, *Journal of Management Development*, 15(3/4), 28–32.

Fang, D., Li, M., Fong, S. and Shen, L. (2004) Risks in Chinese construction market-contractors' perspective, *Journal of Construction Engineering and Management*, 130(6), 853–64.

Fellows, R.F. and Hancock, R. (1994) Conflict resulting from cultural differentiation: An investigation of the new engineering contract, in *Proceedings on Construction Conflict: Management and Resolution*), 259–67. Rotterdam: CIB.

Flanagan, R. and Norman, G. (1993) *Risk Management and Construction*. Oxford: Blackwell Scientific.

Foreign Investment Agency (2010) Foreign direct investment attraction in 2009. Hanoi: Ministry of Planning and Investment. Retrieved on 16 June 10 from http://fia.mpi.gov.vn/Default.aspx?ctl=Article2&TabID=4&mID=52&aID=919/

Gunhan, S. and Arditi, D. (2005) Factors affecting international Construction, *Journal of Construction Engineering and Management*, 131(3), 273–82.

Hampton, J.J. (1993) *Essentials of risk management and insurance*. New York: AMACOM.

Han, S.H., Diekmann, J.E. and Ock, J.H. (2005) Contractor's risk attitudes in the selection of international construction projects, *Journal of Construction Engineering and Management*, 131(3), 283–92.

He, Z. (1995) Risk management for overseas construction projects. *International Journal of Project Management*, 13(4), 231–7.

Hofstede, G. and Hofstede, G.J. (2005) *Culture and organizations: Software of the mind* (2nd edn). New York: McGraw-Hill.

Jain, S.C. (1996) *International marketing management*. Cincinnati: South Western College Press.

Jin, X.H. and Ling, Y.Y. (2005) Model for fostering trust and building relationships in China's construction industry, *Journal of Construction Engineering and Management*, 131(11), 1224–32.

Kartam, N.A. and Kartam, S. (2001) Risk and its management in the Kuwaiti construction industry: a contractor's perspective, *International Journal of Project Management*, 19(6), 325–35.

Lai, X.T. and Truong, Q. (2005) Relational capital and performance of international joint ventures in Vietnam, *Asia Pacific Business Review*, 11(3), 389–410.

Ling, Y.Y., Ang, A.M.H. and Lim, S.S.Y. (2007) Encounters between foreigners and Chinese: perception and management of cultural differences, *Engineering, Construction and Architectural Management*, 14(6), 501–18.

Ling, Y.Y. and Hoang, V.T.P. (2010) Political, economic, and legal risks faced in international projects: case study of Vietnam, *Journal of Professional Issues in Engineering Education and Practice*, 136(3), 156–64.

Ling, Y.Y. and Hoi, L. (2006) Risk faced by Singaporean firms when undertaking construction projects in India, *International Journal of Project Management*, 24(3), 261–70.

Ling, Y.Y. and Lim, H.K. (2007) Foreign firms' financial and economic risk in China, *Engineering, Construction and Architectural Management*, 14(4), 346–62.

Ling, Y.Y., Low, S.P., Wang, S.Q. and Lim, H.H. (2009) Key project management practices affecting Singaporean firms' project performance in China, *International Journal of Project Management*, 27,(1). 59–71.

Long N.D., Ogunlana, S.O. and Lan, D.T.X. (2004a) A study on project success factors in large construction projects in Vietnam, *Engineering, Construction and Architectural Management*, 11(6), 404–13.

Long, N.D., Ogunlana, S., Quang, T. and Lam, K.C. (2004b) Large construction projects in developing countries: a case study from Vietnam, *International Journal of Project Management*, 22(7), 553–61.

Loosemore, M. and Chau, D.W. (2002) Racial discrimination towards Asian operatives in the Australian construction industry, *Construction Management and Economics*, 20, 91–102.

Low, S.P. and Shi, Y.Q. (2002) An exploratory study of Hofstede's cross-cultural dimensions in construction projects, *Management Decision*, 40(1), 7–16.

Luu, T.V., Kim, S.Y., Cao, H.L. and Park, Y.M. (2008) Performance measurement of construction firms in developing countries, *Construction Management and Economics*, 26, 373–86.

Mbachu, J. and Nkado, R. (2007) Factors constraining successful building project implementation in South Africa, *Construction Management and Economics*, 25, 39–54.

Meng, X. (2002) Guarantees for contractors' performance and owners' payment in China, *Journal of Construction Engineering and Management*, 128(3), 232–7.

Ministry of Foreign Affairs, Vietnam (2008a) Vietnam attracts over 57 billion USD in nine months. Retrieved on 7 October 2008 from http://www.mofa.gov.vn/en/nr040807104143/nr040807105039/ns080926083658/

Ngowi, A.B. (1997) Impact of culture on construction procurement, *Journal of Construction Procurement*, 3(1), 3–15.

Ofori, G. (1991) Programmes for improving the performance of contracting firms in developing countries: A review of approaches and appropriate options, *Construction Management and Economics*, 9, 19–38.

Ofori, G. (2000) Challenges of construction industries in developing countries: lessons from various countries, in A.B. Ngowi and J. Ssegawa (eds) *Challenges Facing the Construction Industry in Developing Countries*, Proceedings of 2nd International Conference of CIB Task Group 29, 15–17 November. Gaborone: CIB; 1–11.

Ofori, G. (2007) Construction in developing countries, *Construction Management and Economics*, 25(1), 1–6.

Orman, G.A.E. (1991) New applications of risk analysis in project insurances, *International Journal of Project Management*, 21(7), 537–46.

Petrov, M. (2006) Managing project risk with proper scheduling, *Nielsen-Wurster Communique, 1.5*. Retrieved on 30 July 2008 from http://www.nielsen-wurster.com/Email_Announcements/NW_Communique/NW_Communique_2006_NOV.html/

Rwelamila, P.M.D. (2007) Project management competencies in public sector infrastructure organisations, *Construction Management and Economics*, 25, 55–66.

Sha, K. (2004) Construction business system in China: an institutional transformational perspective, *Building Research and Information*, 32(6), 529–37.

Smith, J., Zheng, B., Love, P.E.D. and Edwards, D.J. (2004) Procurement of construction facilities in Guangdong Province, China: factors influencing the choice of procurement method, *Facilities*, 22(5/6), 141–8.

Swierczek, F.W. (1994) Culture and conflict in joint ventures in Asia, *International Journal of Project Management*, 12(1), 39–47.

Thomas, D.C. and Inkson, K. (2004) Cultivating your cultural intelligence, *Security Management*, 48(8), 30–3.

Thompson, P. and Perry, J. (1992) *Engineering construction risks: a guide to project risk analysis and risk management*. London: Thomas Telford.

Walker, D.H.T. (1998) The contribution of the client representative to the creation and maintenance of a good project inter-team relationships, *Engineering, Construction and Architectural Management*, 5(1), 51–7.

Wells, J. (2007) Informality in the construction sector in developing countries, *Construction Management and Economics*, 25, 87–93.

Wong, J., Wong, P.N.K. and Li, H. (2007) An investigation of leadership styles and relationship cultures of Chinese and expatriate managers in multinational construction companies in Hong Kong, *Construction Management and Economics*, 25, 95–106.

WTO (World Trade Organization) (2005) *Understanding the* WTO. Geneva: WTO. Retrieved on 13 January 2008 from http://www.wto.org/english/thewto_e/whatis_e/tif_e/understanding_e.pdf/

Xenidis, Y. and Angelides, D. (2005) The financial risk in build-operate-transfer projects, *Construction Management and Economics*, 23, 431–41.

Zou, P.X.W., Zhang, G. and Wang, J. (2007) Understanding the key risks in construction projects in China, *International Journal of Project Management*, 25(6), 601–14.

14 A protocol for lean construction in developing countries

Subashini Suresh and Abubakar M. Bashir
University of Wolverhampton, United Kingdom
and
Paul O. Olomolaiye
University of the West of England, United Kingdom

Introduction

The construction sector makes a significant contribution to the gross domestic product (GDP) of both developed and developing countries as it provides the basis for socio-economic development (Ofori, 2007; Kheni et al., 2008). While construction contributes up to 5 percent of GDP in some developing countries (Ahadzie et al., 2009), this figure is up to 8 percent in developed countries such as the UK (Hughes and Ferrett, 2008). The construction industry also provides the nations with essential facilities and physical structures and assets that are necessary for national development. The industry is strongly linked to other sectors of the economy such as manufacturing, wholesale and retail trade, finance and insurance (Callaghan, 2006). Lessons can be learnt from other industries to improve the contribution that construction makes to the GDP.

The description of nations as "developing" and "developed" countries is based on the World Bank's classification determined by the level of per capita national income of that country. According to World Bank (2010), low-income countries have a per capita income of US$975 or less while middle-income countries have a per capita income from US$976 to US$11,905 and high-income countries have a per capita income of US$11,906 or more. The low-income and middle-income countries are collectively referred to as developing countries. However, there are differences among the developing countries in terms of the GDP and many other considerations. For example, while Malta has a GDP (purchasing power parity) of US$9.8 billion, Nigeria and China have GDP of US$357.2 billion and US$8,789 billion respectively (CIA, 2010a). The differences demonstrate the high level of dissimilarity among the countries in terms of socio-economic development and future prospects. Most developing countries are affected by broad factors such as low living standards, inadequate education and training of workers, low level of productivity, high population growth rate, large-scale unemployment, and market imperfections and weaknesses (Lewis, 2006). Other challenges are high levels of poverty, low levels of skills, high urbanisation rates, inadequate institutions and social inequality (Ofori, 2007; du Plessis, 2007).

Several efforts have been made to improve the performance of the construction industries in many developing countries. However, studies by authors such as Dulaimi and Tanamas (2001); Ofori (2005, 2007); Polat and Arditi (2005); Kheni et al. (2008) and Luu et al. (2008) have shown that the construction industries continue to face problems such as: poor cost, time and quality performance; contribution to environmental stress; poor health and safety records; poor professionalism and lack of adherence to statutes; and low productivity, owing to factors such as poor coordination and materials shortages. A number of solutions have been proposed to address these problems. However, these tend to be mostly concerned with the cost of building and completion periods rather than the quality of work and productivity of the workers.

In developed countries, efforts have also been made over the decades to improve quality, time management, procurement systems, health and safety, risk management, sustainable construction practices, and overall client satisfaction. Many of these initiatives were based on the reports of various task forces and committees, many of which drew attention to the merits of lean principles. For example, in the UK, the Egan (1998) report, "Rethinking Construction", recommended the adoption of lean production management principles in the construction sector to eliminate waste and improve productivity and quality of services. Studies have shown that the adoption of lean construction has improved financial performance, provided better products and service to clients, improved workflow, minimised accidents, and promoted safety on construction sites (Mitropoulos et al., 2007; Mossman, 2009; Nahmens and Ikuma, 2009; Thanwadee, 2009).

Ehrenfeld (2008) defined sustainability as a continuous improvement process that involves managing processes in such a way that the environment will continue to support future activities as it does at present. It is suggested that the adoption of sustainable approaches in construction activities should involve rethinking and restructuring the pre-construction, construction and post-construction processes in a way that will improve the economy, protect the environment and improve social responsibility (Shelbourn et al., 2006; Klotz et al., 2007; Rogers et al., 2008) through improved environmental quality, energy efficiency and enhanced health and safety (Lapinski et al., 2006). The principles of lean construction focus on creating a sustainable change by stressing efficient, waste-free and safe flow; storage and handling of materials to minimise cost, energy and resource consumption; and providing value for clients and end users (Mossman, 2009). Hence, lean construction and sustainability share a common goal of eliminating waste in construction activities.

The aim of this chapter is to present a tool that will facilitate the implementation of lean construction strategies in developing countries through the application of a lean protocol developed in the study. The first part of the chapter presents the concept of lean construction. It discusses the origin of lean construction, its principles and tools, and barriers to its implementation. The second part discusses the practice of lean construction in the developed and developing countries. The third part of the chapter discusses case studies of four projects in Nigeria. The research studied the level of leanness of the design and construction process based on interviews conducted among the stakeholders. The penultimate part of the chapter discusses the lean protocol. The final part of the chapter presents recommendations relating to the implementation of lean construction.

Lean construction: an overview

The concept of "lean" was developed by Toyota, a car manufacturing company, under the leadership of Engineer Taichi Ohno in the 1950s as a process to eliminate waste (Womack et al., 1990; Womack and Jones, 1996; Howell, 1999). The company made every effort to work towards 100 percent value-added work with the lowest amount of waste (Abdelhamid, 2004). Ohno considered waste to be any human activity that consumes resources but creates no value. Such an activity could be mistakes requiring rectification, production of unwanted items, unnecessary processing steps, unnecessary movement of workers and goods, waiting for an activity, or goods and services which do not meet the customer requirements (Monden, 1998). Ohno also considered failure to meet the unique requirements of a customer at the first time of doing the work as a waste. Working towards attaining zero waste, he focused on improving the delivery system and designed a production system that would deliver the required product right away without intermediate inventories.

Koskela (1992) discussed the idea of understanding construction as a production process and proposed that the construction industry should consider adopting lean production system to develop the best production process possible in the industry. This led to what is today known as "lean construction".

There have been various definitions of lean construction (Koskela et al., 2002; Pasquire and Connolly, 2002; Abdelhamid and Salem, 2005; Salem and Zimmer, 2004) but the central theme is that lean construction is a way of designing production systems to minimise wastage of materials, time and effort in order to generate the maximum possible amount of value for the client. This is achieved by minimising activities that do not add value to create more time for those that add value to the delivery. Koskela (1992) classified construction delivery processes into conversion and flow activities. While conversion activities produce the output, flow activities are those that support and bind the conversion activities together (Farrar et al., 2004; Senaratne and Wijesiri, 2008). However, both activities consume resources. Lean construction considers conversion processes as value adding activities that need to be made efficient and be continuously improved, whereas flow activities should be minimised to save resources.

Lean construction maintains continuous pursuit of improvement throughout the design, construction, operation and maintenance stages of a construction project and its product to satisfy the client's requirements. It also makes the process more effective, efficient and profitable. In addition, it brings effective value and risk management into construction companies and challenges the belief that cost, time and quality cannot be concurrently pursued (Dulaimi and Tanamas, 2001).

Lean construction focuses on eliminating waste (or *Muda*). Waste exists in different forms. Koskela (1992) identified design errors, defects, variations, rework, cost due to poor safety and excess material consumption as the main causes of waste in construction projects. Alarcon (1995) related construction waste to methods, materials, labour, time, operations, and equipment. Abdelhamid and Everett (2002) considered occupational accidents as waste and non-value-adding events in production systems. In addition to these are delays and wastes resulting from poor organisation by project managers (Song et al., 2008).

Waste can be viewed in two facets. The first one could be non-value-adding waste that, for some reason, is deemed necessary, and usually these forms of waste cannot be eliminated immediately. The second group of waste comprises activities that are non-value-adding and are not necessary; these can be the first targets in any exercise to eliminate waste.

The five principles of lean thinking are: value identification; value stream mapping; value stream flow; achieving customer pull; and striving for perfection and continuous improvement (Womack and Jones, 1996; Mascitelli, 2002; Koskela, 2004). These principles are drivers of continuous improvement and the benefits of lean construction can only be achieved through their holistic implementation (Dulaimi and Tanamas, 2001). To support such holistic implementation, Koskela (1992) identified other principles such as: reducing variability; reducing cycle times; simplicity; benchmarking; increasing output flexibility; and increasing process transparency.

Farrar et al. (2004) considered workflow reliability to be a core principle of lean construction. Furthermore, according to Salem and Zimmer (2004), the principles include waste elimination; focus on client's needs; workplace standardisation; and continuous improvement. However, Seneratne and Wijesiri (2008) suggest that the elimination of waste, flow obstacles and other non-value-adding activities are core principles of lean construction. Based on these principles, construction researchers and professionals have developed and successfully applied several lean construction tools to aid implementation in both simple and complex construction projects.

Lean construction tools

The implementation of lean construction involves developing tools that conform to lean principles, and applying them in project delivery. The fundamental differences between the manufacturing and construction processes in terms of the nature of the operations, planning, and task execution make it difficult for lean production tools to be directly implemented in construction activities. A new set of tools is being developed to suit construction processes. According to Salem et al. (2005), the tools applied in implementing these principles include: last planner system; increased visualisation; daily huddle meetings; first run studies; the 5C/5S; and fail safe for quality (poka-yoke). In a review of the literature, eight key tools used in lean construction projects were identified. These are shown in Table 14.1. A brief description, and an outline of the benefits, of the lean construction tools are presented in the table. It is important to note that these are not the only tools that can be used, and there is also no requirement to use all of them, but the emphasis should always be on using the correct tool at the correct time.

The implementation of lean construction principles and tools has also been suggested as a way of minimising accidents and promoting safety on construction sites (Mitropoulos et al., 2007; Thandawee, 2009; Nahmens and Ikuma, 2009). Lean construction contributes to efforts to realise the goals of sustainable construction by eliminating material waste and promoting health and safety in construction activities (Bae and Kim, 2008; Nahmens, 2009) from the environmental, social and economic perspective. Companies must take advantage of research and technological developments to continually improve efficiency and quality of products and services so as to withstand existing and future market competition (Paton and James, 2008). A change in approach is necessary if construction companies are to be able to continuously meet clients' needs and respond to the global, social and environmental challenges.

Despite the significant benefits that companies can gain by adopting lean principles and their associated tools, the concept does not seem to be generally applied among construction organisations. In the next section, the barriers to the implementation of lean construction are discussed.

Barriers to lean construction practice

Several studies (see, for example, Alarcon et al., 2002; Mossman, 2009; Bashir et al., 2010) have been undertaken in various developed and

Table 14.1 Lean tools

Tools	*Description*	*Benefits*
Last planner system (LPS)	Developed to minimise waste in a system by eliminating barriers to flow through assignment-level planning, control and look-ahead scheduling. Relies on Should-Can-Will analysis to develop a Weekly Work Plan.	Production assignments (works) are planned based on the prevailing site conditions. Empowerment of workers. It provides valuable information for risk management and effective decision-making.
Increased visualisation	Involves effective communication of vital information to the workforce using different signs and labels around the construction site. Use of charts, displayed schedules, painted designated inventory and tool locations.	Enhances communication among project participants. Improves safety.
Offsite fabrication	Simplifies the assembly process through industrialisation, modularisation and standardisation of different elements and units.	Reduces the number of site operations and production steps leading to a for reduction in chances for injuries to occur.
The 5S/5C (House-keeping)	Developed from five Japanese words – Seiso, Seiton, Seiri, Seiketsu and Shitsuke, meaning Clean, Configure, Clear Out, Conformity and Custom. The 5S/5C process (House-keeping or Visual Work Place) makes the site conducive to the flow of value-adding activities by maintaining everything in its right place.	Makes orderliness and standardisation of operations the norm in an organisation. Improves safety by minimising accidents. Reduces time wastage.
Kaizen	Involves continuous improvement activities across the whole organisation. A planned, systematic and innovative approach to reduce waste and inefficiency.	Standards improvement. Educates employees. Improves communications on site.
Poka-yoke (failsafe or error-proofing)	Pokayoke is a Japanese word for failsafe devices or error-proofing. Involves all the measures taken to minimise or prevent defects from occurring. Poka-yoke activities include visual inspection, risk assessment and analysis and any other action that prevents bad outcomes.	Helps to reduce variability in construction. Helps to achieve reliability and maintain improvements.
Daily huddle meetings	A brief daily start-up meeting is conducted to collect reports on the state of the work since the previous meeting.	Ensures a rapid response to problems through continuous open communication and empowerment of workers. Involves workers in problem solving.

Tools	*Description*	*Benefits*
First run studies	Involves studying the assignment, reviewing the work methods, and identifying and reorganising the different functions involved in executing the assignment. The best and simplest approach is illustrated to the workers using video files, pictures, or graphical representations.	Used to plan out and improve the results of crucial assignments. Enables the project manager to redesign critical activities that could otherwise expose workers to high risks and hazards. Reduces interruption and prevents errors.

Table 14.2 Barriers to lean construction practice

Classification	*Barriers*	*References*
Management issues	Delay in decision-making, lack of top management support and commitment, poor project definition, delay in materials delivery, lack of equipment, scarcity of materials, lack of time for innovation, unsuitable organisational structure, weak administration, lack of supply chain integration, poor communication, use of substandard components, lack of steady work engagement, long implementation period, inadequate pre-planning, poor procurement selection strategies, poor planning, inadequate resources, lack of client and supplier involvement, lack of customer focus and absence of long-term planning.	Common et al. (2000); Alarcon et al. (2002); Forbes and Ahmed (2004); Alinaitwe (2009); Abdullah et al. (2009); Mossman (2009)
Educational issues	Lack of understanding, lack of technical skills, high level of illiteracy, lack of training, lack of holistic implementation, inadequate knowledge, lack of project team skills, inadequate exposure to requirements for lean implementation, lack of awareness programmes, difficulty in understanding concepts and lack of information sharing.	Common et al., (2000); Alarcon et al., (2002); Cua et al., (2001); Castka et al., (2004); Alinaitwe (2009); AlSehaimi et al., (2009) ; Abdullah et al., (2009); Mossman (2009)
Financial issues	Corruption, inadequate project funding, inflation, implementation cost, poor salaries of professionals, lack of incentives and motivation, risk aversion.	Common et al., (2000); Olatunji (2008); Mossman (2009)

Classification	Barriers	References
Governmental issues	Government bureaucracy, inconsistency in policies, lack of social amenities and infrastructure, scarcity of materials, unsteady price of commodities.	Olatunji (2008); Alinaitwe (2009)
Technical issues	Lack of buildable designs, incomplete designs, poor performance measurement strategies, fragmented nature of industry, lack of agreed implementation methodology, lack of prefabrication, uncertainty in supply chain, lack of constructability of design, inaccurate and incomplete designs.	Alinaitwe (2009); Mossman (2009)
Human attitudinal issues	Lack of transparency, cultural change, lack of team spirit, lack of self-criticism, lack of teamwork, lack of cooperation, poor house-keeping, poor leadership, leadership conflict, poor understanding of client's brief, misconceptions about lean practice, over enthusiasm, lean is seen as being too complex and feared.	Common et al. (2000); Alarcon et al. (2002); Cua et al. (2001); Castka et al. (2004); Forbes and Ahmed (2004); Alinaitwe (2009)

developing countries to investigate factors that could affect the successful implementation of lean construction. The barriers that have been identified may be grouped into six categories. They are: financial, governmental, educational, human attitudinal, technical and managerial, as shown in Table 14.2. For the efforts to bring about change to be effective, the implementation of lean construction should be well managed and should include specific steps to overcome the barriers highlighted in Table 14.2. The efforts to realise sustainable change also require building trust and establishing a new culture of constant learning, improvement and perfection among employers and employees. A lean construction expert can be consulted to create awareness among clients, contractors, subcontractors, suppliers and consultants on the aims, objectives, goals and benefits of lean construction and its advantages over traditional management approach. The extent and mode of applying the lean protocol should be discussed openly in order to clear misconceptions and misunderstandings.

Lean construction practice

Significant success has been realised in the implementation of lean construction on a range of residential, commercial and institutional projects in various developed and developing countries. In the UK, The Construction Lean Improvement Programme (CLIP) was established in 2003 to support the construction industry in implementing lean construction to improve client satisfaction, efficiency, profits and productivity, by providing guidance to contractors and consultants in the application of lean techniques in the delivery of construction projects.

Table 14.3 presents an outline of studies by various researchers. The outline identifies the lean construction tools, techniques and strategies used to improve project performance in terms of delivering value to the client and eliminating non-value adding activities using the minimum resources. It is evident from the table that lean tools have been applied in some developing countries for more than a decade. The studies in the table identified benefits from the application of lean such as reduction in project duration, improvement in productivity and improved project planning and control. Some of the barriers found in the studies include resistance to cultural change, delay in the delivery of materials, inadequate knowledge of lean concepts, and delays in decision making.

The practice of lean construction has not yet been launched in most developing countries. However, the construction industries in some developing countries have adopted certain methods to minimise and eliminate wastage of resources and maximise value for the client. This is discussed in the next section.

Case studies from Nigeria

The case studies investigate the leanness of the methods used by the client, architect, contractors and sub-contractors in different trades to minimise the wastage of resources such as time, materials and money in the construction of residential and commercial projects owned by private clients in Nigeria.

Nigeria has a population of over 130 million people from more than 250 different ethnic groups (CIA, 2010b). The country covers a land area of 923,768 square kilometres. Although the construction industry accounts for over 50 percent of the national budget, it contributes only 1.72 percent to the nation's GDP (Oladapo, 2007). The Nigerian construction industry is dominated by small- and medium-sized local contractors, mostly engaged in residential projects for private clients (Bashir, 2009). This is the typical situation in developing countries: private individual clients constitute the majority and their projects are mainly residential developments. Although a general picture is depicted in this research, it should be stressed that the exact situation in each country varies. The differences lie in the nature of the construction industries, the problems they face, as well as ways of addressing them.

The research studied the impact of the waste minimising and eliminating methods based on a series of interviews on the application of lean, including the methods used to minimise wastage of resources in the project delivery processes. This enabled the collection of information relating to the respondents' personal experience, understanding and description of the issues in an open and flexible manner.

Table 14.3 Application of lean construction in developing and developed countries

Countries	*Saudi Arabia*	*Brazil*	*Korea*	*Ecuador*	*United Kingdom*	*United States of America*
Project type	Institutional	Commercial	Institutional (tunnelling: heavy civil engineering)	Residential	Residential	Residential
Project nature	a) Faculty of Business and Administrative Sciences b) Classrooms and laboratories	McDonalds restaurant	Seoul and Busan subway construction	102 one-family units	Houses	Industrialised houses
Client	Government	Private	Government	Private	Private	–
Lean tools/ techniques applied	Last planner system	Last planner system	Last planner system	Last planner system	Process mapping, visual management, collaborative planning, workforce training.	Field coordination, lean design management, material and information flow management, Prefabrication.
Benefits	Improved predictability, improved planning and control, improved productivity, reduced uncertainty and improved learning process.	Reduction in project duration from 90 to 83 days, reduction in rework, better resource allocation, and improved site organization.	Improved workflow reliability, workers' empowerment and job satisfaction.	Improved planning and control, satisfaction of workers, higher productivity, empowerment of workers, high level of commitment, and reapplication in future projects.	Reduction in project duration, improved clients' satisfaction, continuously improved quality, cost and performance, improved staff morale and saved cost to both client and contractor, innovative ideas, higher profits and productivity, developed strategies and working culture that can be adopted on future projects.	Minimised site operations, reduced time, simplified work processes and management, satisfaction of workers, enhanced efficiency in production, improved communication and resource optimization.
References	AlSehaimi et al. (2009)	Junior et al. (1998)	Kim and Jang (2005)	Fiallo and Revelo (2002)	BRE (2009)	Chen et al. (2004)

The first section of the interview was intended to obtain information on the characteristics of the participant, including work experience, number of employees, level of involvement and role in the project delivery processes. The second section of the interview focused on the different methods used by the clients, architects, contractors and sub-contractors involved in the project to minimise or eliminate wastage of resources such as materials, time, labour and money. That part also investigated the effectiveness and efficiency of these methods by determining their impact on the quality of the product. The last part of the interviews was on the willingness of the participants to play a role in applying lean construction strategies that will deliver value and eliminate wastage of resources without compromising the quality of work, and the issues they consider as barriers to the successful implementation of lean construction.

Research results

This section discusses the findings from the interviews. The major findings are presented in relation to the aim and objectives of the chapter.

The case studies comprise three residential (R1, R2, R3) and one commercial (C1) building projects in Nigeria. While the commercial project is a three-storey building, the residential buildings are one- and two-storey buildings. All the four projects were funded by private clients. All the four buildings were constructed from sandcrete blocks and reinforced concrete structures. Each of the housing projects has a gross floor area of more than 200 square metres and an entire plot size of over 3000 square metres. Prior to commencing the projects, they were broken up into packages and the labour aspects were awarded to the contractors which subsequently sub-let the packages. These include excavation work, concrete, bending of steel reinforcement bars, carpentry, plumbing, painting, bricklaying, landscaping, tiling, plastering, and so on. The purpose of engaging the contractors is so that they would manage and supervise the sub-contractors. Despite their low level of technical knowledge, the contractors are also responsible for interpreting the design to the sub-contractors who were only engaged to undertake the work on a labour-only basis; the clients supplied all the materials.

The respondents included clients, architects, contractors and sub-contractors. They played various roles. Four clients, four architects, four contractors and 20 sub-contractors were interviewed, making a total of 32 respondents. The architects had an average of 8 years of working experience, while the contractors and sub-contractors had an average of 22 and 17 years of working experience respectively. The clients had an average of 10 years working experience.

While 75 percent of the sub-contractors had less than 10 employees, about 60 percent of both the contractors and architects had an average of less than 3 employees in their organisations. The need to register as a corporate entity or with professional bodies is not considered a necessity among the respondents. Although 75 percent of the architects had the necessary post-graduate qualifications, only one of them is registered with the professional body. Similarly, none of the contractors or the sub-contractors is registered. However, 60 percent of the sub-contractors had formal education up to secondary school level while 25 percent of the contractors had no formal education.

Findings

The interviews revealed the different methods used by the clients, architects, contractors and sub-contractors to minimise wastage of time, materials, money and other resources on the projects. These methods were similar in some trades and differed across others. Some of the methods were applied at the pre-construction stage, at the point of appointing consultants, while some were applied during the construction process. Some of the methods adopted are now discussed.

Limiting the involvement of professionals and consultants

In the first residential project studied (R1), only the architect was appointed to prepare architectural drawings. The clients proceeded to the construction stage without structural drawings, services drawings and bills of quantities, leaving all decisions requiring expert advice to the contractor and relevant sub-contractors. Similarly, only the architect was engaged at the construction stage. However, the client did not engage the architect on a full-time basis; to minimise supervision fees, the architect was paid on a daily basis but restricted to specific working days. As the contractor and sub-contractors lacked the technical knowledge to effectively interpret the designs, many errors were made in the architect's absence (on days on which he was not engaged), resulting in rework and variations. This approach to the engagement of consultants can have a negative impact on the project. In this particular case, it can cause structural failure, electrical faults, fire outbreak and cost overrun.

On the two other residential projects (R2 and R3), the architect's role was extended to supervising the project, procuring materials and making payment to the contractor and sub-contractors. Thus, both the architects were involved during both the design and construction stages.

In the case of the commercial project (C1), various consultants were involved. A complete set of architectural, structural and services drawings was prepared. Furthermore, the quantity surveyor was engaged to prepare a bill of quantities. However, except for the architect, the involvement of the consultants was limited to the design stage to minimise what the client considered as wastage of resources. Thus, there was a risk of misinterpreting their designs. In cases where the architect and contractor were unable to interpret the design and give appropriate instructions, the final decisions were taken by the tradesmen.

This shows that some private clients consider the involvement of professionals other than the architect as adding unnecessary cost to the project. Thus, their non-involvement is seen as a way of reducing the project cost, or minimising "wastage" of resources. Thus, the number of consultants involved in the projects was significantly reduced at both the design and construction stages.

Avoiding contract documentation and signing of agreements

The appointment of a lawyer to prepare the contracts for the projects was seen by all the clients as an unnecessary additional cost in all the case study projects (R1, R2, R3 and C1), thus a "waste" of resources. The projects were executed without written agreements between the parties to the projects. However, there were negotiations on key aspects of the project, leading to a basic common understanding of the rights and obligations of the parties. During construction, variations and their cost implications were also negotiated. These arrangements resulted in several disputes during and after the construction work on the project (R1), which caused delays and additional costs.

Avoiding the appointment of security staff at early stage

At the commencement of two of the residential projects (R1and R3), there were no security personnel on the sites. To minimise cost or "waste" of resources, security personnel were engaged when the work reached the superstructure level when some fixtures such as conduits and electrical wires were installed. At this stage, the clients considered the value of materials and works on site as worth protecting. However, on the other residential project (R2), security staff were engaged on a full-time basis throughout the project. Similarly, in the commercial project (C1), the client built a temporary fence to secure the site. Although no security personnel were employed at the beginning of this project, at the end of every working day the gates were locked to safeguard the site.

Despite the presence of security personnel at a later stage of this project (C1), some bags of cement, timber and reinforcement rods were stolen from the site. However, worse cases were experienced on projects R1 and R3. This shows that avoiding the appointment of security staff at any stage of a project is neither an effective nor an efficient way of minimising cost or a method for minimising the wastage of resources.

Lack of site facilities

The provision of site facilities such as toilets and canteens for the convenience and welfare of the workers were considered by all the private clients, in projects R1, R2, R3 and C1, as another unnecessary additional cost.

Thus, these facilities were not provided on site. At break periods or when the need arose, workers had to leave the site to use public toilets and canteens located at long distances from the sites. This affected the workers' commitment to the projects, resulting in delays, poor workmanship and low productivity.

Price-based award of work packages

On projects R1 and C1, the clients awarded work packages to contractors and sub-contractors on the basis of the lowest bid and earliest completion time. In order not to carry out the work at a loss, some subcontractors employed fewer, cheaper and less qualified workers whose performance could affect the completion period and standard of work. However, on projects R2 and R3, where the architects shouldered this responsibility, besides cost and completion time, factors such as the company's competence, experience and quality of work on previous projects were also considered as criteria for selecting the sub-contractors.

Non-provision of site storage

The construction of a temporary store on-site for building materials used in bulk on the project, such as cement, timber and reinforcement rods, was seen as a waste of space and money by the clients of projects R1 and R3. As a result, materials were stocked under temporary shelters in the early stages. However, as the superstructure was developed, the materials were later stored in the building. On project R2, the materials were stored in uncompleted buildings adjacent to the site. Besides obstructing the workers' movement, this resulted in additional labour cost and time to handle and transport the materials on the site. There were several instances of theft, poor safety and material wastage. In the case of the commercial project C1, a temporary store was built next to the temporary fence. Nevertheless, the absence of security staff at the early stage of the project resulted in the theft of some building materials from the store.

Re-use of building materials

One of the methods used in minimising wastage of resources on all the projects (R1, R2, R3 and C1), especially in carpentry and masonry works, was the continuous reuse of building materials such as timber, broken blocks, cement and nails. During the carpentry works, there was continual reuse of timber from the foundation, through the formwork and scaffolding, and up to the roof truss level. To achieve this high number of times of usage, the clients followed the advice of the sub-contractors to purchase high-quality timber. The cutting of fresh pieces was minimised by making use of smaller pieces already cut. Already used nails were also removed from unwanted pieces and reused. At the end of every stage, the timber pieces were carefully packed to avoid damage. This effort minimised material wastage on the site. Although significant savings in cost were made, the strength of the roof might be affected as the timber might have lost some of its original quality and strength by the time it was re-used in the roof structure.

Non-conformity with building regulations and standards

To minimise costs on the project, the clients generally adopted methods that do not conform to the building regulations in many trades, ranging from the use of sub-standard materials to inappropriate work procedures. The lack of involvement of relevant professionals especially during the construction stage also contributed to this. This was prevalent in concrete, masonry and reinforcement works. For example, each bag of cement was used to make more blocks than the maximum number set by regulations. This was common among all the projects (R1, R2, R3 and C1).

The methods adopted during the projects to save materials and labour cost included reducing the quantity of cement in both the mortar and the concrete, adding more sand to the mix, using larger (cheaper) sizes of aggregates in the concrete mix, increasing the spacing of the reinforcement rods in the lintel, floor slabs, roof beams and other structural elements, reducing the thickness of the floor slab and the size of the column bases specified by the engineer and architect in some cases, manual mixing of concrete, casting and compaction of concrete without a vibrator, and manual ramming of the hardcore.

To save materials at the ground and upper floor levels in project R1, a wire mesh was omitted and the spacing between the links in the lintel and roof beam was increased. In projects R1 and R3, the steel benders used the same size of rods for both primary and secondary beams supporting floor slabs to save costs by not purchasing larger sizes of rods. The thickness of the ground and upper floor slabs was kept to the minimum sizes.

The period for the curing of concrete set by the building regulations was ignored by the clients, contractors and sub-contractors of all the four projects, as they considered such a curing period a waste of time. In this case, bricklaying and concrete curing were done concurrently to save time. This resulted in slight cracks in the walls and the plaster in project R1.

Economical architectural design

At the design stage, all the four clients collaborated with the architects to develop a design that could help to minimise cost and time. There are some differences in the methods used due to the nature of the designs, budgetary limits and functions of the buildings. An example involved aligning the positions of the toilets on the floors in order to avoid unnecessary changes in the alignment of the pipes as well as curves and circular shapes in both the plans and elevations. In all the residential projects, the architects avoided unnecessary recesses in the walls on plan; this reduced cost. However, in the commercial project, where aesthetics were seen as vital, many elements and designs were chosen to facilitate marketing. Although the quantity surveyor liaised with the architect and the client at the design stage to develop specifications based on the client's budget, this specification was partly abandoned during the construction stage.

Barriers to implementing lean construction

Based on the interviews, the major barriers to implementing lean construction in Nigeria are: lack of technical knowledge, poor education, lack of skilled workers, poor communication, poor wages and salaries, poor organisational structure, incomplete brief, clients' uncertainties and inconsistencies, poor management, lack of prefabrication, poor tendering system, and the fragmented nature of the construction industry. Other barriers are: lack of adequate funding, poor government control of standards of materials and lack of enforcement of regulations, instability in government policies, high inflation rates, poor infrastructure, and poor documentation.

An analysis of the interviews showed that the different methods used by the private clients to minimise wastage of time, materials, money and other resources had significant impacts on the project quality, completion period, aesthetics, as well as the building's durability. Moreover, the interviews revealed that, although some of the methods adopted effectively reduced the project cost, most of them had negative impacts on the project. Thus, the implementation of lean concepts is proposed to achieve an effective and efficient minimisation of wastage of resources while delivering a valuable product to the client. A protocol is developed to aid the implementation; this is next discussed.

Implementing lean construction: a protocol

The lean protocol was developed through a review of literature on lean construction and information obtained from the field study. Of the 32 respondents, 62 percent noted that a guide of the principles, or a roadmap or framework for implementing lean, would be helpful. Other issues considered in developing this protocol are the challenges, barriers and pre-requisites of the implementation of lean construction in developing countries. The protocol provides for a continuous process that can be implemented in four phases, as shown in Figure 14.1. These are: Awareness and Enlightenment; Training and Practical sessions; Tools Implementation and Monitoring; and Change Management and Follow-up phases.

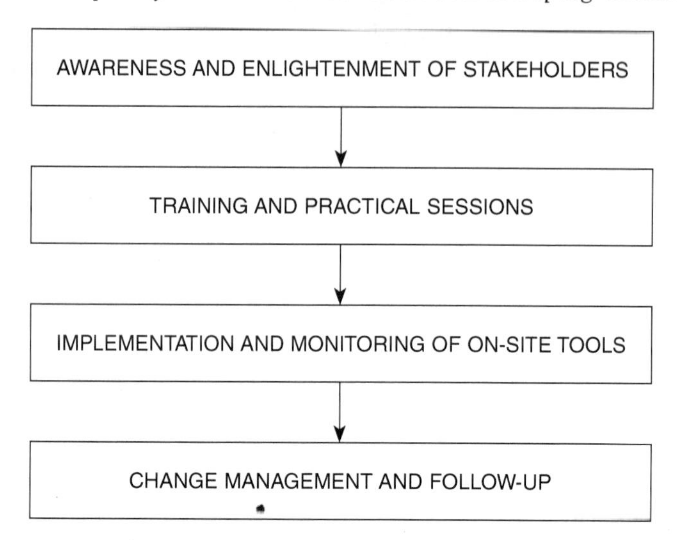

Figure 14.1 Phases of lean implementation protocol

First phase: Awareness and enlightenment of stakeholders

The first phase of the lean implementation protocol involves introducing the concept to stakeholders in the construction sector. Some of the major barriers to lean construction practice are: inadequate knowledge, inadequate exposure to requirements for lean implementation, lack of awareness programmes and lack of information sharing. Findings from the interviews reveal that there is a high level of illiteracy among some private clients, sub-contractors and local contractors in Nigeria. Olatunji (2008) identified inconsistency in government policies and price control among the other barriers. Overcoming these factors is considered to be a prerequisite in the successful implementation of lean concepts. In implementing the protocol, the methods used to minimise waste by both private and public clients and other stakeholders should be discussed at this stage, highlighting their weaknesses and all the negative impacts they have on project performance. The misconception about lean relating to its origin can also be addressed through the enlightenment exercise. Some of the challenges of lean practice include unfamiliarity with the lean techniques among workers, and the risks related to investing in new management concepts. This may discourage some contractors. However, this can also be addressed at this stage of the protocol. A good understanding of the benefits of lean construction practice can generate support and commitment from the government, professional bodies and relevant organisations. In addition, it will expose them to the requirements for its holistic implementation. The government could provide support by formulating policies, and providing the national or provincial level facilities needed.

In various developed countries such as the UK, the US, Japan and Denmark, the creation of awareness and enlightenment is handled by chapters of the Lean Construction Institute (LCI). Similar chapters can be formed in the developing countries. Collaborative efforts and support from the professional institutions, as well as the governments, would also help. A lean construction expert can create awareness among clients, contractors, sub-contractors, suppliers and consultants on the aims, objectives, goals and benefits of lean construction. The achievements made by users of the concept in other countries should be highlighted in such an exercise.

The awareness and enlightenment programme should cover the roles of various elements of the project such as professionals, building standards and contract agreements in adding value to projects. The second part of this phase should involve enlightening the stakeholders on how to manage change in their organisations in order to achieve a sustainable lean construction practice. The relevance of "Kaizen" can be discussed to promote change management. Kaizen is a lean tool that emphasises the involvement of all the members of the project team, in a continuous improvement process throughout the project (Senaratne and Wijesiri, 2008). Kaizen leads to improvement in the quality of work, site safety and labour productivity. It gives the client, consultants and contractor the responsibility for allocating resources and developing strategies and procedures that will eliminate wastage of resources. The site engineer is responsible for monitoring the implementation of these strategies. The site engineer should also ensure that the workers are well trained to achieve effective implementation. The sub-contractors should effectively supervise the workforce and improve communication with, and among them.

Second phase: Training and practical sessions

The second phase of the application of the protocol involves training the consultants and contractors on how to apply lean construction tools during the construction processes. Training is the most crucial part of the lean implementation process (Conner, 2001). Most of the tools require the users to be trained in order to be effectively applied. Training improves the efficiency of managers and workers and eliminates barriers to behavioural changes (Harmon, 2007; Paton and James, 2008).

Several authors identified lack of understanding among workers and employers, lack of technical skills and lack of training, as barriers to lean construction practice in some countries. Training is an effective and efficient way that enables the stakeholders to collaborate in identifying and eliminating waste during the design and construction processes. Similarly, lean practice demands that there should be continuous improvement across all aspects of the project. Some workers could find the consistent and continuous application of some lean techniques such as the 5S difficult owing to the traditional working culture. Certain strategies can be developed to aid consistency and continuity in their practice. Furthermore, some of the demerits of lean practice that are often highlighted include the cost of employing a lean consultant as well as the risks associated with new methods. Effective training can save the organisations the cost of employing a consultant to guide implementation in the organisations. This could also minimise the risk that may be encountered. The difficulty in understanding the concepts also suggests that engaging the stakeholders in practical sessions could significantly add value to the training sessions.

The first part of the training programme should involve a discussion of the principles of lean construction. This should guide the stakeholders on how to identify what is valuable to the client and how to map out value streams so as to eliminate non-value-adding (waste) activities. Value is any activity that the client is willing to pay for (Mascitelli, 2002). The organisations in the supply chain should only engage themselves in activities that add value to the project without compromising the quality, at both the design and construction stages. The client should be provided with a product of highest possible performance, quality and reliability within the cost budget without exceeding the client's needs and requirements. However, the simplest approaches should be adopted. As much as possible, the approaches should be waste-free so that the client will be convinced that so much value is given for the money spent. Thus, they should consume the least resources in terms of time, capital, materials and labour; any unnecessary additional service that the client does not consider valuable and is unwilling to pay for should be considered as a potential waste, and avoided. To ensure effective understanding and appreciation of the lean construction tools, the training should be immediately followed by some practical application of every tool.

Third phase: On-site tools implementation and monitoring

The third phase of the application of the protocol is practical implementation. The initial part involves developing policies and strategies to support lean construction practice in all the projects involving the organisation. A reasonable time frame for the full transformation should be designed as a guide for the organisation.

The second part of the implementation and monitoring phase involves the practical application of the tools to improve the entire delivery process. Although several lean construction tools have been successfully implemented in developed countries (Abdelhamid and Salem, 2005), the level of impact they make depends on the environment and ability of the workforce to participate in the implementation process. The level of literacy and skills of the workers determines how effectively some tools can be adopted. Barriers such as high level of illiteracy, lack of equipment, unsuitable organisational structure, weak administration, use of substandard components, lack of social amenities and infrastructure, poor procurement selection strategies, non-availability of materials and inadequate resources suggest that not all lean tools can easily be applied in some developing countries. The interviews identified a high level of illiteracy among some contractors and sub-contractors in Nigeria that could also affect the application of certain tools. The involvement of lean consultants may be necessary on projects where there is a low level of knowledge and poor understanding of lean concepts among the contactors, sub-contractors and the clients.

Tools required for lean practice

Some basic tools which are instrumental and fundamental to the implementation of lean construction practice, and related issues in the context of developing countries, are now discussed.

Last Planner System (LPS) LPS is a lean tool that improves productivity and minimises wastage through production planning and scheduling based on the workers' ability and prevailing site conditions. Last planners are usually trades foremen who decide what work is to be done the following day (Song et al., 2008), based on the workers' capabilities and prevailing site conditions. The LPS develops a Weekly Work Plan using a *Should-Can-Will* analysis (Ballard, 2000; Song et al., 2008). "Should" indicates all the work that is required to be carried out. However, various constraints put a limit to the work that "Can" be carried out. Thus, the last planners make a commitment to the work that "Will" be carried out (Salem et al., 2005). This enables improvement in workflow reliability, predictability, health and safety, quality, and speed of delivery. A good understanding of the tasks to be carried out is necessary among the last planners for this tool to be implemented. In situations where the foremen are illiterate, the tasks need to be well interpreted to them by the site manager or supervisor.

Productive meetings Meetings are essential to the successful delivery of every project. They are a route for encouraging cultural change and reminding all participants to focus on value-creating activities. However, a meeting is considered a waste of time if it does not lead to a decision or an action. The project stakeholders should, when necessary, hold meetings to ensure the coordination of activities on the project in order to facilitate the execution of the project plan. Meetings could also be held with the workers to achieve a rapid response to problems and eliminate possible barriers (Abdelhamid and Salem, 2005).

Prior to the meeting, the purpose should be communicated to all those expected to attend to achieve a proper plan. During meetings, the supply team should share ideas and knowledge, weigh alternatives, relate decisions to the client's needs, and solve all technical problems. Resources could also be reallocated to maximise efficiency and eliminate waste.

Although a meeting takes the workers away from performing their real tasks on the project, it reduces the chance of making errors that could result in variations and redesign (Mascitelli, 2002). It also encourages teamwork and builds emotional commitment among the participants.

Increased visualisation Owing to the low level of literacy among the sub-contractors and their workers in developing countries, some vital information can be graphically communicated using signs, charts and photographs (Salem et al., 2005). This tool, known as increased visualisation, can be used to alert the workforce about the required quality, appearance, safety and target completion date. Besides guiding the workers, the information helps them to judge their productivity and work progress (Pasquire and Connolly, 2002). In some cases, the tool can serve as a guide to the workers with respect to the location and movement of materials on site to minimise wastage of time (Abdelhamid and Salem, 2005). Furthermore, it reduces the chances for errors and mistakes that can lead to accidents (Sacks et al., 2009). When applying this tool, the use of words and numbers should be minimised.

Off-site prefabrication The wall panels, beams, columns and other relevant components of the building can be manufactured off-site and transported to the site for assembly. In this way, the time spent building up the elements on site is minimised. This has an additional advantage of saving the client from poor workmanship and sub-standard mix ratios. It also gives the client some certainty about the price and project duration. The major advantages of off-site prefabrication include quality enhancement, safety, efficiency and time saving (Pasquire and Connolly, 2002). It also minimises materials wastage, and time loss through the movement of goods and accidents on site. However, to apply this tool effectively, the workforce has to be trained and their contributions well coordinated during the assembly on site.

Despite the benefits associated with off-site fabrication, its application may be affected by factors including poor transportation routes to the sites, poor technical skills among designers and constructors, and availability of manufacturing, transportation, handling and installation facilities, which are problems experienced in many developing countries.

5S/5C (House-keeping) House-keeping is a lean tool that minimises accidents, theft and wastage of resources by ensuring that all working tools and building materials are well maintained and kept in their places (Abdelhamid and Salem, 2005). It makes the working environment tidy and orderly. It also makes movement within the site safe and convenient (Salem et al., 2007). The site environment should be clear of all unwanted materials. The neat arrangement of tools and materials in their appropriate location makes it easy to identify them when required. A good house-keeping culture helps in eliminating trips, falls and exposure to hazards which could result in accidents and affect the production process (Nahmens and Ikuma, 2009). The major advantages of this tool are improvements in productivity, quality and safety.

To improve safety and prevent accidents caused by obstructing building materials, as found in the interviews, the application of 5S and Increased Visualisation are key parts of the protocol. However, adequate awareness and motivational incentives are necessary to achieve continuity and consistency in their application.

Mistake-proofing/poka-yoke Mistake-proofing involves all the necessary strategies adopted to identify defects at early stages before they result in any failure. This could be in the form of visual inspection, or risk assessment and management. According to Koskela (1992), the strategy increases reliability and predictability, and reduces variability. In promoting workers' health and safety, poka-yoke devices involve the use of safeguards, personal protective equipment and other measures that prevent workers from coming too close to unsafe conditions (Saurin et al., 2005). It also protects the workers from falling objects, and excess heat and noise (Saurin et al., 2006). The application of this tool may involve some financial support from the government, and training.

Root cause analysis/Five "Whys" It is suggested that the repetition of "why" five times can help to trace the roots of a problem (Nicholas, 1998). By tracing the roots, the problem can be better handled and solved. This may involve revising the architectural, structural and services drawings to achieve a permanent solution to the problem. For example, if a crack develops in a wall, its cause should be traced rather than the defect being made good. This approach improves the quality and reliability of the design. Besides continuously improving the efficiency, it helps to minimise errors on the project. Similarly, if an accident occurs on site, its root causes can be investigated using this tool.

First Run Studies Some critical tasks may require special training on how to execute them. First Run Studies is a technique where the workforce is assembled and given comprehensive training on how to carry out a particular task. The technique involves using videos, pictures and other graphical representations (Abdelhamid and Salem, 2005). The technique improves the workers' efficiency, performance, productivity and quality of work. It also saves time and minimises costly errors (Mitropoulos et al., 2007).

Just-in-time Just-in-time is a management philosophy aimed at eliminating wastes by providing only the required quantity of resources at the right place at the right time. It challenges traditional methods of executing tasks and encourages the use of the latest technology (Conner, 2001). The continuous modification of construction methods is a key to continuous improvement in quality and productivity. Although it may be difficult to adopt, it could help to eliminate waste. The supply team should ensure an efficient flow of information, materials, equipment and labour at both the design and construction stages within a minimum duration. Its effective implementation requires much commitment and a cordial relationship between the supply team and the client.

Findings from the interviews show that lack of site welfare facilities such as toilets, canteen and project office also contribute to low productivity and delay in work processes. There should be adequate provision of site facilities and security personnel. A project office shall serve as a place where all documents related to the project are kept. This will minimise or eliminate the time wasted in retrieving past documents and transferring or communicating information to other team members. Provision of site security is necessary to avoid theft of building materials and vandalism. The site should be fenced and the number of access points into the site should be minimised. The building materials should be well arranged inside a storage facility to avoid wastage, theft and exposure to the elements (Bicheno, 2000).

One of the major principles of lean construction is reduction in variability. This is necessary to achieve reliability and transparency in the workflow. The architectural, structural and services design should be very detailed. The aim of detailing the design is to reduce variability, deliver value and improve predictability and workflow (Ballard et al., 2001). To further minimise waste, the construction processes must be integrated into the detailed design so as to deliver value to the client with minimal consumption of resources.

Fourth phase: Change management and follow-up

The last phase of the protocol involves change management and following up to measure the level of implementation and the impact made in delivering value to the client. Although change is necessary for the construction industry to continuously meet clients' needs and respond to the global, social and environmental challenges, its implementation has to be well managed for it to be effective and sustainable.

Managing change is a major challenge in project management. Due to lack of transparency, difficulty in effecting a change in culture, lack of self-criticism, poor house-keeping, leadership conflict and over-enthusiasm, change management and follow-ups constitute a core part of the lean implementation protocol. The interviews also reveal poor wages, incentives and motivation among both consultants and subcontractors. This can significantly affect consistency and continuity in applying tools such as the 5S and "kaizen" activities. Change management and follow-up are necessary to address these sorts of problems. Success in change management requires commitment from the top management of the company (Salem et al., 2005; Hudson, 2007). Moreover, success in implementing lean lies in their behaviour and attitudes towards change. Although change is behavioural and psychological, it requires an holistic long-term approach. The management must motivate the people and provide the necessary support. This may also involve re-organising the management structure to support the implementation of lean construction (Harmon, 2007). Sustainable change requires building trust and establishing a new culture of constant learning and improvement among employers and employees.

Open communication is an integral part of change management (Hudson, 2007). This can be achieved by facilitating actions, recognising and rewarding contributions, providing and requesting feedback, reviewing progress, providing staff support and development, and giving exemplary leadership. The organisation should welcome mistakes based on trust so that their root causes can be dealt with to achieve lasting solutions.

The problems of, and barriers to, effective implementation of the different tools should be identified during this phase so that necessary measures will be taken to avoid them on future projects.

Lean construction is still in its early stage of development. The concept is new to many developed and developing countries. The implementation protocol developed in this research will be beneficial to construction industries in developing countries. It would aid the clients and the project delivery team to implement the concept in order to effectively and efficiently eliminate wastage of resources throughout the construction process, and help to achieve improvement in the workers' productivity, and the companies efficiency and profitability. Considering the benefits of lean construction in promoting quality, safety and sustainable construction, a government policy should be developed to support its implementation. Its practice could be centrally guided and monitored by a Lean Construction Institute. The policy and implementation phases could be gradually reviewed based on feedback from the industry.

Conclusions and recommendations

Lean construction seeks to attain the improvement of value-adding activities and the minimisation of non-value-adding activities, termed as waste, from the construction processes as a way of minimising resource consumption and increasing productivity and quality of outputs. Significant benefits have been realised from the implementation of lean concepts in many countries. However, in many developing countries such as Nigeria, clients, consultants and contractors adopt certain methods to minimise wastage of resources that have negative impacts on the projects and are hence inconsistent with the goals of lean construction. It is likely that similar practices are adopted on many of the construction projects in developing countries, as attempts are made to minimise wastage in their delivery. In order to achieve an effective and efficient minimisation or elimination of wastage of resources, a protocol has been proposed in this chapter.

The protocol was developed with due consideration of prerequisites for, barriers to, and demerits of, lean practice as well as the challenges of its implementation in developing countries. It is believed that following this protocol would lead to an effective and sustainable implementation of lean concepts in developing countries, leading to the minimisation of the wastage of resources throughout the construction process.

Considering the benefits of lean construction in promoting quality, safety and sustainability in the construction process, governments in developing countries should support its implementation. Its practice could be centrally guided and monitored. The policies and their implementation should be continually reviewed, based on feedback from the industry.

References

Abdelhamid, T. (2004) Lean Production Paradigms in the Housing Industry. *Proceedings of the NSF Housing Research Agenda Workshop*, Feb. 12–14, Orlando, Florida.

Abdelhamid, T.S. and Everett, J.G. (2002) Physical Demands of Construction Work: A Source of Workflow Unreliability. *Proceedings of the 10th Annual Conference of the International Group for Lean Construction*. 6–8 August, Gramado, Brazil.

Abdelhamid, T. and Salem, S. (2005) *Lean Construction: A New Paradigm For Managing Construction Projects*. http://www.hbrc.edu.eg/ehbrc/workshop/contents/CMN02.pdf/ (accessed on 17 December 2009): 1–4.

Abdullah, S., Abdul-Razak, A., Abubakar, A. and Mohammad, I.S. (2009) Towards Producing Best Practice in the Malaysian Construction Industry: The Barriers in Implementing the Lean Construction Approach construction work. Paper presented at International Conference on Construction Industry, Padang, Indonesia.

Ahadzie, D.K., Proverbs, D.G., Olomolaiye, P.O. and Ankrah, N. (2009) Towards Developing Competency-based Measures for Project Managers in Mass House Building Projects in Developing Countries. *Construction Management and Economics*, 27, 89–102.

Alarcon, L.F. (1995) Training Field Personnel to Identify Waste and Improvement Opportunities in Construction. In L.F. Alarcon (ed.) *Lean Construction*. Rotterdam: A.A. Balkema.

Alarcon, L.F. and Calderon, R. (2003) Implementing Lean Production Strategies in Construction Companies. In K.R. Molenaar and P. S. Chinowsky (eds) *Proceedings: International Conference on Integration and Innovation of Construction*. 19–21 March, Honolulu: American Society Civil Engineers.

Alarcon, L.F., Diethelm, S. and Rojo, O. (2002) Collaborative Implementation of Lean Planning Systems in Chilean Construction Companies. *Proceedings of the 10th Annual Conference of the International Group for Lean Construction*. 6–8 August, Gramado, Brazil.

Alinaitwe, H.M. (2009) Prioritizing Lean Construction Barriers in Uganda's Construction Industry. *Journal of Construction in Developing Countries*. 14(1), 15–30.

AlSehaimi, A., Tzortzopoulos, P. and Koskela, L. (2009) Last Planner System: Experiences from Pilot Implementation in the Middle East. *Proceedings of 17th Annual Conference on Lean Construction*, July, Taiwan.

Bae, J. and Kim, Y. (2008) Sustainable Value on Construction Projects and Lean Construction. *Journal of Green Building*, 3(1), 156–67.

Ballard, G. (2000) *The Last Planner System of Production Control*. PhD dissertation, Birmingham University.

Ballard, G. (2008) The Lean Project Delivery System: An Update. *Lean Construction Journal*, 4, 1–19

Ballard, G. and Howell, G. (1998) Shielding Production: An Essential Step in Production Control, *Journal of Construction Engineering and Management*, 124(1), 11–17.

Ballard G., Koskela, L., Howell, G. and Zabelle, T. (2001) Production System Design in Construction. *Proceedings 15th Annual Conference of the International Group for Lean Construction*, Singapore.

Bashir, A.M. (2009) *Developing a Framework for Lean Construction Practice in Developing Countries*. Unpublished M.Sc. Dissertation, University of Wolverhampton.

Bashir, A.M., Suresh, S., Proverbs, D.G. and Gameson, R. (2010) Barriers Towards the Sustainable Implementation of Lean Construction in the United Kingdom Construction Organisations. *Proceedings of the Association of Researchers in Construction Management (ARCOM) Doctoral Workshop*, Wolverhampton.

Bertelsen, S. (2004) Lean Construction: Where are we and how to proceed? *Lean Construction Journal*, 1(1), 46–64.

Bertselen, S. and Koskela, L. (2002) Managing the Three Aspects of Production In Construction. *Proceedings of the 10th Conference of the International Group for Lean Construction*, 6–8 August, Gramado, Brazil.

Bicheno, J. (2000) *The Lean Toolbox*. 2nd edn. Buckingham: PICSIE.

BRE (2009) *Constructing Excellence*. http://www.bre.co.uk/filelibrary/CLIP/Vol_1__CLIP_Case_Study_Booklet_28-09-04.pdf / (accessed 22 December 2009): 1–5.

BRE (2010) *Construction Lean Implementation Programme*. http://www.bre.co.uk/page.jsp?id=355/ (accessed 21 May 2010).

Callaghan, E. (2006) An Investigation of the Productivity of Concrete Operations in the Australian Construction Industry. Unpublished M.Sc. Dissertation, University of Wolverhampton.

Castka, P., Bamber, C. and Sharp, J. (2004) Benchmarking Intangible Assets: Enhancing Teamwork Performance using Self Assessment. *Benchmarking*, 11(6), 571–83.

Chen, Q., Wakefield, R.R. and O'Brien, M. (2004) Lean Applications on Residential Construction Site, in *Proc. of ASCE Specialty Conference on Leadership and Management in Construction*, 24–26 March, Hilton Head, SC: 145–57.

Central Intelligence Agency (2010a) *Country GDP Comparison*. https://www.cia.gov/library/publications/the-world-factbook/rankorder/2001rank.html?countryName=Nigeria&countryCode=ni®ionCode=af&rank=32#ni/ (accessed 22 May 2010).

Central Intelligence Agency (2010b) *The World Factbook: Nigeria*. https://www.cia.gov/library/publications/the-world-factbook/geos/ni.html/ (accessed 22 May 2010).

Common, G., Johansen, E. and Greenwood, D. (2000) A Survey of the Take-Up of Lean Concepts Among UK Construction Companies. *Proceedings of the 8th International Group for Lean Construction Annual Conference*. Brighton.

Conner, G. (2001) *Lean Manufacturing for the Small Shop*. Dearborn, MI: Society of Manufacturing Engineers.

Cua, K.O., McKone, K.E. and Schroeder, R.G. (2001) Relationships Between Implementation of TQM, JIT and TPM and Manufacturing Performance. *Journal of Operations Management,* 19(6), 675–94.

Dulaimi, M.F. and Tanamas, C. (2001) The Principles and Applications of Lean Construction in Singapore. *Proceedings, 9th Annual Conference of the International Group for Lean Construction*, Singapore.

du Plessis, C. (2007) A Strategic Framework for Sustainable Construction in Developing Countries. *Construction Management and Economics*, 23, 67–76.

Egan, J. (1998) *Rethinking Construction: The Report of the Construction Task Force.* London: Department of the Environment, Transport and the Regions.

Ehrenfeld, J.R. (2008) Sustainability Needs to be Attained Not Managed. *Sustainability: Science, Practice and Policy*, 4(2), 1–3.

Farrar, J.M., AbouRizk, S.M. and Mao, X. (2004) Generic Implementation of Lean Concepts in Simulation Models. *Lean Construction Journal*, 1(1), 1–23.

Fiallo, C. and Revelo, V. (2002) Applying LPS to a Construction Project: A Case Study in Quito, Equador, *Proceedings of the 10th IGLC Conference*, Gramado, Brazil.

Forbes, L. and Ahmed, S. (2004) Adapting Lean Construction Methods for Developing Nations. *2nd International Latin America and Caribbean Conference for Engineering and Technology*, Florida, June.

Harmon, P. (2007) *Business Process Change: A Guide for Business Managers and BPM and Six Sigma Professionals*. 2nd edn. London: Sage.

Hook, M. and Stehn, L. (2008) Lean Principles in Industrialised Housing Production: the need for a cultural change. *Lean Construction Journal*, 4, 20–33.

Howell, G.A. (1999) What is Lean Construction? *Proceedings IGLC-7*, Berkeley.

Howell, G. and Ballard, G. (1994) Lean Production Theory: Moving Beyond "Can-Do". *Proceedings of the 2nd Annual Conference of International Group of Lean Construction.* http://www.leanconstruction.org/lcj/V2_N2/LCJ_05_009.pdf/

Hudson, M. (2007) *Managing Without Profit: The Art of Managing Third-sector Organizations*. 2nd edn. London: Directory of Social Change.

Hughes, P. and Ferrett, E. (2008) *Introduction to Health and Safety in Construction.* 3rd edn. Oxford: Butterworth-Heinemann.

Junior, A., Scola, A. and Conte, A. (1998) Last Planner as a Site Operations Tool, *Proceedings of the 6th IGLC Conference*, Guaruja, Brazil.

Kheni, N.A., Dainty, A.R.J. and Gibb, A. (2008) Health and Safety Management in Developing Countries: A Study of Construction SMEs in Ghana. *Construction Management and Economics*, 26, 1159–69.

Klotz, L., Horman, M. and Bodenschatz, M. (2007) A Lean Modeling Protocol for Evaluating Green Project Delivery, *Lean Construction Journal*, 3(1), 46–64.

Kim, Y. and Jang, J. (2005) Case Study: Application of Last Planner to Heavy Civil Construction in Korea, *Proceedings of the 13th IGLC Conference*, Sydney.

Koskela, L. (1992) *Application of New Production Theory in Construction.* Technical Report No. 72, CIFE, Stanford University, CA.

Koskela, L. (1993) Lean Production in Construction. *Proceedings of the first Annual Conference of the International Group for Lean Construction (IGLC-1)*, Espoo, Finland.

Koskela, L. (2000) *An Exploration Towards a Production Theory and its Application to Construction*, Espoo, Finland: VTT Building Publications. http://www.inf.vtt.fi/pdf/publications/2000/P408.pdf/

Koskela, L. (2004) Moving On – Beyond Lean Thinking. *Lean Construction Journal*, (1)1, 24–37.

Koskela, L., Howell, G., Ballard, G. and Tommelein, I. (2002) The Foundations of Lean Construction. In R. Best and G. de Valence (eds) *Design and Construction: Building in Value*, Oxford: Butterworth-Heinemann, Elsevier.

Lamming, R. (1993) *Beyond Partnership: Strategies for innovation and lean supply*. London: Prentice-Hall

Lapinski, A.R., Horman, M.J. and David, R.R. (2006) Lean Processes for Sustainable Project Delivery. *Journal of Construction Engineering and Management*, 132(10), 1083–91.

Lewis, T.M. (2006) Impact of Globalization on the Construction Sector in Developing Countries. *Construction Management and Economics*, 25, 7–23.

Luu, T.V., Kim, S.Y., Cao, H.L. and Park, Y.M. (2008) Performance Measurement of Construction Firms in Developing Countries. *Construction Management and Economics*, 26, 373–86.

Mascitelli, R. (2002) *Building a Project-driven Enterprise: How to slash waste and boost profits*. Northridge: Technology Perspectives.

Mastroianni, R. and Abdelhamid, T.S. (2003) The Challenge: The Impetus for Change to Lean Project Delivery. *Proceedings: Annual Conference for Lean Construction*, 22–24 July, Blacksburg, VA: 610–21.

Mitropoulos, P., Cupido, G. and Namboodiri, M. (2007) Safety as an Emergent Property Of the Production System: How Lean Practices Reduce the Likelihood Of Accidents. *15th International Group for Lean Construction Annual Conference*, East Lansing, MI, July.

Mitropoulos, P., Howell, G.A. and Reiser, P. (2003) Workers at the Edge: Hazard Recognition and Action. *Proceedings 11th Annual Conference of the International Group for Lean Construction*, Blacksburg, VA.

Monden, Y. (1998) *Toyota Production System: An Integrated Approach to Just-in-Time*. 3rd edn. London: Chapman and Hall.

Mossman, A. (2009) Why isn't the UK Construction Industry going Lean with Gusto? *Lean Construction Journal*, 5(1), 24–36.

Nahmens, I. (2009) From Lean to Green Construction: A Natural Extension. Conference Proceeding Paper. *Building a Sustainable Future. Proceedings of the 2009 Construction Research Congress*: 1058–67.

Nahmens, I. and Ikuma, L.H. (2009) An Empirical Examination of the Relationship between Lean Construction and Safety in the Industrialized Housing Industry. *Lean Construction Journal*, 5(1), 1–12.

Narang, P. and Abdelhamid, T.S. (2006) Quantifying Workers' Hazard Identification Ability Using Fuzzy Signal Detection Theory. *Proceedings 14th Annual Conference of the International Group for Lean Construction*, Santiago.

Nicholas, J.M. (1998) *Competitive Manufacturing Management: continuous improvement, lean production, customer-focused quality*. Boston: Irwin/McGraw-Hill.

Ofori, G. (1993) Research in Construction Industry Development at the Crossroads. *Construction Management and Economics,* 11, 175–85

Ofori, G. (2005) Revaluing Construction in Developing Countries: a research agenda. *Journal of Construction in Developing Countries*, 11(1), 1–16.

Ofori, G. (2007) Construction in Developing Countries. *Construction Management and Economics,* 21, 1–6.

Oladapo, L. (2007) *An Investigation into the Use of ICT in the Nigerian Construction Industry.* http://www.itcon.org/cgi-bin/works/Show?2007_18/ (accessed 25 May 2011).

Olatunji, J. (2008). Lean in Nigerian Construction: State, Barriers, Strategies and Go-to-gemba Approach. *Proceedings 16th Annual Conference of the International Group for Lean Construction*, Manchester.

Pasquire, C.L. and Connolly, G.E. (2002) Leaner Construction through Off-Site Manufacturing. *Proceedings IGLC-10*, August, Gramado. http://www6.ufrgs.br/norie/iglc10/papers/64-Pasquire&Connolly.pdf/

Paton, R. and James, M. (2008) *Change Management: A Guide to Effective Implementation*. 3rd edn. London: Sage.

Polat, G. and Arditi, D. (2005) The JIT Materials Management System in Developing Countries. *Construction Management and Economics*, 23. 697–712.

Rogers, P., Jalal, K. and Boyd, J. (2008) *An Introduction to Sustainable Development.* United Kingdom: Earthscan.

Sacks, R., Rozenfeld, O. and Rosenfeld, Y. (2009) Spatial and Temporal Exposure to Safety Hazards in Construction. *Journal of Construction Engineering and Management*, 135, 726–36.

Salem, O., Lothlikar, H., Genaidy, A. and Abdelhamid, T. (2007) A Behaviour-Based Safety Approach for Construction Projects. *Proceedings 15th Annual Conference of the International Group for Lean Construction*, Dearborn, MI.

Salem, O., Solomon, J., Genaidy, A. and Luegring, M. (2005) Site Implementation and Assessment of Lean Construction Techniques. *Lean Construction Journal,* 2(2), 5–29.

Salem, O. and Zimmer, E. (2004) Application of Lean Manufacturing Principles to Construction. *Lean Construction Journal,* 2, 51–4

Saurin, T.A., Formoso, C.T. and Cambria, F.B. (2005) Analysis of a Safety Planning and Control Model from the Human Error Perspective. *Engineering, Construction and Architectural Management* 12(3), pp. 283–298.

Saurin, T.A., Formoso, C.T. and Cambraia, F.B. (2006) Towards A Common Language Between Lean Production and Safety Management. *Proceedings 14th Annual Conference of the International Group for Lean Construction*, Santiago.

Schonberger, R.J. (1982) *Japanese Manufacturing Techniques: Nine hidden lessons in simplicity*. London: Collier Macmillan.

Senaratne, S. and Wijesiri, D. (2008) Lean Construction as a Strategic Option: Testing its Suitability and Accessibility in Sri Lanka. *Lean Construction Journal,* 4, 34–48.

Shelbourn, M.A., Bouchlaghem, M.D., Anumba, J.C., Carillo, P.M., Khalfan, M.M.A. and Glass, J. (2006) *Managing Knowledge In The Context Of Sustainable Construction.* http://www.itcon.org/data/works/att/2006_4.content.07629.pdf/ (accessed 29 April 2010).

Song, L., Liang, D. and Javkhedkar, A. (2008) *A Case Study On Applying Lean Construction To Concrete Construction Projects*. University of Birmingham.

http://ascpro0.ascweb.org/archives/cd/2008/paper/CPGT201002008.pdf/ (accessed 16 December 2009).

Thanwadee, C. (2009) The Use Of System Dynamics Modelling In Improving Construction Safety. *The 17th International Group for Lean Construction Annual Conference*, 13–19 July, Taiwan.

Thomassen, M.A., Sander, D., Barnes, K.A. and Nielsen, A. (2003) Experience and Results from implementing Lean Construction in a Large Danish Contracting Firm. *Proceedings of 11th Annual Conference on Lean Construction*, 22–24 July, Blacksburg, VA: 644–55.

Womack, J.P. and Jones, D.T. (1996) *Lean Thinking: Banish waste and create wealth in your corporation*. New York: Simon and Schuster.

Womack, J.P., Jones, D.T. and Roos, D. (1990) *The Machine that Changed the World*. New York: Rawson Associates.

World Bank Report (2010) *Country Classifications*, http://data.worldbank.org/about/country-classifications/ (accessed 24 June 2010).

Part VI

Industrial developement

15 Green jobs in construction

Edmundo Werna
International Labour Organization

Introduction

Green jobs are important because they bring together improvements in environmental protection and new opportunities for employment. Furthermore, those enterprises and workers trained in green techniques are likely to become champions of environmental protection, as they learn about its importance and also have a motivation to apply the knowledge that they have acquired. Construction has been recognised as having a significant impact on the environment. One notable example relates to climate change, through emissions of global warming gases (GWG). New construction as well as refurbishment give the opportunity to improve the environment in many ways, for instance, reducing CO_2 emissions and energy consumption, thus encouraging the development of new professional skills, leading to employment opportunities. The "greening" of the construction industry requires the development and implementation of new technologies aimed at reducing the negative impact of construction on the environment, and the delivery of enhanced performance by infrastructure. This green technology development requires new skill sets, new training methodologies and new entrepreneurs.

This chapter discusses the challenges of, as well as the potential for, green jobs in construction, with particular attention to developing countries. The first section gives a background to the environmental impact of construction and the potential for greening the industry. This is not exhaustive, because the focus of the chapter is on labour issues. The next section presents the "world of labour", using the overall concept of decent work promoted by the International Labour Organization (ILO). It analyses the state of decent work in construction in developing countries. The following section analyses issues related to green jobs specifically, building upon what was presented in the previous sections. The chapter concludes with policy implications for green jobs in construction in developing countries. As a background to all the sections, the remainder of the introduction presents general features of green jobs.

Green jobs

Green jobs have become an emblem of a more sustainable economy and society that preserve the environment for present and future generations and are more equitable and inclusive of all people and all countries. Green jobs reduce the environmental impact of enterprises and economic sectors, ultimately to levels that are sustainable. Specifically, but not exclusively, this includes jobs that: help to protect ecosystems and biodiversity; reduce energy, materials and water consumption through high-efficiency strategies; de-carbonise the economy; and minimise or altogether avoid the generation of all forms of waste and pollution. Green jobs in emerging economies and developing countries include opportunities for managers, scientists and technicians, but the bulk of the job opportunities can benefit a broad cross-section of the population which needs them most: youth, women, poor workers.

However, many jobs that are green in principle are not green in practice because of the environmental damage caused by inappropriate practices. Thus, the notion of a green job is not absolute, but there are "shades" of green, and the notion will evolve over time. Moreover, the evidence shows that green jobs do not automatically constitute decent work. Many of these jobs are "dirty, dangerous and difficult". Employment in construction and some other industries such as recycling and waste management, as well as biomass energy, tends to be precarious, and incomes from such jobs are low. If green jobs are to be a bridge to a truly sustainable future, this needs to change. Therefore, green jobs need to comprise decent work. Decent, green jobs effectively link Millennium Development Goal (MDG) 1 (poverty reduction) and MDG 7 (protecting the environment) and make them mutually supportive rather than conflicting.

The United Nations Environment Programme (UNEP), the International Labour Organization (ILO), the International Organization of Employers (IOE) and the International Trade Union Confederation (ITUC) launched a Green Jobs Initiative to assess, analyse and promote the creation of decent jobs as a consequence of the needed environmental policies. It supports a concerted effort by governments, employers and trade unions to promote environmentally sustainable jobs and development in a climate-challenged world. Construction was the first specific sector of the economy to be addressed in the Green Jobs Initiative. This process started with a global study (Keivani et al., 2010), with special focus on retrofitting of buildings (and comparisons with new construction). This study has set the scene for a number of studies in specific developing countries, with an aim of feeding policy making. The European Union, the United States government and other actors have also promoted green jobs in construction and other sectors.

There is a need to better understand the technical and economic dynamics of the construction industry as well as the relationship between the technological changes towards green construction and labour issues. It is important to gain an appreciation of how to use the shift to environment-friendly technologies to improve employment and decent work opportunities. How is a possible win-win situation forecast and, especially, encouraged? These are issues addressed in this chapter. Before delving into the specific implications regarding labour issues, an overview of the environmental impact of construction and the greening of the industry is presented.

1 The construction industry and the environment

The environmental impacts of the construction industry have already been well documented (see, for example, The Environment Agency in England and Wales et al., no date; Gallego Pinol et al., 2009; Gangolells et al., 2009; Keivani et al., 2010; Khatib, 2009; Moavenzadeh, 1994; Sarsby and Meggyes, 2009).

There are impacts related to the choice of sites for construction, to the construction process and the choice of building materials and equipment, as well as to the products of the industry (such as buildings). The purpose of this section is to present an overview of such issues based on the work of the authors mentioned above, as a way of discussing how the construction industry can, and should, improve its practices to the benefit of the environment. This is important in order to understand the potential for green jobs (this is analysed later in the chapter).

The choice of the site for a construction project has an impact on the environment when there is land contamination (i.e. if contamination is not treated). Subsequently, examples of impacts of the construction process include: air emissions (such as dust from earthworks or emissions from plant and equipment); noise pollution (for example, from operating plant and equipment); waste disposal (such as spoil, offcuts and other building materials); and water discharges (such as those from the dewatering of excavations and testing of pipes).

The choice of building materials also has a significant impact on the environment, depending on the characteristics of the extraction of raw material (thus linking back to mining); whether a given material is renewable or not; the characteristics of its manufacturing process (for example, consumption of energy, air and noise pollution, waste and water discharges); and how it contributes to an environmentally-sound product (this is explained below). The manufacturing of construction equipment also has an impact on the environment in similar ways to that of materials.

An examination of buildings provides a good picture of the environmental impact of the products of the construction industry. These are related to the interior of the buildings as well as to the impact of the buildings on the wider environment.

The interior of buildings can have environmental health impacts, which have been referred to as the “sick building syndrome”. The problematic materials used in construction include volatile organic compounds (that may be adhesives, plastics or paints that emit gases, sometimes over a long period, that may affect breathing or be irritants), and a range of materials such as asbestos which have possible carcinogenic effects. Dampness within buildings can also produce moulds, the spores of which affect the health of the occupants of the buildings. There are also psychological health concerns. For example, failure to provide adequate natural lighting might lead to the creation of claustrophobic spaces and feelings among occupants of being trapped or being too enclosed.

In regard to the impact of buildings on the wider environment, energy use is of the utmost concern (this is analysed in detail later in this section). Another important aspect is the use of water, given the increasing scarcity of this natural resource. Buildings also need to be durable – it is important to bear in mind the alarming rapidity with which many modern buildings decay and need to be demolished. The early decay of an existing building speeds up the need for a new building to replace it, with all its possible environmental implications. When an existing building does need to be torn down, proper waste management and recycling may reduce the environmental impact of demolition.

Finally, the clustering of buildings and infrastructure in a given area (the urban fabric) has a significant impact on microclimates, ecosystems and local biodiversity.

While all the different types of environmental impact of the construction industry are important and need to be addressed, there has been a preponderant worldwide attention on energy use and its impact on climate change, given the risk it implies for the survival of the planet. The construction industry (and buildings in particular) is the sector of the economy with the largest impact on climate change. The remainder of this section provides more information on this issue. It brings together selected parts of the ILO's global study on green jobs in construction (Keivani et al., 2010).

According to the Inter-Governmental Panel on Climate Change (IPCC) (2007) report, the building industry was responsible for 8.6 $GtCO_2$ in 2004. The report presents two scenarios which project these emissions to 11.4 $GtCO_2$ (the scenario of lower economic growth) and 15.6 $GtCO_2$ (the scenario of rapid economic growth, particularly in developing nations) emissions in 2030, representing an approximately 30 percent share of total CO_2 emissions. In member countries of the Organisation for Economic Cooperation and Development (OECD), buildings account for 35–40 percent of national CO_2 emissions from the use of fossil fuels. In developing countries, coal and biomass are significant sources of energy for heating in buildings, invariably with adverse effects on the occupants. The fluctuations in extreme temperatures, with increases in the intensity of tropical storms or heavy rainfall events, will directly affect buildings. Although heating energy use will decrease in cold climates, the demand for cooling will increase. At the same time, many of the passive- and low-energy techniques for cooling buildings that are needed to reduce their contribution to GHG emissions (such as evaporative cooling, or night ventilation) will become less effective as heat waves become more intense and longer-lasting (Urge-Vorsatz et al., 2007). Therefore, urgent mitigation action is needed to reduce energy use in buildings – both in the design of new buildings and through the refurbishment of existing buildings that, together, provide the greatest low-cost mitigation opportunities among all economic sectors (IPCC, 2007). Thus, the construction industry is fundamentally important in any climate change mitigation effort.

A study by Urge-Vorsatz and Novikova (2008) indicates that the greatest economic potential (at net negative costs) for mitigating CO_2 emissions in buildings lies in developing countries. This is because many of the low-cost opportunities for CO_2 abatement have already been captured in the more developed economies due to the progressive policies that are in place or in the pipeline. Urge-Vorsatz and Novikova (2008) estimate that, by 2020, globally, 29 percent of the projected baseline emissions can be avoided cost-effectively through mitigation measures in buildings. However, developing countries have the largest cost-effective potential abatement with up to 52 percent of the total reduction, transition economies with up to 37 percent, and developed countries up to 25 percent. It is also worth noting that, in developing countries, the largest CO_2 mitigation potential results from savings in the use of electrical appliances, whereas in developed countries this saving is gained from measures oriented to space and water heating. The former are considered less complicated in terms of switching to more efficient alternatives with quick payback periods, while the latter involve shell retrofitting and fuel switching that are often more expensive and require longer payback periods.

In order to assess, with the available information, the potential areas of energy-efficient construction or refurbishment of buildings, it is important to identify the different kinds of energy end-uses in the residential and commercial sectors in different climates. Urge-Vorsatz et al. (2007) present a breakdown of energy end-use in residential and commercial buildings in the US, Canada and the EU, whilst the IPCC (2007) presents a similar breakdown of energy end-use in the US, based on more recent sources of information. These combined figures are reflected in Figure 15.1, representing typical end-uses in developed countries.

Space heating represents the single largest use of energy in residential buildings in these industrialised regions, followed by water heating. Space heating also represents the single largest use of energy in commercial buildings in both Canada and the EU, accounting for up to two-thirds of total energy use. Lighting is the largest single use of electricity in commercial buildings in the US. Water heating is not significant in commercial buildings in the developed countries.

Figure 15.2 depicts the breakdown of energy end-use in the residential and commercial sectors in China, an economy in transition. The single largest end-use of energy in residential buildings in China is space heating, followed by water heating and electric appliances. The largest energy end-use in commercial buildings in China is space heating, followed by water heating, lighting and cooling.

In regard to developing countries, Figure 15.3 reflects data for Brazil from Poole and Geller (1997) and for South Africa (residential sector only) from the *Energy Outlook for South Africa* report (Department of Minerals and Energy et al., 2002). The largest end-use in the residential sector in Brazil is refrigeration, followed by lighting and water heating. Cooking is the largest end-use in the residential sector in South Africa, followed by space heating. Lighting is by far the single largest use in the commercial sector in Brazil, followed by air conditioning and refrigeration. The differences in hot and cold climates in these two countries account for the disparity in end-uses.

The differences in the energy consumption patterns among representative developed and developing countries can best be compared by ranking the end-uses, as shown in Table 15.1. The consumption patterns are different. Space heating and water heating predominate in the US, Canada and the

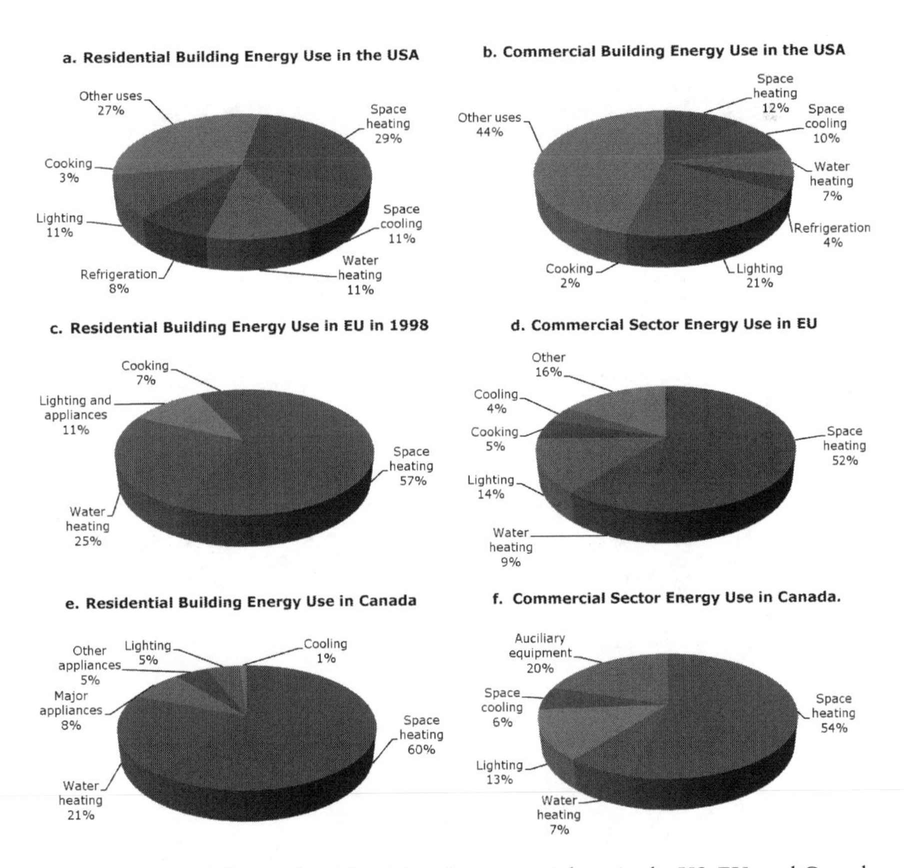

Figure 15.1 Breakdown of residential and commercial use in the US, EU, and Canada (sources: a, b: Keivani et al. (2010) after IPCC (2007); and c to e: Keivani et al. (2010) after Urge-Vorsatz et al. (2007))

EU. However, energy used in cooking predominates in South Africa and Mexico, whereas refrigeration energy consumption predominates in Brazil. Overall, no clear pattern emerges for household energy consumption in developing countries. This is largely due to the disparity in climatic and weather factors that necessitate the partitioning of a country into smaller units in order to obtain a clearer picture.

The choice between renovating existing buildings and constructing new ones depends largely on individual cases in different regions. However, supporting evidence points to a huge potential in energy savings from the renovation or refurbishing of existing buildings rather than the construction of new ones. As noted above, in the near to medium-term future (at least up to 2020), new buildings will comprise only 15 percent of the total housing stock, inevitably focusing attention on refurbishment of existing stock as the main route for addressing targets on CO_2 abatement. In addition, studies carried out by the European Alliance of Companies for Energy Efficiency in Buildings (EuroACE)

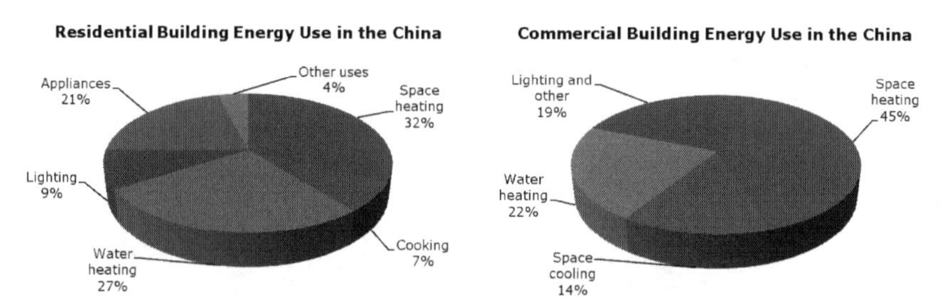

Figure 15.2 Breakdown of residential and commercial use in China (source: IPCC (2007))

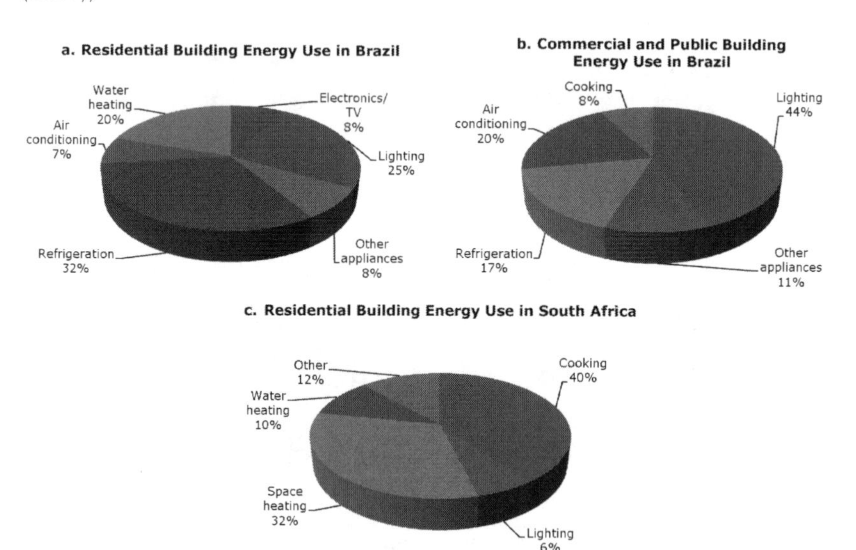

Figure 15.3 Breakdown of residential and commercial use in Brazil and South Africa (source: a, b: IPCC (2007); and c: Keivani et al. (2010) after Department of Minerals and Energy et al. (2001))

demonstrate an energy saving potential of over 50 percent, while Danish studies (Bach, 2006) indicate an energy saving potential of 40–60 percent if energy efficiency measures are implemented. A review of the relevant literature reveals that a raft of improvement measures for sustainable refurbishment is being recommended for both residential and non-residential buildings (EBN, 2007; RICS, 2007a, 2007b; and Smith, 2004) in hot and cold climates worldwide. A summary of typical measures is presented in Table 15.2.

In sum, construction has a significant impact on the environment. Consequently, the greening of the construction industry has a major potential to help address the environmental challenges currently faced worldwide. In order to implement changes in the construction industry, and indeed to consider related changes in jobs in the industry, it is important to discuss the challenges faced by the labour force as the workers are the ones who will implement the greening of the industry in practice.

2 Labour aspects of construction

The concept of decent work is used as a framework to analyse the labour aspects of construction. Decent work is a concept used by the ILO and other organisations in the "world of labour" to analyse and address the different aspects of work. It has four elements, namely: employment generation, social protection, workers' rights and social dialogue. These elements

Table 15.1 Residential energy consumption pattern in selected countries

	Developing countries			*[1]Developed countries*		
	[2]Mexico	*[3]Brazil*	*[4]South Africa*	*USA*	*Canada*	*EU*
Cooking	1	///	1	8	///	4
Space heating	///	///	2	1	1	1
Water heating	2	3	3	2	2	2
Lighting	2	2	4	5	4	3
Electronics		4				
Refrigeration		1		3		
Air conditioner	3	5				
Space cooling				4	5	
TV				6		
Clothes dryers				7		
Furnace fans				9		
Appliances	3				3	
Washing machines						

Source: Keivani et al. (2010), after: [1]Ürge-Vorsatz et al. (2007), [2]Olivia (2007), [3]Poole and Geller (1997), [4]Department of Minerals and Energy et al. (2002).

Table 15.2 Global improvement measures for sustainable building refurbishment

Category	*Improvement measures*
Energy – Building Fabric and Envelope	Air-seal foundations Moisture-proof basement/insulate walls Draught proofing/air-seal building Building fabric – external wall insulation enhancement Upgrade windows/enhanced glazing insulation Modify windows/shading devices to reduce heat gain and solar controls Insulate floors Insulate roofs Make roof reflective
Energy – Mechanical and Electrical Systems	Install energy-efficient lighting and appliances Install energy-efficient heating Install energy-efficient ventilation Optimize pipe sizes Optimize ducts Upgrade pumps Upgrade chillers Install energy-efficient appliances
Energy – Renewable Energy Sources	Install solar water heating Install combined heat and power (CHP) Install ground source heating and cooling pumps (GSHP) Install micro wind turbines Install photovoltaic
Water Efficiency Measures	Install water consumption meters Install water management plan Install rainwater recycling Install dual flush toilets Install reed bed water treatment Install water efficient devices/fittings Drainage irrigation
Waste Reduction Measures	Select carpets with high level of recycled content Select vinyl flooring with high level of recycled content Selection of carpet underlay with up to 100 percent recycled content Use plasterboard containing high levels of recycled material Use recycled building materials Use recycled internal features Use external recycled features Recycle complete structures
Sustainable Facility Management	Install building energy management system (BEMS) Implement efficient maintenance strategies Implement energy-efficient policies and staff awareness and training

Source: Keivani et al. (2010).

are analysed in turn in relation to construction. The section concludes with some specific considerations for green jobs.

The construction sector generates 5 to 15 percent of gross domestic product (GDP) at the national level, approximately 10 percent of global GDP, and approximately 7 percent of the world's workplaces. This accounts for 5 to 10 percent of employment at the national level, amounting to over 111 million people directly employed worldwide, with 75 percent in developing countries and 90 percent in micro firms (entities with less than 10 employees) (Keivani et al., 2010). However, aggregate figures hide problems in the conditions of employment, which are explained next.

The following analysis brings together selected parts of the report of the latest comprehensive meeting of the ILO on labour in construction (ILO, 2001). The characteristics and trends of the industry noted in such a report still prevail.

The products of the construction industry are fixed in space, so production takes place on a project-by-project basis with the production site constantly moving. This implies that the labour force has to be mobile. The construction industry has a long tradition of employing migrant labour. During the process of economic development, work in construction provides a traditional point of entry to the labour force for migrant workers from the countryside. Construction work is often the only significant alternative to farm labour for those without any particular skill or education and it has special importance for the landless.

When the pool of surplus labour from the countryside has dried up, or there is a shortage of local labour for other reasons, labour may be recruited from overseas. Migrant construction workers are generally from less developed and lower wage economies with labour surpluses. Despite migration of some of the workers to richer countries, the majority of construction workers are still in developing countries. However, as many construction workers in these countries are informally employed and therefore not counted in official data, the real numbers are much higher than the published data.

The reason for the greater employment-generating potential of construction activity in the developing countries can be traced to differences in technology. There is a very wide choice of technology available for most types of construction and the technology adopted tends to reflect the relative cost of labour and capital. In the richer countries, machines have largely replaced workers in many of the tasks involved in new construction (although repair and maintenance is still very labour intensive). In developing countries, the majority of tasks are undertaken by manual methods.

Despite the employment-generation potential, instability of work is one of the major problems facing the construction industry. Fluctuations in demand, the project-based nature of construction and the widespread use of the contracting system all conspire to make it difficult for contractors to obtain a steady flow of work which would allow them to provide continuity of employment. Hence there is a constant friction between the need of employers for "flexibility" and the need of workers for stable jobs. It has become the norm for construction workers to be employed on a short-term basis, for the duration of the whole or part of a project, with no guarantee of future work. The number of casual as well as informal workers has greatly increased, including in developed countries.

Due to the kind of "triangular employment relationship" (contractors – sub-contractors and labour agents – workers) and related casualisation, there are considerable barriers to training in the industry. The heightened division of labour into ever more specialised trades, which is implicit in sub-contracting, limits the range of skills that can be acquired in any one enterprise. This means that all-round craftsmen and general supervisory workers are difficult to train. In many countries, the public sector used to provide stable employment and a good training ground, but its role in training has diminished as public-sector units have been disbanded.

In addition to the implications for employment generation, the characteristics and trends of the construction industry also have an impact on the three remaining elements of decent work.

In regard to social protection, occupational health and safety deserves special attention, as construction is one of the most hazardous occupations. Data from a number of industrialised countries show that construction workers are three to four times more likely than workers in other sectors to die from accidents at work. Many more suffer and die from occupational diseases arising from past exposure to dangerous substances such as asbestos. In the developing world, the risks associated with construction work are much greater: available data would suggest three to six times greater. This is explained by the fact that sub-contracting, on a piecework basis, intensifies the pressure to produce while increasing the difficulties of coordinating work and ensuring site safety. It is estimated that 95 percent of serious accidents involve workers employed by sub-contractors. Most of these workers are on temporary contracts that, in a context of fluctuating demand, encourages them to work long hours in order to make the most of work while it lasts. They are also less likely than workers in permanent contracts to gain the training and experience required to work safely in a dangerous working environment, and they are in a weaker position to refuse to work in unsafe conditions. A construction worker with a fixed-term contract is three times more likely to suffer an occupational accident than one with a permanent contract. Informal workers are particularly vulnerable. Inappropriate conditions of health and safety are not only a question of workers' rights, but they also have a negative impact on their productivity. For instance, Figure 15.4 shows that there is a clear inverse correlation between the rank of a country in terms of competitiveness and the number of fatal occupational accidents.

Still related to social protection, there is evidence from many countries that employers do not pay contributions into social security funds for

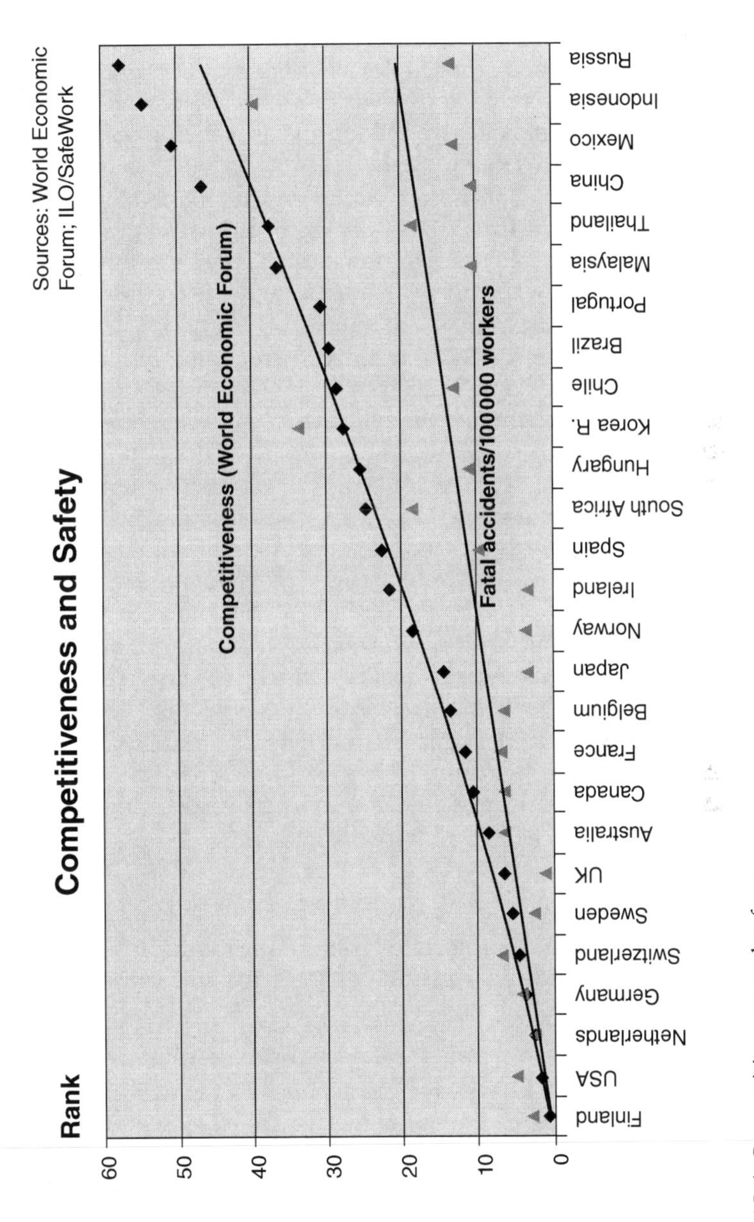

Figure 15.4 Competitiveness and safety

workers who are on temporary contracts. Hence, the workers who are most in need receive no healthcare, no holiday pay, and no protection against loss of pay when they are unable to work due to unemployment, ill health, accidents or old age. Many of the self-employed are believed to be actually working for employers, but to be doing so without social security contributions. Similarly, most workers on temporary contracts are not entitled to benefits during periods of unemployment between contracts.

There are also decent work deficits related to rights. In this section, the "triangular employment relationship" in construction (contractors – sub-contractors and labour agents – workers) and related casualisation has been explained. Under such circumstances, workers' rights are often unclear and workers may enjoy less protection from the law than those who are directly employed. The same applies for the informal sector. In regard to migrants, while many countries have recognised their dependence on such workers and have attempted to regularise and control the process of migration for work in the construction industry, there are still many foreigners working illegally in many countries. These workers are vulnerable to exploitation. Even when they are legally employed, migrant construction workers are sometimes deprived of trade union membership and other rights. The recruitment process is also very often exploitative.

Finally, there are also implications for social dialogue. Social dialogue can include various forms of negotiation, consultation or simply exchange of information between representatives of governments, employers and workers, on issues of common interest relating to economic and social policy. Social dialogue with employers – and also with governments – has traditionally been a powerful means for workers to collectively bargain for better wages and better conditions of work. Box 15.1 presents one example of how social dialogue led to benefits for both (unionised) workers and employers in construction. However, nowadays, the vast number of temporary, casual, informal and unemployed workers find it difficult to organise themselves to engage in social dialogue.

Box 15.1 Social dialogue for job creation and improvement of working conditions in the construction industry in Hong Kong

The construction industry was severely hit by the financial crisis throughout the world, and Hong Kong was no exception. Unemployment rose, forcing many construction workers including some who had worked in the industry for more than 20 years to look for employment in other sectors. Throughout the world, it was estimated that at least 5 million construction workers were laid off during 2008, with a similar estimation for 2009. Building and Wood

Workers International (BWI) reported that, in August 2009, while the official unemployment rate in Hong Kong was at 5 percent, the rate for construction workers was much higher, at 12.7 percent. In addition, in a trade union survey, 50 percent of those interviewed stated that they were underemployed and at least 20 percent stated that they have been out of work for a month or longer.

In an effort to address the crisis, the Concrete Industry Trade Union (CITU) held dialogues with the Mass Transit Railway Corporation (MTRC) which was established in 1975. The sole shareholder was the Hong Kong Government. The company was re-established in June 2000 after the government sold 23 percent of its issued share capital to private investors. MTRC shares were listed on the Stock Exchange of Hong Kong on 5 October 2000. The operations of the other government-owned rail operator, the Kowloon–Canton Railway Corporation, were merged into the MTRC on 2 December 2007.

Afterwards, workers (CITU) and employers (MTRC) together approached the Hong Kong Government to begin construction of infrastructure projects that it had proposed earlier. According to BWI, CITU and MTRC successfully lobbied the Hong Kong government to approve HK$12 billion for the construction of the West Island Line project. This new project, which began in August 2009, plans to create 5000 new jobs for the construction industry. There are currently 10 project sites under construction with 2000 workers already employed by sub-contractors of MTRC. According to MTRC, the breakdown of the total of 5000 jobs is as follows:

- Bar benders: 7 percent;
- Carpenters: 8 percent;
- Concrete workers: 7 percent;
- Plant operators: 11 percent;
- Skilled workers (building facilities and electrical devices): 21 percent;
- Skilled workers (civil) and general workers: 46 percent.

The project will also provide a further 1600 jobs for more supervisory and technical positions, such as:

- Director/project manager: (3 percent);
- Assistant project manager/Senior engineer: (15 percent);
- Engineer/Assistant engineer: (39 percent);
- Site supervisor (technical): (43 percent).

Furthermore, in early November 2009, MTRC invited the Hong Kong unions to conduct occupational health and safety training on the sites. The unions were scheduled to conduct other rounds in 2010, when MTRC employs more workers in the project sites. In a recent meeting between MTRC and the unions, other issues such as back wages and the sub-contract system were also discussed.

Source: Information provided in November 2009 by the Asia and Pacific Office of the Building and Wood Workers International (BWI).[1]

Specific considerations for green jobs

In sum, the construction industry faces a set of challenges related to the different elements of decent work. It is important to bear them in mind in the discussion of green jobs specifically. It is necessary to promote green jobs hand-in-hand with improvements in decent work, particularly in developing countries where the conditions of workers are worse than in industrialised countries.

Green jobs entail changes in the production process, with implications for working conditions, which also need to be addressed. One example relates to occupational health and safety. Green jobs include activities involved in the production of new (green) building materials and the assembly of new equipment (such as solar panels and wind generators). They require specific techniques to protect the workers, which need to be taken into account. Cases of accidents involving workers in activities such as assembling wind generators have been reported (see, for example, O'Neill, 2009). Box 15.2 presents a summary of information from articles comparing health and safety in existing green and non-green construction, noting that problems remain in existing green construction.

Training is also particularly important in developing countries, where green jobs are still incipient. The creation of green jobs entails an opportunity to provide training for specific target groups, such as women, who have had limited openings in construction. Young people can also be targeted, as they often come to the construction industry without much in terms of skills – therefore, training in green jobs can give them specialised skills to enter the market. Specialised training for migrant workers can also give them more power to negotiate better working conditions.

1 The Asia and Pacific Office of BWI confirmed and updated information for two articles on BWI's website: http://www.bwint.org/default.asp?index=2383&Language=EN and http://www.bwint.org/default.asp?index=2360&Language=EN

Box 15.2 Green construction is no safer

Green construction is no less hazardous for workers than the less environmentally concerned alternative, a US study has found.

Researchers from Oregon State University and East Carolina University compared reportable injury and illness rates at building sites to the highly sought after green Leadership in Energy and Environmental Design (LEED) standard with comparable conventional sites (Rajendran et al., 2009).

The study concluded: "There appears to be little or no difference between green and non-green projects in terms of construction worker safety and health. With both green and non-green buildings having the same safety performance, a question arises as to whether LEED buildings should be labelled as sustainable buildings."

In a separate paper in the same issue of the *Journal of Construction Engineering and Management*, the same authors call for a sustainable construction safety and health (SCSH) rating system to "rate projects based on the importance given to construction worker safety and health and the degree of implementation of safety and health elements" (Rajendran and Gambatese, 2009).

The public health blog The Pump Handle backs the approach. A 15 April blog entry notes: "What a great idea! This takes a LEED-like approach to rating worker safety and health.

"Now, how do we roll out this good idea with LEED-like flare and fanfare such that occupational safety and health becomes as attractive and interesting to the masses as the concept of green buildings?"

Source: O'Neill (2009)

Informal workers constitute another significant target group, given their large numbers and the decent work deficits that they face. Training in green construction may open opportunities for bringing informal workers into the mainstream of the economy to work as employees or to provide services to enterprises attending to the growing demand for green construction. At the same time, it is worth noting that informal construction workers play a crucial role in the building and upgrading of low-income neighbourhoods throughout the developing world. Formal construction enterprises have had at best a limited penetration in such neighbourhoods. The low-income families and community-based organisations usually rely on local informal workers to undertake construction activities. Even in so-called "self-help" construction, such workers are frequently hired to carry out at least the most technically demanding activities. Therefore, the training of informal workers may be a conduit for green practices to be applied in low-income neighbourhoods, which constitute a sizeable share of the built environment in developing countries (see also Box 15.3).

Box 15.3 Avoiding a green/brown divide

With few notable exceptions, the existing practice in green construction (and following on, green jobs in construction) in developing countries tends by and large to focus on more sophisticated buildings located in higher-income neighbourhoods. While this may be the beginning of a trend that could scale up to other neighbourhoods, it is important to make sure that this does take place. Otherwise, there is the risk that cities could increasingly have a green/brown divide: green construction for the rich and non-green for the poor, thereby exacerbating intra-urban differentials that already exist in developing countries. While the greening of low-income neighbourhoods encompasses many factors – such as decision-making for investments and the combat of poverty – the training of those who actually build the low-income neighbourhoods is an important part of the picture.

Source: Werna (2000)

Having discussed some of the major issues pertaining to construction labour and some implications for green jobs, the chapter now analyses the potential for the creation of such jobs.

3 Employment generation and green jobs

Employment in the green building sector directly accounts for 5–10 percent of the people employed in the construction industry around the world (UNEP, 2008).

According to UNEP (2011), considering the different types of environmental impact of the construction industry noted above, there are many channels through which green construction (and green buildings in particular) generates employment:

- the construction process (air emissions, noise pollution, waste disposal, water discharges);
- construction of new green buildings;
- construction of durable buildings (postponing demolition);
- retrofitting old buildings;

- increased production of green building materials, products, appliances and components;
- energy-efficient operation and maintenance of existing buildings;
- expansion of renewable energy in the total energy mix;
- waste management and recycling in demolitions.

It is pertinent to note that contaminated sites, while related to construction as noted above, are not included here because their clearance generates green employment outside the construction industry. Some examples of the potential under discussion here are found, for instance, in UNEP's *Green Economy Report* (UNEP, 2011). The following paragraphs in the present section are based on preparatory work for this report (see "acknowledgements").

New building construction creates moderate employment growth. An analysis of new green jobs in construction in the US evaluated the types of jobs that are expected to increase or decrease as conventional building construction is replaced by green building (Booz Allen Hamilton, 2009). The study predicts increases for:

- construction of new commercial, healthcare and manufacturing buildings;
- construction of other new non-residential buildings;
- construction of new residential single- and multi-family buildings; and
- maintenance and repair of non-residential structures.

Job categories that will contract due to utility, operations and maintenance savings are:

- maintenance and repair of residential and non-residential buildings;
- waste management and remediation services;
- water, sewage, and other water treatment systems; and
- electric power generation, transmission and distribution.

The data for LEED buildings show incremental job creation of about 60 jobs per million square feet of green building constructed, including direct, indirect and induced jobs. Green building engineers and architects, and other green building consultants, will benefit from this growth. There are currently over 133,500 LEED-accredited professionals in the US, another 1500 in India, 900 Green Star professionals in Australia and 1197 BRE Environmental Assessment Method (BREEAM)-licensed professionals in the UK. These figures have steadily risen over the years and are expected to grow dramatically.

Existing building retrofits, in particular, show huge potential for employment growth. Several studies have estimated that every US$1 million invested in building efficiency retrofits would create 10–14 direct jobs and 3–4 indirect jobs. Using a value of 12.5 jobs per US$1 million invested, one study (Hendricks et al., 2009) calculates the jobs that would be created if 40 percent of the American building stock (50 million buildings) were to be renovated by 2020 with an average investment of US$10,000 per retrofit. This would result in a US$500 billion market that would lead to 6,250,000 jobs over a period of 10 years.

Retrofitting programmes across the world are beginning to address this opportunity. For example, a massive retrofitting project undertaken by the German Alliance for Work and Environment led to around 140,000 new or saved jobs, and a study of ten new EU member states projected that 50,000–185,000 jobs would be created from the retrofitting of the existing residential building stock (UNEP, 2008). The growing interest in the opportunities for retrofitting is also increasingly providing evidence that training is of crucial importance in the conversion to green. Indeed, lack of skilled and certified professionals is proving to be a barrier to the adoption of green buildings, especially in developing countries.

In addition to employment arising within the construction industry, important additional employment opportunities are also generated from the manufacture of green materials and products, and from renewable energy. For example, research by the US Department of Energy estimates that adopting standards for washing machines, water heaters and fluorescent lamps alone would create 120,000 jobs through to 2020. The Apollo Alliance (UNEP and ILO) also estimates that an investment of US$3.5 billion to modernise appliance standards would result in 29,876 jobs. In India, the introduction of a single type of appliance, the fuel-efficient bio-mass cooking stove, to replace the traditional stoves in 9 million households could produce 150,000 jobs with an added benefit of reducing negative health impacts due to poor indoor air quality.

Specific information on green employment in construction in selected developing countries, including China, is presented in the case studies that follow.

Case study 1: China

The building industry in China is massive and has tremendous potential for creating green and decent jobs across a range of professions, including engineering and architecture, project management and auditing, various technical trades (such as plumbing, installing heating, ventilation and air conditioning (HVAC) installations and lighting), and direct construction work on site. According to the World Energy Outlook (IEA, 2007), China is experiencing an unprecedented construction boom with 2 billion square metres of new building stock being added every year.

Building on its previous efforts in reducing energy consumption in buildings (since 1986 the Chinese Ministry of Construction has adopted a number of codes and standards emphasising energy conservation in buildings), the Chinese government brought the debate on energy efficiency in buildings to the next level by adopting the *Evaluation Standard for Green Building GB/T50378* in 2006. This code is the green building evaluation system for China and is combined with the other two relevant codes, the *Management Methods for Green Building Evaluation and Certification* and *Technical Code for Evaluating Green Buildings*, both adopted in 2007. Together with different subsidies and tax incentives, these codes are designed to promote energy efficiency in buildings and fuel the growth of the green building sector. Since 2007, the Chinese government has provided funding for energy efficiency in buildings in five areas:

1 Financial support for retrofitting of existing buildings;
2 Tax refunds for companies promoting energy-efficient building;
3 Financial support for five model cities: Shanghai, Beijing, Tianjin, and Shenzhen;
4 Funding for development of regulations in public buildings;
5 Subsidies for model projects that meet the 65 percent energy saving target.

Both new building construction and retrofitting of existing buildings provide opportunities for creating new green jobs, with the emphasis frequently being put on the latter, according to some sources (for example, the IPCC maintains that retrofitting and replacing equipment in buildings has the largest potential within the building sector to reduce greenhouse gas emissions by 2030). Given this viewpoint, retrofitting the existing stock of buildings in China increases employment opportunities and fuels local and regional economies. The Chinese government has recognised this potential and has issued a series of regulations and guidelines to push for retrofitting of existing buildings, although there are no minimum retrofit standards in place yet. According to the Ministry of Housing and Urban–Rural Development, at least one third of the total of 42 billion square metres of existing buildings in China – that is, about 14 billion square meters – needs retrofitting. If the retrofitting cost is 200 Yuan per square metre, the market for the retrofit of existing buildings would be 2.6 trillion Yuan.

Apart from the big sums of money that will be brought to bear on the retrofitting of existing buildings in China, the large number of jobs that will be created due to the labour-intensive nature of such activities will contribute to employment growth in the construction industry. According to some estimates, around 12.6 million new jobs will be created due to the effect of energy conservation measures in the construction industry in China. Most of these jobs will go to migrant workers, as the construction industry in China relies almost entirely on people who leave their home towns to work on sites in cities. This trend will raise labour-related issues, including provision of minimum wage, rights, safety, and health benefits, as migrant workers are less educated and lack the necessary means to protect their interests at work. Moreover, although far from being perfect, due to the lack of necessary education and skills among the migrant workers, the government has prioritised and created training programmes that address capacity building in the building industry, therefore preparing the groundwork for the creation of long-term green jobs in the industry. For example, the Shandong provincial government has established 69 training bases and 91 certification agencies to cover 17 cities and 70 counties, which can train over 60,000 migrant workers each year.

Source: ILO(2010)

Case study 2: Bangladesh

By employing 1.52 million people and contributing around 267.9 billion Taka* to Bangladesh's GDP every year, the construction industry is one of the major sectors of the country's economy. The size of the labour force engaged in construction activities has grown significantly since the 1990s and is projected to grow further in the coming decade. It increased from 525,000 people in 1991 to 1,524,000 in 2006 and then to 2,024,673 in 2009 (indicating an average annual growth rate of 7.36 percent). Based on this growth rate, it is projected that the construction labour force will grow to 2,887,803 people in 2014 and eventually to 3,328,530 in 2020. Green building construction and creating green jobs in the sector are new phenomena that increasingly receive attention in the private and public sectors, mainly due to concerns over the protection of the environment. Although currently there is no specific policy that addresses green jobs in the construction industry, there are some policies in place that have green building and green jobs implications. For example, both Environmental Conservation Act (1995) and Environmental Conservation Rules (1997) of Bangladesh emphasise the importance of environmental impacts of buildings, and a number of regulations adopted by the Public Works Department require incorporation of environmental measures in construction work. Similarly, the Real Estate and Housing Association of Bangladesh prohibits the use of firewood for burning bricks used for construction purposes, and the National Land and Transport Policy requires that Environmental Impact Assessments be undertaken and cleaner energy in the construction industry promoted.

The ILO's global report on green jobs in construction (Keivani et al., 2010) indicates that 66 percent of real estate companies consider incorporating green building criteria in site selection, choice of construction materials, electrical appliances and equipment, and utility service system design, and half of them use up to 80 percent locally produced building materials such as bricks, concrete, sand and cement. Following this, if 66 percent of real estate companies in Bangladesh consider incorporating green building technology and construction materials in their projects – and considering that the construction industry as a whole employed 2.025 million people (2009) – there were as many as 1.34 million workers engaged in work

*US$1.00 = Bangladeshi Taka (BDT) 74.80.

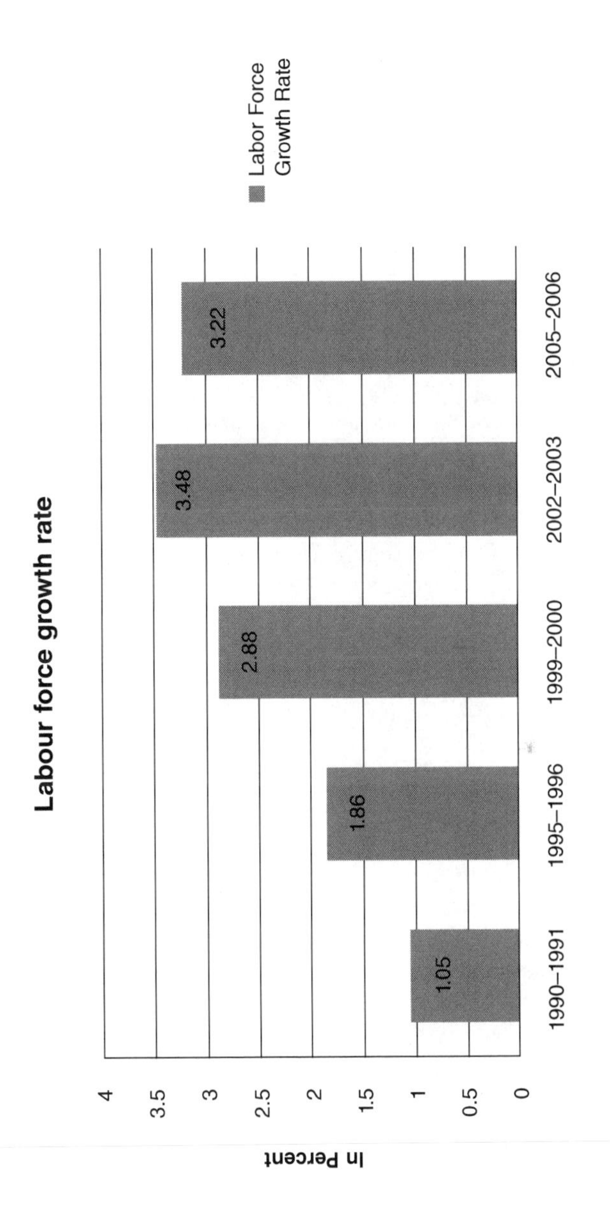

Figure 15.5 Labour force growth rate in construction in Bangladesh, 1990–2006

that adopted or at least considered green jobs requirements in 2009. Based on the observed 7.36 percent growth rate in the construction industry, it is projected that green employment in the construction industry will grow to between 1.91 million and 2.927 million people in 2014 and 2020 respectively. Based on the same assumption that 66 percent of real estate companies incorporate green building technologies and materials on their projects, and that the calculated growth rate in GDP of the construction industry has been 6.7 percent during the period 1996–2009, the net contribution of green employment GDP will increase from 299,372 million Taka in 2009 to 499, 951 million Taka in 2020.

However, there are a number of barriers to the promotion of green construction practices and creation of green jobs in the construction industry in Bangladesh. These barriers entail a range of issues including policy, institutional, technical, financial and informational. The lack of clear policy for modernising the construction industry, or providing financial incentives to encourage energy efficiency, has been a major constraint. Institutional barriers range from lack of coordination among relevant government bodies, through lack of interest in improving design and planning standards, to lack of adequate knowledge about green construction technologies among the parties concerned. Technical barriers include lack of knowledge of energy efficiency in buildings, lack of expertise in the use of green construction materials, and deficiency of green technologies available in the country. Financial barriers include resource limitations, and information barriers range from lack of awareness of environmental sustainability, through lack of information about green construction technologies, to lack of awareness of quality standards in the production of different construction materials.

Source: Waste Concern Consultants (2010)

Case study 3: South Africa

With a share of 8.28 percent of the total workforce in South Africa (as of June 2008), the construction industry is one of the country's major economic sectors. With the current need for housing and the government's declared commitment to invest in infrastructure development, the industry is expected to grow significantly over the coming years. The South African government has stated that it will spend R400 billion on infrastructure projects in the coming years. There is also a housing backlog of approximately 2.1 million units (ILO, 2009). Creating jobs and improving energy efficiency in the building industry is an issue that has been receiving increasing attention from several government bodies since 2005. Although the absence of solid legislative measures remains a concern, the government's various departments, institutions and programmes for the development and growth of the building industry strongly support energy conservation and energy efficiency as part of their larger sustainable development endeavours. The

National Energy Efficiency Strategy and the *Intensive Multimedia Energy Efficiency Campaign* are some of the initiatives of the Department of Minerals and Energy that address energy efficiency in buildings. Similarly, the Department of Public Works, a major property owner and developer of public infrastructure and buildings, is integrating energy efficiency in buildings into its programmes in fulfilling the infrastructure and building needs of the country. The Department of Environment and Tourism has also developed a national framework for sustainable development that provides support for programmes to realise energy conservation in buildings.

With respect to green jobs, the South African government has several policies and campaigns to transform the response of the construction industry to the prevailing environmental concerns. One of these measures taken by the government is the retrofitting of the existing stock of public buildings. Over 100 buildings in Pretoria, the Western Cape and the Free State have already been retrofitted and there are plans to retrofit another 106,000 government buildings throughout the country in the near future. Retrofitting provides an excellent opportunity for creating new green jobs, and can be supplemented by jobs generated through building new infrastructure facilities in the country, something that South Africa needs. Calculations made by the ILO (2009) in its "Employment Aspects of Energy-Related Improvements in Construction in South Africa" put the total number of estimated new jobs due to greening buildings at around 13,833 between 2009 and 2020, or 1153 new jobs per year. However, the ILO acknowledges that, in South Africa, the relationship between technological changes in energy efficiency in buildings and employment creation from such changes is quantitatively weak (that is to say, the total number of jobs created will not rise significantly) in so far as the required skill sets to a large extent already exist. On the other hand, this relationship is qualitatively strong in that up-skilling of the existing skill sets is needed in order to improve productivity and create decent jobs in the building industry.

There are a number of barriers and challenges in creating green jobs in the building industry in South Africa. These challenges mostly relate to the industry itself, and ultimately translate into barriers to the growth of the green building industry and job creation. The building industry in South Africa is not well developed and there is a lack of knowledge on the part of the developers which hinders meaningful energy-related improvements. There is a shortage of skills and capacity in the construction companies to undertake sophisticated energy improvements in buildings, and there is a common understanding that the procurement practices and conventions currently used by client bodies deter quality improvement work. Moreover, the perceived additional cost necessary for improving energy efficiency in buildings is a major constraint for developers and affects the growth of the green construction industry significantly.

Source: Van Wyk et al. (2009)

Case study 4: Malaysia

Malaysia's building industry is at an early stage of adopting energy efficiency and green building construction techniques. However, it has the potential for growth and creation of green jobs. The construction industry is a major employer in Malaysia (Department of Statistics Malaysia, 2002, 2004, 2006, 2009). The Construction Industry Development Board (CIDB) of the Malaysian government indicates that the construction industry's contribution to GDP during the period 2000–8 ranged between 2.9 percent and 4.0 percent, with an average of 3.4 percent. The Department of Statistics of Malaysia puts the contribution of the construction industry to total employment in the country at around 9 percent (nearly 1 million people), making it one of the largest sectors of the economy in terms of the size of its workforce. The issue of energy efficiency in buildings has received increasing attention from both the public and private sectors. Since the beginning of 2009, several policy initiatives have been formulated and enforced by the government and the private sector to initiate and encourage energy efficiency improvements in buildings. Although currently there is no policy that mandates building green, the government has invested in showcase buildings and has provided a range of tax incentives to encourage investment in energy efficiency in buildings in the country.

So far, no study has been undertaken by any organisation in Malaysia to estimate the number of green jobs that will be created in the building industry. Although the term "green jobs" is fairly new to the industry, the potential

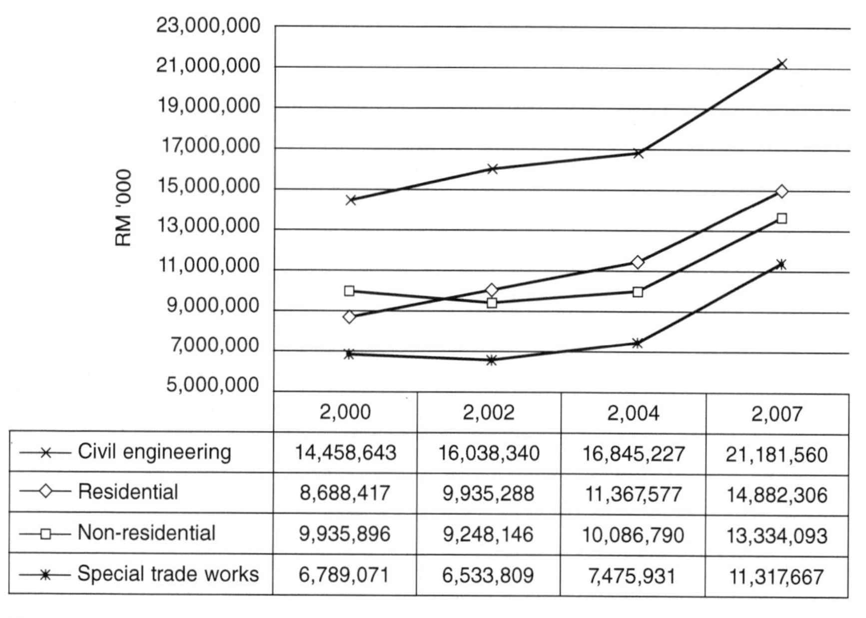

	2,000	2,002	2,004	2,007
Civil engineering	14,458,643	16,038,340	16,845,227	21,181,560
Residential	8,688,417	9,935,288	11,367,577	14,882,306
Non-residential	9,935,896	9,248,146	10,086,790	13,334,093
Special trade works	6,789,071	6,533,809	7,475,931	11,317,667

Figure 15.6 Construction output in Malaysia, 2000–7 (source: Department of Statistics Malaysia (2002, 2004, 2006, 2009))

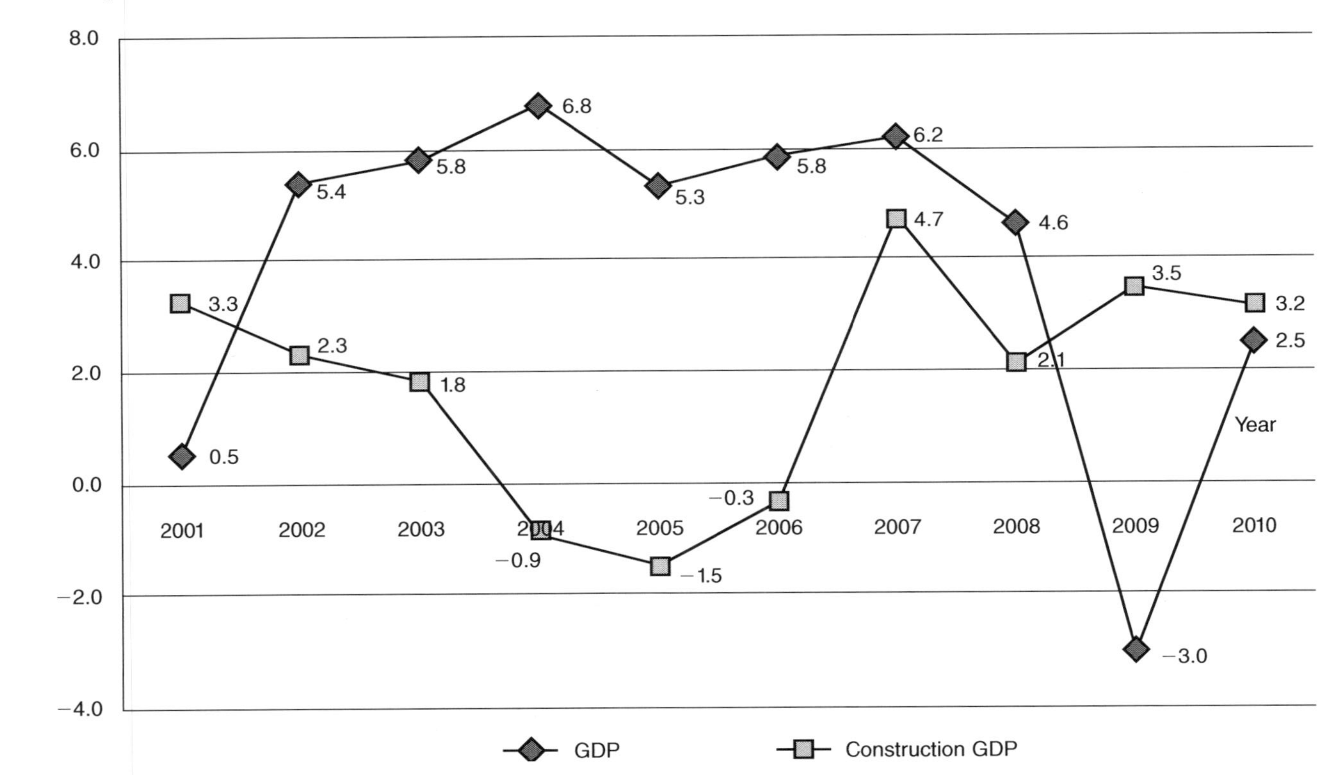

Figure 15.7 Growth rates of national GDP and construction in Malaysia, 2001–10 (source: Ofori and Abdul-Aziz, 2010)

for creating such jobs exists. The ILO's (2010) report, "Energy Efficiency and Green jobs in the Building Industry in Malaysia", which was based on interviews with key stakeholders in the construction industry, indicates that applying energy efficiency measures on new building projects as well as retrofitting the existing stock of buildings will lead to the creation of new jobs in the industry. It anticipates that, with more emphasis on green building technologies and energy efficiency, there will be more pressure on the real estate companies and building owners to invest in retrofitting activities and energy efficiency improvements in buildings – which will ultimately lead to the readjustment of the existing skill sets in the industry and creation of new green professions. In fact, as the ILO study suggests, this has already started to happen to some extent, but more effort and time (and perhaps a new job creation programme) is needed to bring it to a significant new level. With time and training, new positions such as energy simulation and monitoring specialists, building assessment professionals, Green Building Index (Malaysia's green building assessment and certification programme) certifiers and facilitators will emerge. These positions will be new additions to the building construction workforce and will complement the work of more conventional building professionals such as architects, engineers, quantity surveyors and other specialists.

There are a number of obstacles and barriers that need to be overcome in order to promote green jobs in the building industry in Malaysia. These include: limited awareness of the importance of environmental sustainability in the country; the existence of substantial subsidies for fossil fuels; the additional cost of employing energy efficiency technologies in buildings; and the absence of a critical mass of energy efficiency experts.

Labour conditions in the construction industry, including wage, safety and health conditions for construction workers, are far from ideal. Construction work in Malaysia is not perceived to constitute decent work, and this is one of the reasons why the industry has a hard time in recruiting local employees. Wages are often low and inequitable (among local and foreign workers), safety measures are poor, and health conditions hazardous. Furthermore, there are no minimum wage requirements in the industry, despite the government's declared commitment to decent work.

Source: Ofori and Abdul Aziz (2010)

4 Conclusion and recommendations

It is important to discuss ways to address the problems of construction workers, taking into consideration that the characteristics and trends of the construction industry which have generated most of the problems are likely to continue, even if green practices are applied in the industry.

There are many opportunities for green employment in construction. The magnitude of such openings depends on the growth of green construction, which, in turn, depends on the willingness of different actors to invest in it. At the same time, the training of workers and employers in green techniques is likely to motivate them to apply the knowledge they acquire. In this context, workers and enterprises will become a push factor for green construction.

In addition, there is evidence that, for some types of infrastructure, the employment-generating potential of investment in construction may not have been fully realised. This is due to constraints in the planning and procurement of projects, as well as to lack of capacity in the local construction industries, particularly in developing countries. Also, in many such countries, investment in construction is at a very low level. The way forward is to expand the volume of output and employment in the industry, for example through the development of public–private partnerships and an appropriate choice of technology (ILO, 2001). This is compatible with green practices.

Investments in cultural heritage may also create green jobs. There are many documented examples of traditional construction materials and building designs that are environment-friendly and have a low impact on the consumption of energy – and perform better than their modern counterparts. Therefore, the current focus on green construction represents an opportunity to look into such old good practices (Werna, 2009a).

Regarding the limitations to the training of workers, innovative solutions should be considered. For instance, the training of master craftsmen as trainers will improve the on-the-job apprenticeship system that is found in several countries. Another way to supplement skills acquired through the apprenticeship system is to issue target groups with training vouchers, which they can spend as necessary. The merits of this have been demonstrated in a project in Kenya. Also, the involvement of sub-contractors, labour contractors and intermediaries in joint training schemes, with cost reimbursement, seems to be essential if these schemes are to be effective in meeting the real skill needs of the industry (ILO, 2001). These general recommendations could be used for the development of green construction.

It is important to include green techniques in the syllabuses of training courses and/or to have specific green courses, depending on the potential in each country. One example of such potential is the "green plumbers" initiative that is in place in some countries such as Australia and South Africa. Green plumbers have specific knowledge on how to save water, for instance with techniques to collect rainwater and to use grey water to flush toilets. This attracts clients with a concern for the environment as well as those who are interested in saving expenditure on water consumption, therefore generating business and employment for the plumbers.

Training can be used to target specific groups of workers who need further support. The suggestions in this chapter could be used in these regards.

Occupational health and safety still needs attention in construction, and these issues have specific implications for green jobs. Therefore, it is also important to discuss improvements related to this theme.

In some countries it is necessary to update the laws for construction in general and green construction in particular (see, for example: Keivani et al., 2010; Lawrence and Werna, 2009). Moreover, in many other countries, although appropriate laws are already in place, there is a problem of enforcement. There are never enough inspectors to police even the big sites, let alone the myriad of small ones. Corruption is also a problem in many places. The way forward is to change the role of labour inspectors to one of education and prevention, as opposed to inspection and prosecution. This is already being adopted in a number of countries, but should also be considered in the rest of the developing countries (see, for example, ILO, 2001). The greening of the industry brings the opportunity to create synergies between inspections relating to the environmental, and the health and safety components of construction.

There have also been some experiments with safety cards. For example, the Construction Industry Development Board (CIDB) of Malaysia has pioneered a scheme to make every construction worker undergo a safety and health induction course, after which the worker is issued with a green card. Those without the card would be barred from entering work sites, while contractors that fail to send their workers to undergo the training have been threatened with blacklisting. Contractors with major projects are also required to send their management staff to attend the training course. A similar scheme is in operation in a number of developed countries, such as Australia, Ireland and the US (ILO, 2001). Lawrence and Werna (2009) provide examples of further training schemes.

However, securing real improvement in occupational health and safety requires more than advice and training. To meet this challenge, many insurance initiatives have been introduced in developed countries that offer financial incentives to encourage employers to implement accident prevention strategies. Examples of such schemes are a lower annual premium if claims are reduced, and a surcharge on excessively high levels of claims. Switzerland and Germany are examples of countries that use such schemes; the insurance companies provide significant advice and support to employers. If insurance initiatives take up the opportunity to ensure the environmental aspects of construction are addressed, they could devise schemes to integrate the provisions with safety and health. An alternative that might be more appropriate in developing countries – where insurance schemes are not well developed – is to take the costs of health and safety measures and those addressing environmental considerations out of competition by including them in the prime costs of a competitively tendered contract (ILO, 2001).

Regarding social protection, where there are state insurance schemes that apply to permanent workers, attempts can be made to extend them to all workers (see Lawrence and Werna, 2009, for a review). However, in many countries, a new approach might be required, with schemes specifically tailored to the needs of construction workers. In Australia, unions in the construction industry have responded to the prevalence of short-term employment by developing collective industry agreements at the state level for portable benefit systems. These schemes allow construction workers to accrue benefits on the basis of length of service in the industry, rather than with a specific firm. The government of the Republic of Korea has recognised the special needs of construction workers and has introduced a law that is a mutual aid project for retirement allowances. Some states in India have been operating a Construction Workers Welfare Board, which is funded by a tax on all building works (ILO, 2001).

Regarding social dialogue, it is important to seek new roles for trade unions and other actors (see, for example, Van Empel and Werna, 2008). Where there are legal restrictions on the rights of sections of the workforce to organise, trade unions can campaign for their removal. It is also important for trade unions to secure positive improvements in collaboration with employers. For example, in Canada, there have been joint activities to raise the level of safety, quality and productivity. While trade unions are adopting new roles, new organisations are joining in to campaign for the welfare of workers. For example, in India, the National Campaign Committee on Central Legislation for Construction Workers has campaigned to realise better legislation to protect the workers in the industry (ILO, 2001). There are also cases of informal construction workers who have organised themselves and receive formal recognition, such as in Dar es Salaam, Tanzania (Jason, 2008).

The greening of construction may provide a new impetus for social dialogue. Many employers and government authorities have shown enthusiasm regarding green construction. This may open a new door for them to sit together with workers to discuss labour issues in connection with the greening of the industry.

The suggestions offered in this chapter will help to improve the labour conditions in construction, and in green construction in particular. However, not all the problems of construction stem from the characteristics of the industry. For example, much depends also on economic growth and a stable political environment in each country (ILO, 2001). At any rate, the different actors within the construction industry can, and should, make a difference, ideally through joint efforts where appropriate.

The advocacy for green construction is burgeoning. Examples include the work of the Building and Climate Change Initiative of the UNEP and the certification of buildings through green construction councils in a growing number of countries. The case for green construction is often argued based on the notion of sustainability. Considering this, it is important to bear in mind that sustainability entails not only environmental conservation and economic growth, but, equally, social development, in which labour plays a major role. Sustainability is linked to improved quality of jobs; equal access to, and benefits from, generated opportunities; and compensation and retraining of workers displaced in the greening process (Werna, 2009b). In other words, green jobs must also be decent jobs.

Acknowledgements

This chapter was compiled from contributions from Edmundo Werna, but also from other authors mentioned below. Keivani et al. (2010) produced the first ILO study on green jobs in construction. The latter part of Section 1 of the chapter is heavily based on the analysis of green construction produced by these authors. In 2001 Jill Wells produced a comprehensive study of labour in the construction industry (ILO, 2001). By and large the characteristics and trends of the industry noted in that study are still valid today. The data on labour in construction in general included in Section 2 and the recommendations for change included in Section 4 are heavily based on the work of Jill Wells. At the same time, the implications for green jobs in section 2 and the recommendations for green jobs in section 4 are from Edmundo Werna, based on his own experience as well as drawing from other ILO studies on green jobs in construction (referred to in the chapter). The main paragraphs of Section 3 are based on preparatory work for the buildings chapter of UNEP's "Green Economy Report" (Philipp Rode is the principal author of the chapter). In addition, Adrian Atkinson provided useful insights on a comprehensive approach to environment-friendly construction. Abdul Sabor summarised the case studies in the chapter, which are based on ILO studies (referred to in the cases). The work of all these individuals and organisations is here acknowledged. They have the gratitude of Edmundo Werna, who remains solely responsible for the final text.

References

Bach, P. (2006) *How can Governments Support Improved Building Energy Performance? Good practice examples*. Copenhagen: Danish Energy Authority, Ministry of Transport.

Booz Allen Hamilton (2009) *Green Jobs Study*. Prepared for the U.S. Green Building Council, October.

Department of Minerals and Energy, ESKOM, Energy Research Institute and University of Cape Town (2002) *Energy Outlook for South Africa 2002* [Online]. http://www.info.gov.za/view/DownloadFileAction?id=124706

Department of Statistics Malaysia (2002) *Report on Survey of Construction Industries, Malaysia 2001*, Kuala Lumpur.

Department of Statistics Malaysia (2004) *Report on Survey of Construction Industries Malaysia 2003*, Kuala Lumpur.

Department of Statistics Malaysia (2006) *Report on Survey of Construction Industries, Malaysia 2005*, Kuala Lumpur.

Department of Statistics Malaysia (2009) *Report on Survey of Construction Industries, Malaysia 2008*, Kuala Lumpur.

EBN (*Environmental Building News*) (2007) The challenge of existing homes: Retrofitting for dramatic energy savings. *Environmental Building News*, 16(7), 18–19.

Gallego Pinol, E., Roca Mussons, F.J., Perales Lorente, J.F. and Guardino, X. (2009) Determining indoor air quality and identifying the origin of odour episodes in indoor environments. *Journal of Environmental Sciences (China)*, 21(3), 333–9.

Gangolells, M., Casals, M., Gassó, S., Forcada, N., Roca, X. and Fuertes, A. (2009) A methodology for predicting the severity of environmental impacts related to the construction process of residential buildings. *Building and Environment*, 44(3), 558–71.

Hendricks, B., Goldsterin, B. Detchon, R. and Shickman, K. (2009) *Rebuilding America: A national policy framework for investment in energy efficiency retrofits*. Washington, D.C., Center for American Progress and Energy Future Coalition.

International Energy Agency (IEA) (2007) World Energy Outlook (2007). Vienna: IEA.

Intergovernmental Panel on Climate Change (IPCC) (2007) *Climate Change 2007: Synthesis Report*, http://www.ipcc.ch/pdf/assessment-report/ar4/syr/ar4_syr_spm.pdf/

International Labour Organization (ILO) (2009) *Employment Aspects of Energy-Related Improvements in Construction in South Africa*. Geneva: ILO.

International Labour Organization (ILO) (2001) *The Construction Industry in the Twenty-First Century: Its Image, Employment Prospects and Skill Requirements*. Report for an ILO Tripartite Meeting on the Construction Industry. Geneva ILO.

International Labour Organization (ILO) (2010) *Study on Energy Efficiency in Existing Buildings and Green Jobs Initiative in China*. Geneva: ILO.

Jason, A. (2008) Organising informal workers in the urban economy: The case of the construction industry in Dar es Salaam, Tanzania. *Habitat International*, 32(2), 192–202.

Khatib, J. (ed.) (2009) *Sustainability of Construction Materials*. Basingstoke: CRC Press.

Keivani, R., Tah, J.H.M., Kurul, E. and Abanda, H. (2010) *Green Jobs Creation Through Sustainable Refurbishment in the Developing Countries*. ILO Sectoral Activities Department Working Paper 275, Geneva: ILO.

Lawrence, R. and Werna, E. (eds) (2009) *Labour Conditions for Construction: Building cities, decent work & the role of local authorities*. Oxford: Wiley-Blackwell.

Moavenzadeh, F. (1994) *Global Construction and the Environment: Strategies and Opportunities*. Oxford: Wiley.

Ofori, G. and Abdul Aziz, A.R. (2010) *Energy Efficiency and Green Jobs in the Building Industry in Malaysia*. Report prepared for the ILO.

Olivia, G.S. (2007) Comparative Analysis of Resources Consumption in Dutch and Mexican Housing. *ENHR (European Network for Housing Research) International Conference*, 25–28 June, Rotterdam.

O'Neill, R. (2009) Maintenance worker dies at UK wind farm. *Hazards Magazine*, 17 September. http://www.hazards.org/greenjobs/blog/2009/09/17/maintenance-worker-dies-at-wind-farm/

O'Neill, R. (2010) Green construction is no safer. *Hazards Magazine*, 22 April. http://www.hazards.org/greenjobs/blog/2010/04/22/green-construction-is-no-safer/

Poole, A.D. and Geller, H. (1997) The Emerging ESCO Industry in Brazil, Washington, D.C.: Amercian Council for an Energy-Efficient Economy. http://www.inee.org.br/down_loads/escos/esco_mkt.pdf/

Rajendran, S. and Gambatese, J.A. (2009) Development and initial validation of Sustainable Construction Safety and Health Rating System. *Journal of Construction Engineering and Management*, 135(10), 1067–75.

Rajendran, S., Gambatese, J.A. and Behm, M.G. (2009) Impact of green building design and construction on worker safety and health. *Journal of Construction Engineering and Management*, 135(10), 1058–66.

Royal Institution of Chartered Surveyors (RICS) (2007a) Transforming existing buildings: The green building challenge. London: Royal Institution of Chartered Surveyors.

Royal Institution of Chartered Surveyors (RICS) (2007b). Transforming existing buildings: The green building challenge – Appendix b. London: Royal Institution of Chartered Surveyors.

Sarsby, R. and Meggyes, T. (eds) (2009)*Construction for a Sustainable Environment*. Basingstoke: CRC Press.

Smith, P.F. (2004) *Eco-refurbishment: a guide to saving and producing energy in the home*. Oxford: Elsevier.

StatsSA (2008) *Labour Force Survey, 2007*. Pretoria: StatsSA.

The Environment Agency in England and Wales, SEPA in Scotland and the Northern Ireland Environment Agency (NIEA) in Northern Ireland (no date). *How can construction and building trades affect the environment?* http://www.netregs.gov.uk/netregs/businesses/construction/62311.aspx/ (accessed 10 Sept 2010).

United Nations Environment Programme (UNEP) (2008) *Green Jobs: Towards decent work in a sustainable low-carbon world*. Study commissioned by the ILO. http://www.ilo.org/global/What_we_do/Publications/Newreleases/lang--en/docName--WCMS_098503/index.htm/

United Nations Environment Programme (UNEP) (2011) *Green Economy Report*. UNEP.

Ürge-Vorsatz, D., Harvey L.D.D., Mirasgedis, S. and Levine, M.D. (2007) Mitigating CO2 emissions from energy use in the world's buildings. *Building Research & Information*. 35(4), 379–98.

Ürge-Vorsatz, D. and Novikova, A. (2008) Potentials and costs of carbon dioxide mitigation in the world's buildings, *Energy Policy*, 36, 642–61.

Van Empel, C. and Werna, E. (2008) Labour-oriented participatory approaches as instruments of urban governance. Paper presented at the 12th EADI (European Association of Development Research and Training Institutes) General Conference, 24–28 June, Geneva. http://www.eadi.org/fileadmin/Documents/Events/General_Conference/2008/Paper_Werna.pdf/

Van Wyk, L., Kolev, M., Osburn, L., de Villiers, A. and Kimmie, Z. (2009) *Research on the Employment Aspects of Energy-Related Improvements in Construction in South Africa*, Report of the Council for Scientific and Industrial Research (South Africa) prepared for the ILO.

Waste Concern Consultants (2010) *Assessment of Green Jobs in Construction Sector*, Report prepared for the ILO.

Werna, E. (2000) *Combating Urban Inequalities: Challenges for Managing Cities in the. Developing World*, Cheltenham: Edward Elgar.

Werna, E. (2009a) *Cultural Heritage, Economic Development and the Built Environment: Considerations about the Role of Employment*. Paper presented at a hearing of the European Parliament on Cultural Heritage and the Economy, 5 March, Brussels. http://www.ilo.org/public/english/dialogue/sector/papers/construction/culturalheritage.pdf/

Werna, E. (2009b) Green jobs for sustainable cities. *Vital Spaces*, Newsletter of the UNECE Committee on Housing and Land Management and the Working Party on Land Administration, 2(4), 6.

World Economic Forum and ILO/Safework (no date) Competitiveness Influenced by Occupational Health and Safety. Occupational Health Authority Media Release on 03 August 2008, http://www.ohsa.org.mt/docs/comp.pdf/

16 Case studies of construction technology development and innovation in developing countries

Emilia van Egmond
Eindhoven University of Technology, The Netherlands

Introduction

At the 2005 World Summit, the United Nations member states reaffirmed their commitment to the Millenium Goals for a Better World in the 21st Century (UN, 2010). Although there have been some advances since the turn of the century, by 2005 some 36 percent of the urban population in developing countries were living in slums lacking at least one of four basic amenities: durable housing; adequate living space; improved sanitation; and clean water. Sub-Saharan Africa remains the region with the highest prevalence of slums. Some 63 percent of the population live in such settlements.

Advancements still have to be made to safeguard the natural environment and resource base, given the evidence of deforestation, water scarcity (in a number of countries), waste production, and the greenhouse gas emissions from human activity. If the MDG targets are to be met by 2015, more efforts are needed to improve the living circumstances of a large number of the members of the world's population, living particularly in the developing countries. This poses challenges to the construction industry, which can be regarded as opportunities for innovation and the development of more efficient (green) technologies that will contribute to sustainable development.

In this chapter, case studies of construction technology development and innovation that have occurred by means of international technology transfer in some developing countries are presented. The issues considered include: how, and in which form, they are accomplished and applied; and why their uptake fails even though they promise superior performance compared to incumbent technologies.

The methodology applied in the evaluation of the cases is built on the main principles of the innovation system theories. The first two cases, in Ghana and Tanzania, describe technology and knowledge flows (TKFs) in collaboration between foreign and local contractors on construction projects. The cases that follow focus on TKFs in the innovation development and implementation phases in Tanzania, Ecuador, Costa Rica and Bangladesh.

The comparison of the cases enabled some conclusions to be made regarding technology development and innovation in construction related to the technological regime and capability in the innovation system of construction in developing countries. The findings reflect the way ahead with respect to technology development and innovation in construction in developing countries in order for construction to contribute to the achievement of the MDGs.

Technology, innovation, development and transfer

New technologies and knowledge are not necessarily generated by in-house research and development (R&D); they can also be obtained through simple acquisition of the technology and knowledge from elsewhere in the country or abroad, or acquisition of the parts of the needed technology and knowledge components that are missing, and which can be combined with the in-house development and production process. This implies that technology and knowledge are transmitted over space, across borders, between institutions or within the same institution (Dunning, 1981).

Technologies and knowledge flow from source to recipient through various mechanisms. These can be placed in two main groups: the conventional and the non-conventional channels (Radosevic, 1999). The conventional channels are based on the type of contractual agreement. They include: foreign direct investment (FDI); licensing; joint ventures; franchising; marketing contracts; technical services contracts; turnkey contracts; and international sub-contracting. Examples of the non-conventional channels are reverse engineering and reverse brain-drain.

An advantage of the acquisition of new technologies and knowledge through these mechanisms is that the enterprises can improve their production just by copying or borrowing existing technologies. Thus, technology and knowledge flows are seen as important mechanisms to foster learning, capability building and innovation. They also lead to the stimulation of technological development, the formation and growth of entrepreneurial start-ups, and, therefore, the creating of new jobs and, ultimately, economic development. However, evidence from many studies shows that several factors, some of which are presented in Table 16.1, make their uptake fail although they might promise superior performance compared to existing technologies.

The partners in the technology and knowledge transfer process – the recipient firm, recipient country and supplier – often have different and sometimes opposing objectives. Table 16.2 provides examples of these objectives.

Construction in developing countries

Construction activity in developing countries has been dynamic in the recent years. The growth rates in China and India have been higher than 8 percent per annum, and the industries in Brazil, Mexico and the United Arab Emirates have also grown substantially. The economic growth in China and India has provided opportunities for the construction industry. It is predicted that the strongest growth will be in the Asia–Pacific region, due to the population growth, especially in China which is estimated to become the leader of the global construction market in 2018 (Betts et al., 2009). India and Brazil will also see high levels of growth in construction activity, particularly in the residential sector. This is reinforced by labour

Table 16.1 Factors influencing up-take of new technology and knowledge

Level of influence	*Factor*
Country level	1. National policies, orientation, regulations, subsidies, protection 2. Economic situation 3. Available human resources skills and knowledge 4. State of industrialisation 5. Physical infrastructure 6. Available natural resources
Industry level	1. Market demand: structure and culture 2. Market supply and competition 3. (Local) supply of raw materials and intermediate products 4. Technological capabilities 5. Technological infrastructure: institutional relations
Enterprise level	1. Technology status 2. Technology needs: type, costs, time limits 3. Enterprise backgrounds: type, age, location, scale 4. Enterprise management: orientation, scope, targets 5. Enterprise production performance
Technology level	1. Products: range, type, specification, age, quality 2. Process: type, complexity, age, scale of production 3. Need for improvement
Transfer agreement level	1. Source of the transferred technology 2. Relationship between receiver and supplier 3. Nature of the transferred technology 4. Conditions and costs of the technology transaction 5. Modality in which the technology is transferred 6. Advantages and risks

Sources: Egmond and Kumaraswamy, 2003; Stewart, 1979.

mobility and demand for higher living standards. Residential construction will continue to be the largest market (40 percent of the total global construction market by 2020) but infrastructure construction will also continue growing in the emerging markets over the next decade.

There is a contrast between the fast growing output and employment in the construction industries of the emerging countries and the poorer ones, such as the developing countries in Africa. The period of relatively high African growth came to a sudden end during the global economic crisis in 2008 (ADB, 2010). Although the African countries were not really as adversely affected by the global meltdown as the developed countries, Africa remains the poorest and least developed region in the world, with a high rate of urbanisation and an acute shortage of infrastructure, productive facilities and housing, hampering the region's economic growth and its global competitiveness. The structural economic adjustment programmes which were introduced at the end of the 20th century in most African countries led to a sharp decline in the volume of public-sector building and infrastructure projects (Mashamba, 2002; Wells, 2001). Moreover, many

Table 16.2 Objectives and risks in international technology transactions

Recipient firm	*Recipient country*	*Supplier*
Objectives • obtain profitable new product • save R&D costs and risks • improve skills and knowledge of the workforce • develop new products • establish new contacts	Objectives • enable import substitution • realise savings in foreign exchange • improve balance of trade • increase employment • stimulate foreign investment • accelerate industrialisation • improve technological infrastructure • attain economic growth • realise socio-economic development	Objectives • increase income and profit • realise product diversification and improvement • increase sales of tied-in products • develop spin-offs • enter new markets • reduce costs • gain access to protected markets
Risks • dependence on supplier • no technological capability building • technology failures without warranty • possible incompetence of supplier • poor contract conditions	Risks • environmental damage • excessive capital investments • pressure on stock of foreign exchange • double importation of technology	Risks • royalty period is too short • creation of strong competitor • possible damage to firm's good name • lack of feedback • possible publication of production secrets

Source: Egmond, 2006, adapted from UNESCAP, 1992

local construction companies cannot bid for infrastructure projects because they lack adequate financing and working capital and the necessary plant and equipment (IFC, 2010). On the other hand, there has been an increase in the volume of private-sector (particularly informal) building activities undertaken by the small enterprises and individual self-employed workers in many countries such as Tanzania and Kenya (Wells, 2001).

Developing countries in the international construction market

Construction output, by value, had previously been heavily concentrated in the rich, developed world. Also, the overwhelming majority of construction firms operating in the international market were from developed countries. During the last decades of the 20th century, up to 70 percent of the projects undertaken by these firms overseas, primarily infrastructure projects (measured by the size of contracts), were in developing countries (UNCTAD, 2000). The ranking of the top 200 firms in the Architecture and Engineering Services sector shows only 15 developing country firms (UNCTAD, 2000).

The ranking of firms in a list of top 20 international contractors (ENR, 2008) showed that German and French contractors took the lead, followed by some other firms from Europe, the US and Japan. This ranking is based on revenue from projects outside of each company's home country in 2007. The figures include the value of installed equipment when a firm has prime responsibility for specifying and procuring it within the scope of a construction contract.

The ranking of the top global contractors by their total construction contracting revenue in 2007, both at home and abroad, also showed a dominance of European firms (with French firms in the top two positions), followed by a US and a Japanese firm (ENR, 2008). The figures include prime contracts, shares of joint ventures, sub-contracts, design–construct contracts and construction management "at-risk" contracts when a firm's risks are similar to those of a general contractor, as well as the value of installed equipment when a firm has prime responsibility for specifying and procuring it within the scope of a construction contract.

However the lists also show that some companies from emerging developing countries (especially China and Brazil) have entered the top 20, which indicates their increasing capability to compete in the international market (ENR, 2008).

Many developing countries still depend on a sustained acquisition of technologies and knowledge from abroad while making little use of the existing domestic technology stock. Moreover, there is limited chance for them to successfully penetrate the markets of developed countries. Although the share of developing countries in global technology and knowledge inflows and outflows has increased in recent years, the bulk of technology flows are received by a handful of relatively more developed and technologically more dynamic countries.

The observations made above gave rise to the execution of case studies in a number of developing countries to investigate what construction technology development and innovation has taken place; how it took place; and in which form it took place. The results of these studies are discussed in this chapter after the methodology used is presented.

Theoretical perspective

The methodology used in the case studies builds on the innovation system theories, which stress the following principles and hypotheses. Figure 16.1 illustrates the theoretical framework for the discussion in this chapter.

Innovation is the result of interactive learning – technology and knowledge exchange. Knowledge has been recognised as a key component of the capabilities for technology development and innovation competitiveness in the world economy. It is argued that new technologies are based on both know-how (craftsmanship, skills to carry out a job) and knowledge (know-why), and if the new technology is not supported by specific knowledge, then it will collapse sooner or later (Stiglitz, 1999).

Technology development and innovation are undertaken by individuals and organisations in innovation systems. An innovation system is embedded

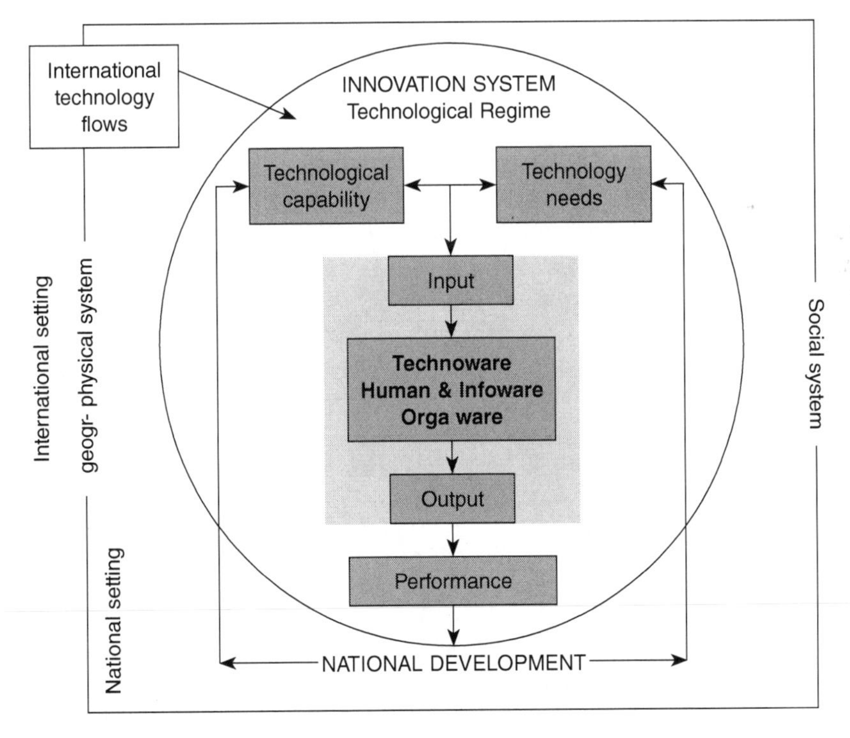

Figure 16.1 Theoretical framework (source: Egmond, 2006)

in a wide macro-level social context (Carlsson and Stankiewicz, 1991; Breschi and Malerba, 2002). Such a system is composed of three dynamic basic building blocks: a network of actors interacting in a specific economic area; following a particular technological regime; and creating and applying a set of competitive technologies, products, processes and services that fulfil specific needs in the markets.

The actor network not only refers to linkages among firms but also the linkages between firms and other institutions and public organisations in the innovation system (such as universities and research institutes) which are increasingly important as sources of knowledge (Von Hippel, 1988; Lundvall et al., 2006).

In production and decision-making, firms rely on routines (Teece, 1997). Routines are standard procedures or regularly followed courses of action that are built up over time and exist in different social environments such as industry, science, politics and the market (Geels, 2004). Routines are based on technological regimes which are coherent collections of rules, regulations collective memories (experience, explicit and tacit knowledge), conventions, consensual expectations, assumptions, or thinking shared by stakeholders in an innovation system. The technological regime amongst the actors determines their behaviour, characterises professional practice, and supports and guides technology development, innovation and continuous learning to increase knowledge, skills and capabilities. It influences the competitiveness and growth rates of firms and industries (Nelson and Winter, 1982; Dosi et al., 2000; Malerba, 2002).

A new technology serving a particular limited domain of application forms a niche in a particular set of technologies and exists alongside other technologies, forming a technological network in an innovation system. The new technology may be promising, but under-supplied in the market, because of high uncertainty, high up-front costs, or because its technical and social benefits are insufficiently valued in the market. The properties of the technology determine whether it fits the technological regime of the innovation system. Any one of the different actors in the innovation system (policy-makers, a regulatory agency, local authorities, a citizen group, private company, an industry organisation, or a special interest group) may act as a niche manager. This manager takes action to promote, and to bring about a change in regime and stimulate the diffusion of a new technology (Schot and Rip, 1996).

The technology climate (considering the national and international setting) is important as a facilitating framework to create a favourable climate for technology development and innovation, competitive production and societal development (UNESCAP, 1994). The major indicators of the national setting are classified in social system characteristics and geographical physical characteristics of the natural and man-made environment. Indicators for the international setting are the features of the international technology market. Yet, whether developing countries can take advantage of the globalisation process, particularly to increase their technological capabilities through technology flows from the ever growing rich pool of global knowledge, depends on their position in the global market.

The first two case studies presented in the following sections, which are from Ghana and Tanzania, describe TKFs in project collaboration between foreign and local contractors. The following case studies focus on TKFs in the development and implementation phases of innovation in construction in Tanzania, Ecuador, Costa Rica and Bangladesh respectively.

Case 1. International technology transfer in Ghanaian construction

Data collection.

A survey was undertaken in 2003 among 40 Class 1 contractors (30 percent of the total population) whose details were derived from the list of general building contractors (1998, 2002), as well as a list of building and civil engineering contractors in the Greater Accra region. First, ten foreign contractors on the list of building and civil engineering contractors were interviewed to obtain information about the Ghanaian contractors with which they had been collaborating during the period 1999–2003. The Ghanaian contractors can be divided in two groups: (1) Ghanaian contractors with Ghanaian owners; and (2) mixed contractors (MCs) which had a foreign owner, but did not have any ties with a company abroad. The survey population finally included five foreign contractors, five MCs, eight Ghanaian contractors which had collaborated with foreign contractors, and 22 contractors which had never been involved in any joint foreign contractor–Ghanaian contractor project execution: 17 Ghanaian contractors, three foreign contractors, and two MCs (Egmond and Vullink, 2005).

National setting

Ghana has a relatively reasonable position for attracting FDI and international joint ventures, compared to other African countries, due to its relatively predictable political and social stability as well as the size of its domestic market. The country is well endowed with natural resources and has a large pool of inexpensive, unskilled labour. It is estimated that about 44 percent of the labour force has no formal education (GSS, 2000). The physical infrastructure – transport, information and communication technology network, electricity and water supply – is not yet up to international standard (CIA, 2002).

Ghanaian construction innovation system

The Ghanaian construction industry contributed 3.1 percent to GDP and was responsible for 2.3 percent of overall employment in 2002. The total number of formally registered construction firms was 7095. Ninety percent of them which were small contractors. Infrastructure improvement was accorded a high priority on the political agenda. The housing needs in Ghana were of a more qualitative than quantitative nature. Capability building in construction through training in formal vocational education and training institutes is negligible; training mainly takes place on an informal basis in apprenticeships (Egmond and Erkelens, 2005).

The major construction projects are awarded to the very few large firms (approximately 10 in 2002), which are mostly foreign contractors. Approximately 75–90 percent of the construction activities take place in the informal sector. About 60 percent of the foreign contractors are fully-owned subsidiaries of European companies. Others are from Israel and China. Of the fully-owned foreign contractors, 40 percent operate as main contractors and 60 percent in joint ventures with Ghanaian contractors. Sixty percent of the foreign contractors have executed more projects in Ghana. Most Ghanaians working for foreign contractors' subsidiaries are artisans and labourers (they number some 100–500 per company, depending on the demand during the execution phase of the project). They are either employed directly by the foreign firm, or engaged through sub-contracting. The foreign personnel are managers, professionals such as engineers, and plant and heavy machinery operators. Appropriately qualified Ghanaians who have proven themselves on the job become site engineers, plant and equipment operators and foremen.

Most of the joint ventures are set up by the foreign contractor and its local counterpart only for the duration of a project. Technology and knowledge are complementary. In this study, only one of the Ghanaian contractors in the sample was involved in a joint venture; the others were MCs with foreign owners. MCs operate either as sub-contractors (40 percent), in a joint venture (20 percent), or as main contractors (40 percent). The foreign–local joint venture that was considered in the study was set up with 40–60 percent equity ownership; the foreign contractor is responsible for the project management and the Ghanaian contractor for the site work.

The Ghanaian contractors are mainly sub-contractors (75 percent) or plant and equipment suppliers (25 percent). Sub-contracts are assigned to Ghanaian contractors which show evidence of their ability to carry out the job, have the equipment in-house and are familiar with the techniques that will be used. For about 85 percent of the Ghanaian contractors it was not the first time they were working as sub-contractors on joint foreign contractor–Ghanaian contractor projects. Thus, through working with foreign contractors on successive projects, the Ghanaian contractors have better learning opportunities and gain more tacit knowledge. This is especially the case with the small group of permanent personnel. Thus, the newly acquired technology and knowledge embodied in the labour force partly stays within the firm, but some of it is lost when the temporary employees, mainly artisans and labourers, leave the firm. Again, where plant and equipment are supplied together with the operators, there are no direct international technology and knowledge flows (ITKFs) to the Ghanaian contractor, since the equipment and the operator return when the project is completed.

Initial capabilities of Ghanaian contractors

To be registered in a particular category, the contractor should meet certain requirements regarding the number and level of education of permanent employees as well as its plant and equipment holdings. All the Class 1 Ghanaian contractors in the study met the requirements for plant and equipment holding as well as level of education of personnel. Nevertheless, some have found it beneficial to collaborate with foreign contractors.

The owners, works managers, quantity surveyors, site agents and engineers of the Ghanaian contractors generally had a university or polytechnic education. The lowest level of education of the labourers of Ghanaian contractors which collaborated with foreign contractors is senior secondary school or vocational training institute. Those firms that did not collaborate with foreign contractors had a labour force with slightly lower qualifications; they were generally only middle school leavers with the exception of the electricians. Masonry, carpentry and painting are the trades that are generally learned in informal apprenticeships.

Although class 1 contractors must have a minimum of 29 full-time "core workers" even during periods of low construction activity, 12.5 percent of the Ghanaian contractors which participated in joint foreign contractor–Ghanaian contractor projects did not meet this requirement; the comparative figure for the 17 Ghanaian contractors which had never worked with a foreign contractor was 40 percent. All contractors also hired occasional temporary workers on their projects. Some 49 percent of the contractors hired more than 50 temporary workers, and 21 percent engaged more than 100 temporary workers. These workers were generally not always hired again for subsequent projects by the same contractor. Only 40 percent of the contractors indicated that they work with the same temporary employees on different projects.

The number of years of registration was taken as a proxy for the experience and related technological capabilities of the Ghanaian contractors. The local contractors which collaborated with foreign contractors had been in business for at least 11 years; the oldest company had been in business for 56 years. Most of the 17 Ghanaian contractors which had never participated in joint foreign contractor–Ghanaian contractor projects were established after the 1980s; one of them had existed for 26 years, three of them were 11 years old, and the youngest D1 Ghanaian contractor was six years old. The size of the Ghanaian contractors depends largely on their project portfolios at a particular time.

Technology and knowledge flows in joint foreign contractor–Ghanaian contractor projects

New technology and knowledge was introduced in 72 percent of the joint foreign contractor-Ghanaian contractor project execution cases. In 22 percent of the cases, only a new material or piece of equipment was introduced, which did not involve radically changing the traditional construction process on site. Learning to use the new materials and equipment on site involved only an adjustment of the existing technological capabilities. New management systems such as those for cost, quality and safety, personnel and logistics management were introduced in 17 percent of the projects. A combination of new management systems, materials and equipment was introduced in 33 percent of the projects.

To investigate whether the Ghanaian contractors and their workers had indeed gained the technology and knowledge, the following proxies were used: continued use of new technology and knowledge; and mastery of the new technology and knowledge (knowing how, why, where, when, by whom) and independence in using it. On 76 percent of the projects, both Ghanaian and foreign personnel worked with the new technology and knowledge; whereas on 16 percent of the projects, only the foreigners, and on 8 percent, only Ghanaians worked with the new technology and knowledge.

Some 7 percent of the Ghanaian contractors did not continue to use the new building materials, whilst 23 percent of the Ghanaian contractors would need assistance on projects where they would use them again. Some 70 percent of the Ghanaian contractors indicated that they would either employ skilled labour or train their employees, and 23 percent of the Ghanaian contractors indicated that they would employ skilled labour to use the new material. About 92 percent of the Ghanaian contractors and their workforce continued to use the labour-substituting mechanical equipment that was introduced on the joint foreign contractor–Ghanaian contractor projects. Some 33 percent of the Ghanaian contractors mentioned that they would need assistance on a future project to use the equipment. Thus, not all Ghanaian contractors and their employees acquired all the necessary technology and knowledge to use the new equipment independently, possibly because they employed temporary workers. The firm loses the technological capabilities it had gained when the workers leave. It is also pertinent to note that knowing how to work with the new materials and equipment does not directly mean that one has a profound understanding of its maintenance, adjustment or repair. Acquiring in-depth and tacit technology and knowledge requires more than collaboration and learning by doing and using. It requires dedicated investment in training. In the present economic situation in Ghana, this is difficult for many private enterprises to undertake.

Case 2. International technology transfer in Tanzanian construction

Data collection

The research population of the survey included all the 10 foreign contractors in the most up-to-date list of the Contractors Registration Board (CRB) of Tanzania at the time of the study (March 2004) and 12 of their local partners (48 percent of the total) which had been involved in completed construction projects executed in the Dar es Salaam region in Tanzania between 2000 and 2004. Table 16.3 shows the profile of the foreign companies. Included in the survey were (a) the local employees of foreign contractors; (b) the Tanzanian contractors which collaborated with the foreign contractors; and (c) approximately 13 percent (129 people) of the Tanzanian contractors' workers. These persons were traced through the interviews with the foreign contractors (Egmond et al. 2007).

National setting

The Tanzanian national setting at the time of the study in 2004 was similar to that of Ghana. The economic situation and policies were generally favourable to business and to attracting FDI and international joint ventures. In 2003, 372 investment projects were approved by the Tanzanian Investment Centre (TIC), of which 29.3 percent were FDI and 29 percent were by joint ventures; 13 percent of these were in construction.

Table 16.3 Survey results for foreign contractors in Tanzania in 2003

	Origin	*Ownership*		*% Tzn empl*	*Type of company*		*Years in industry*	*Projects Tanzania*
		For	*Mix*		*Establishment*	*Public/ Private*		
1	South Africa	X		26–50	Subsidiary	private	6–10	>15
2	Japan	X		51–75	Subsidiary	private	11–15	>15
3	Norway	X		76–100	Subsidiary	private	>15	>15
4	China	X		0–25	Subsidiary	public	11–15	>15
5	Kuwait	X		76–100	Foreign investor	private	0–5	0–5
6	China		X	0–25	Foreign investor	private	6–10	0–5
7	Denmark	X		76–100	Subsidiary	private	>15	>15
8	China	X		51–75	Foreign investor	private	6–10	11–15
9	Italy	X		76–100	Foreign investor	private	>15	>15
10	China	X		0–25	Subsidiary	private	6–10	6–10

Tanzanian construction innovation system

The list of the CRB included 2911 contractors, classified under building (40 percent), civil engineering works, mechanical, electrical and specialist contractors. Large contractors (classes 1, 2 and 3) represented 13 percent, medium-sized contractors (classes 4 and 5) were 13.7 percent, and small contractors (classes 6 and 7) 73.3 percent. Foreign contractors (3.6 percent of all registered contractors) carried out 70 percent of the large- and medium-sized projects. A Tanzanian contractor is defined as "a contracting firm whose majority shares are owned by citizens of the United Republic of Tanzania"; otherwise, the company is listed as foreign. The number of registered building contractors increased in ten years to 1176 in 2004. Based on the annual returns, foreign building contractors, approximately 1.8 percent of all building contractors, accounted for about 44 percent of the reported building works (CRB, 2004). A significant portion (more than 75 percent) of the construction activities in Tanzania take place in the informal sector.

Joint foreign contractor–Tanzanian contractor project execution occurred solely in the form of sub-contracting of work by the foreign contractor to Tanzanian contractors. Almost all foreign contractors were fully foreign owned. Chinese companies dominated the group of foreign contractors in Tanzania in terms of number (40 percent), but they employed only a small number of Tanzanian staff (0–25 percent). Some 50 percent of the larger contractors were Chinese. Of the companies which were solely registered as foreign building contractors, 60 percent were Chinese, which employ very few Tanzanians. All the European companies (30 percent of the total) employ many Tanzanians, who make up at least 75 percent of their permanent employees. Large contractors, in terms of turnover, do not necessarily have more employees (whether permanent or temporary), and also they do not employ fewer or more Tanzanian employees than smaller foreign contractors. Of the foreign contractors involved in the project case studies, only one is registered solely as a foreign building contractor.

Some 70 percent of the foreign contractors indicated that they only sub-contract part of their projects, mainly the specialist works. A disadvantage of sub-contracting work, rather than forming joint ventures, is that local companies are only involved in a (small) part of the construction. The hierarchy of contractors is also the main reason for the unbalanced role division between the Tanzanian and foreign employees. In general, foreigners dominate project management and supervision, and the Tanzanian artisans and labourers execute practically all the construction work.

Initial capabilities of Tanzanian contractors

The Tanzanian contractors which collaborated with the foreign contractors can be classified into: specialist companies, such as those which undertake plumbing works, lift and air conditioning installations and electrical works;

and "all-round" building contractors, to which some of the main contractors can sub-contract practically all building activities. These companies employ an almost entirely Tanzanian workforce, with regard to artisans and labourers. They are rather small companies, in terms of both permanent employees and turnover. The Tanzanian contractors that work with the foreign companies are quite experienced; it was clear that the foreign contractors choose experienced Tanzanian contractors to collaborate with, and the same Tanzanian contractors work on different joint foreign contractor–Tanzanian contractor projects.

The main source of skills of the majority of the Tanzanian employees of the foreign contractors is vocational training and apprenticeships (see Figure 16.2). Many of the workers underwent this apprenticeship at the firms they were working for. Slightly more than 50 percent of these workers had been involved in four or more projects with foreign firms. The Tanzanian labourers of the foreign contractors were mostly male (96 percent). The permanent Tanzanian employees of the foreign contractors can be divided into two groups. First, 50 percent of the workers had higher technical education, and were mainly engineers and quantity surveyors who were involved in project management and supervision. They had little experience beforehand, and obtained most of their skills in the foreign contractor's organisation; around half of them had been involved in four or more projects for the foreign companies. The second group comprised skilled workers employed on-site, most of whom had completed vocational training, and had been employed by the foreign contractor owing to their specific skills. Most of the employees obtained their experience through working with the foreign contractors (some 67 percent of them had worked on four or more projects involving foreign contractors).

Table 16.4 Technological capabilities of Tanzanian contractors involved in joint foreign contractor–Tanzanian contractor project execution

	Number of years industry	*Number of projects Tanzania*	*Number of foreign projects*	*Employees Permanent*	*Employees Temporary*	*Turnover US$*
1	6–10	>15	>4	>100	>100	>400,000
2	>15	>15	>4	<10	<10	40,000–160,000
3	11–15	>15	–	<10	<10	<40,000
4	11–15	>15	1	<10	>100	<40,000
5	>15	>15	>4	<10	26–50	40,000–160,000
6	0–5	0–5	1	<10	26–50	40,000–160,000
7	6–10	6–10	2	<10	10–25	40,000–160,000
8	>15	>15	>4	>100	>100	40,000–160,000
9	>15	>15	>4	10–25	<10	40,000–160,000
10	>15	>15	>4	<10	10–25	<40,000
11	>15	>15	4	<10	<10	<40,000
12	>15	>15	4	76–100	51–75	2000–3500

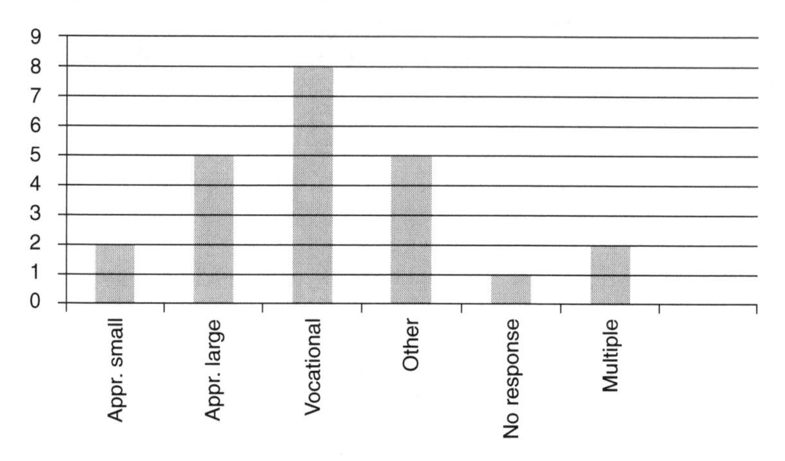

Figure 16.2 Tanzanian personnel of foreign contractor

The Tanzanian contractors worked primarily as labour-only sub-contractors. This is reflected in the percentages of skilled (47 percent) and unskilled workers (11 percent) of the Tanzanian contractors. About 47 percent of the labourers had not finished secondary school; their main sources of skills were vocational training (15 percent), apprenticeship in a small firm (12 percent), or independent learning (13 percent). The workers of the Tanzanian contractors had specialisation in certain trades – over 94 percent of the employees indicated that they had one specialisation for which they were employed on the joint foreign contractor–Tanzanian contractor projects. The painters had primary or secondary school (75 percent), or no education (25 percent). They obtained their skills in informal-sector apprenticeships. The workers in the other trades – carpenters, masons, plumbers – had at least secondary school or technical school education. The electricians generally had at least vocational training. The foremen had at least vocational training (17 percent), most of them had secondary-technical education (33 percent), 17 percent finished polytechnic, and 22 percent were university graduates. The percentage of engineers and quantity surveyors is small (7 percent).

All the engineers had a university education; the quantity surveyors finished either university or polytechnic. The Tanzanian contractors themselves did not have a high education. Most of them had completed secondary school (37.5 percent); this was topped up with an apprenticeship with a small Tanzanian contractor; 25 percent finished secondary-technical school, and 25 percent a polytechnic. About 30 percent of the employees could not read and write English. Amongst these were those with no education or only primary education, and 5 percent had finished secondary school. Whilst 43 percent of the employees had been working in the construction industry for five years maximum and were involved in no more than five projects, 35 percent were involved in four or more joint foreign contractor–Tanzanian contractor projects, through which they gained their experience and skills.

Technology and knowledge flows in joint foreign contractor–Tanzanian contractor projects

Capability building was viewed as the increase of technology (equipment and tools), knowledge (know why, where, when, by whom) and skills (know how) through TKFs from the foreign contractor to the Tanzanian contractor and Tanzanian workers in case studies of four joint foreign contractor–Tanzanian contractor projects. The novelty of technology and knowledge components to the Tanzanian contractors and the influence of the TKFs on the capabilities were also investigated. The sample of four residential and/or commercial building projects was taken from the contracts of six foreign contractors which executed 95 percent of the total value of projects during 2000–04. The projects were comparable in terms of size and type of building system. Table 16.5 shows some information on the projects in the case studies; and Figure 16.3 presents photographs of the buildings constructed.

Table 16.5 Overview of cases of joint foreign contractor–Tanzanian contractor projects

Project case	*Principal*	*Main contractor*	*Permanent employees*	*Value (US$)*	*Project duration*
1. Residence of Japanese Ambassador	Min. of Foreign Affairs Government of Japan, Japanese Embassy	Konoike – Japan; Foreign affiliation 100% Foreign owned 100% Foreign Management 1992 in Tanzania	51–75% Tanzanian	7,000,000	Aug '01 – Dec '02
2. British High Commission	Foreign and Commonwealth Office	Noremco – Norway 100% Foreign owned subsidiary Mixed Foreign–Tanzanian Management 1984 in Tanzania 1950 in Africa	76–100%. Tanzanian	20,000,000	Jan '01– Dec '03
3. Mabibo University Students Hostel	National Social Security Fund (public sector)	China Civil/ Railway Branch office Foreign Public owned subsidiary 1994 in Tanzania	0–25% Tanzanian	15,000,000	Aug '00 – Mar '02
4. The Millennium Towers	Local Pension Fund (public sector)	Kharafi – Kuwait Regional branch Foreign owned subsidiary 1997 in Tanzania	76–100% Tanzanian	12,500,000	Sep '00 – Jul '03

Figure 16.3 Buildings constructed through foreign contractor–Tanzanian contractor project collaboration

Case 2.1: Japanese contractor The only Japanese building contractor in Tanzania carried out the construction of the new residence for the Japanese Ambassador in Dar es Salaam. The company had already executed over 15 projects in Tanzania by 2003, mostly civil works. Its average annual turnover was about US$7 million. The project involved the construction of a two-storey residential building (floor area of 1570 m²) with a reinforced concrete structure on a strip foundation. The project was completed in December 2002 (a 16-month construction period) on time, within budget and to the great satisfaction of the client.

Case 2.2: Norwegian contractor A Norwegian contractor constructed the British High Commission complex. The company has a good reputation on all types of building and civil works, and had executed some 15 projects in Tanzania. Its average annual turnover was US$20 million. The project comprises an administrative and commercial building (total gross internal floor area 6982 m²), providing offices for the British High Commission and a number of other embassies. The six-storey structure was a heavy, in-situ reinforced concrete frame on a raft foundation. The project was completed in August 2002 (19 months), within time and budget and to a world class standard.

Case 2.3: Chinese contractor A university hostel project was executed by a branch of a Chinese state-owned company which had been working as a Class 1 building contractor in Tanzania since 1994. The company had already executed over 15 projects in Tanzania. Its average annual turnover was US$15 million. The hostel – a complex to accommodate about 4200 university students – consists of six hostel blocks of three or four storeys, a canteen, shops, a police post and a sports complex. The construction was a mixture of on-site work and prefabrication, and the main structure was made of reinforced concrete. The project was completed in March 2002 (19 months later), on time, but over budget.

Case 2.4: Kuwaiti contractor The Millennium Towers project was assigned to a Class 1 building and civil works contractor which was a subsidiary of a foreign firm in Tanzania owned by a Kuwaiti company. It had entered Tanzania only in 1997, and had done fewer than five projects at the time of the study. The project comprises hotel facilities, offices, apartments, a swimming pool, and shopping and other commercial spaces. It is a high-rise building with a pile foundation and a concrete structure, completely constructed on-site. The project was completed in July 2003, within budget and on time.

Plant and equipment was brought in by the foreign contractors in all cases, and retained by them after the project was completed. Therefore, the local companies and workers did not gain any equipment by working on the projects. In the case of the Chinese contractor, the equipment was sold to the Tanzanian contractors if there were no new work opportunities on which they could be used.

Transfer of knowledge components (why, where, by whom, when) to the Tanzanian contractors and workers took place in all cases. In most cases this occurred on-site and on-the-job. The Norwegian contractor organised classroom sessions for the local workers. About 86 percent of the Tanzanian workers of the Japanese contractor and those of the local contractors that it had collaborated with reported that they had gained from knowledge transfer. Over 93 percent of the workers of the Norwegian contractor indicated that they had obtained new knowledge as well as information on health and safety on-site; and a similar indication was given by 50 percent of the workers of the Tanzanian contractors. The Chinese contractor's workers indicated that they had gained knowledge only on construction methods. This was not in line with the statement of the foreign contractor, which stated that they had provided much more. In the Millennium Tower project, all the workers indicated that they had gained new knowledge in executing their work, as well as on other areas such as building contracts, materials databases, organisation and health and safety. Not much knowledge regarding the project organisation was transferred in the cases. Although the Japanese contractor indicated that they had transferred knowledge on planning aspects, this was not confirmed by the Tanzanian contractors. Again, the foreign contractor indicated they had provided knowledge regarding project management and English, but that was not confirmed by the labourers; 50 percent of them indicated that they had obtained information on working schedules and regulations.

Transfer of skills (know how) took place in various ways (see Table 16.6). The Japanese contractor provided training on-site; this was confirmed by all of its own workers and 86 percent of the employees of the Tanzanian contractors. The employees of the Japanese contractor obtained training in more areas than those of the Tanzanian contractor, who were only trained on construction activities while the former also received training in the area of documentation. According to the Japanese contractor, training on site planning, health and safety was provided, but the employees did not confirm this. All the Japanese firm's employees and 86 percent of the sub-contractors' workers indicated that they had learned skills "by doing". Practical sessions on-site and skills training abroad were offered to the workers on the British High Commission project. Some 87 percent of the employees of the main contractors reported that they had gained skills through these means, but only 33 percent of the sub-contractors' employees indicated likewise. The training provided to the foreign contractors' employees was limited to construction, but the workers of the sub-contractors also indicated that they had received training in some other areas. Moreover, the main contractor stated that it had provided training on the use of databases and on management, but the employees did not confirm this. Some 93 percent of the employees of the foreign firm and 33 percent of the workers of the local contractors indicated that they had learned skills by doing.

The Chinese contractor indicated that it had provided training on-the-job, but only half of the employees reported that they had received any. The main contractor believed it had trained its employees in many areas, but the workers confirmed only three areas: construction method, new applications, and organisation and management. All the workers indicated that they had learned skills by doing on the project, but only in one area: operation of construction equipment. The foreign contractor of the Millennium Tower sent some workers to workshops and then those people had to teach others on-site. The training was mainly on construction activities, health and safety and organisation. Some 93 percent of the workers indicated that they had learned skills by doing in the same areas as those in which they were trained.

Table 16.6 Types of skill flows

Kind of Skills	*1*		*2*		*3*	*4*	*1*		*2*		*3*	*4*
	F	*T*	*F*	*T*	*F*	*T*	*F*	*T*	*F*	*T*	*F*	*T*
	Training						*'By doing'*					
Construction activities												
Methods	X	X	X		X	X	X	X	X			X
New applications	X	X	X		X	X	X	X	X			X
Operation of equipment	X		X	X		X			X	X	X	X
Documentation												
Building contracts	X						X		X			
Working schedules	X			X			X		X			
Equipment databases	X											
Materials databases	X						X					
Organisation process												
Organisation/management					X	X						X
Financing												
Planning												
Employment procedures				X								
Regulation				X						X		
Health/safety				X		X				X		X
Training												
Other												

Case 3. International technology transfer during the establishment of a building materials company in Tanzania

Data collection

Interviews, on-site observation and reference to documents were used to investigate the Tanzanian ITKF case which took place in the building materials industry, on a project to establish, run and maintain a private-sector cement-based roofing materials enterprise in Tanzania (Egmond and Kumaraswamy, 2003).

The national setting

The national setting in Tanzania at the time the transfers took place (1990s) was not favourable, although liberalisation and privatisation policies offered opportunities for private investment. Local financing of investments by enterprises was subject to stringent loan conditions, which were often difficult for the entrepreneurs to meet. The physical infrastructure, mainly the road network, and supply of electricity and water, formed a constraint on entrepreneurial actions.

Tanzanian construction innovation system, technology needs and capabilities

The proposed enterprise was a small-scale family business (with less than 10 employees) located in Dar es Salaam, and with a focus on production for the local urban market. Linkages between construction firms and various other agents such as governmental, financial, R&D and educational institutes were rather weak, loose or even non-existent. Investment studies revealed a local market demand for the envisaged building material; there was only one local competitor. The studies also identified production technologies and knowledge components that needed to be imported.

Technological capabilities were not very strong. The local production and supply of capital goods for the intended production process was negligible. Moreover, the spare parts and some of the intermediate products needed in the production, such as pigments for building materials, had to be continuously imported. Human resources (to be employed as construction as well as production workers) were abundant, but had low educational and skills levels. Natural resources such as sand and water needed for the production of the building material were abundant but supply was constrained by insufficient exploitation, infrastructure and transport facilities. During the period of the study, the electricity supply was interrupted several times for rather long periods. Consequently, there were also problems with cement delivery and this affected the continuity of the production of the building materials.

Technology and knowledge flows, regime and performance

Table 16.7 shows the nature of the components of technology and knowledge that were transferred to Tanzania, their sources and the particular mode through which the components were transferred.

The production performance in terms of the quality of the output of building materials was basically good at the start. The materials met the quality requirements of the market. However, the quantity of production did not keep pace with the actual market demand, and the process components that were introduced did not fit the prevailing technological regime.

The equipment that was imported for the production line was easy to maintain and repair. However, the labour force did not always have the skills and knowledge to solve any problems with the imported machinery. The majority of employees belonged to the (extended) family of the director of the enterprise. Production and maintenance skills needed to be upgraded through training by foreign suppliers. Information and documentation on the production process and maintenance had to be imported.

Table 16.7 ITKFs in the building materials industry in Tanzania

Nature of technology and knowledge	*Source*	*Mode*
Execution of feasibility studies	Europe	Consultancy
Selection and negotiation of technologies	Europe	Consultancy
Production equipment and machines	Europe	Direct purchase
Production knowledge and skills	Europe	Training
Maintenance knowledge and skills	Europe	Training
Plant information and documentation	Europe	Included in purchase of prod. line
Management and organisation	Europe	Consultancy
Product knowledge	Europe	Training
Raw materials, components, spare parts	Europe	Continuous purchase

Management capacities were initially considered adequate to run the company smoothly. However, this component formed a major bottleneck. The Tanzanian partner and workers were not familiar with the European business model and management principles that were introduced. The socio-cultural background of the personnel was prominent in all decision-making, and this was usually in conflict with the European approach to organisational management. This issue could not be addressed with simple TKFs.

Commercialisation of the production did not succeed. To a large extent, this can be attributed to the limited and weak linkages of the enterprise with other agents in the innovation system of Tanzanian construction. Unfortunately, the enterprise collapsed.

This case study shows that the technological regime (social-cultural aspects including knowledge, values, norms, expectations, attitudes and motivations, and socio-economic background) adversely affected the management and production performance. Moreover, the national setting (conditions of the local financing institutions when granting credits, and deficiencies in the supply of inputs) and international issues (dependence on foreign consultants) were detrimental to the performance of the production entity.

During the pre-investment and investment phases, no insurmountable problems were identified by the foreign consultants. They carried out a traditional economic feasibility study. This revealed opportunities and advantages of ITKFs for the establishment of the enterprise. Employment could be created in a neighbourhood of poor, low-skilled in Dar es Salaam, while the hope was that the acquired skills in manufacturing and sales of roofing materials could be passed on within the sector over time. However, it is clear that the investment studies did not recognise the potential constraints.

Case 4. International technology transfer for residential construction in Ecuador

Data collection

Bamboo has been used to a substantial extent in the construction of dwellings in locations where it grows naturally. Bamboo grows rather fast, and can be a substitute for timber or imported materials. Thus, it helps to reduce foreign exchange problems, deforestation, land degradation and carbon dioxide emissions. Bamboo plantations can be established on sites near fast-expanding urban areas where large numbers of dwellings are required. This case study is based on a literature review to gain insights into the factors that influenced the success of the development and diffusion of innovative bamboo construction technologies for low-income households in Ecuador (Egmond and de Vries, 2002).

National setting

The Ecuadorian development situation appeared to be a direct result of its history, social structure, less favourable economic conditions and policy-making. The country faces deficiencies in the physical infrastructure, as well as a shortage of housing, in particular for the lowest income households.

Construction innovation system, needs and capabilities in Ecuador

In Ecuador, the majority of the formal residential construction is undertaken with concrete blocks. There is a large need for housing for low-income households; these members of the population cannot afford the concrete block houses. They generally rely on the output of informal construction activities. This informal sector has a limited availability of technological capabilities and very weak linkages with the agents in the formal construction innovation system. The clients and providers of informal residential construction have to rely mainly on their own resources; they receive no support from governmental, financial, educational or R&D organisations. The technologies used in the informal construction of houses are limited to predominantly bamboo technologies. The knowledge and skills of most of those involved in the actual execution of the informal construction projects appears to be limited. This is apparent in the traditional use of the bamboo technologies and the limited availability, and use, of tools and equipment. The bamboo structures are often left uncovered, which makes them vulnerable to deterioration at a rapid rate. The natural resources seem to be sufficient, but a problem in bamboo supply can be foreseen in the near future owing to a lack of appropriate bamboo forest management.

Technology and knowledge flows, regime and performance

Through intervention by a non-governmental organisation (NGO) in Guayaquil, the "Viviendas de Hogar de Cristo" (VHC), technologies and knowledge available in the formal Ecuadorian construction sector could be drawn upon for incremental development of the technology for low-cost housing construction. This NGO has formal relations with the agents in the construction industry, through which the non-existent linkage between the formal and informal residential construction sector in the country could be overcome. The VHC developed a design and a prefabricated bamboo construction system for the lowest income households based on the traditional rural bamboo houses.

The prefabrication in a serial production process and on-site assembly of the building elements by the households with tools provided by the VHC helped to overcome their lack of capabilities to carry out the on-site construction activities. Moreover, VHC offers employment to the households it has trained to work in the production of the elements for, and construction of, the basic housing units. The VHC also offered soft loans for the construction of the houses whilst it managed the processes. Figure 16.4 illustrates an innovative bamboo house in Ecuador.

The materials and physical properties of the houses match both the Technology Regime and the national setting in Ecuador. The appearance of the houses blends into the Ecuadorian landscape. The design, construction system and costs of the innovative housing technology correspond well

Figure 16.4 Innovative bamboo construction in Ecuador
Source: de Vries, 2002.

with the requirements of the target group. Therefore, social acceptance of the innovative building technology appeared to be good. The construction process is developed in such a way that the costs of the houses can be kept very low. On-site construction with the system requires limited process technology components. A disadvantage of the houses is the lack of facilities, such as private water, electricity and cooking, and sanitation facilities that are not provided during the first construction stage but can be added when the owner has repaid the loan for the basic house.

This case study shows that the TKF in terms of the diffusion of the locally developed innovative, albeit basic, technology can be considered successful. Ecuador's problems of financing of public housing for the lowest income households are partially alleviated with this innovative bamboo technology development and additional technical and financial support of the NGO.

Case 5. International technology transfer for residential construction in Costa Rica

Data collection

Literature studies, field observations and interviews with the experts involved in the project were carried out to gain insights into the factors that influenced the success of the technology development and knowledge flows of innovative bamboo technologies for residential construction for low-income households in Costa Rica. The innovations took place within the Proyecto Nacional de Bamboo (PNB), a government-initiated project that formed part of the National Housing Programme which was launched in 1985 in Costa Rica (Egmond, 2000).

National setting

Costa Rica is a small country with approximately 500,000 households; it is a middle-income country; nevertheless, it faces deficiencies in the physical infrastructure and a shortage of housing, particularly for the lowest income households. Since the early 1950s, the national policy orientation has changed nearly every four years from liberal to social-democratic and then back again after elections, thereby changing the scope of the government's financial support for various national programmes. This forms a constraint on the continuation of development projects in a number of cases. During the 1980s, an economic crisis struck Costa Rica that manifested itself in various sectors of the economy, including the construction industry. The accelerated population growth and the poor condition of approximately 32 percent of the housing stock stimulated the government at that time to give first priority in the national development plans to solutions to the housing problems.

Residential construction Innovation system, needs and capabilities

In Costa Rica, residential construction is dominated by the concrete building materials suppliers and contractors that form a tight actor network, and so the country predominantly builds with concrete-based systems. The skills and knowledge of the construction workers were developed mostly to work with these concrete systems. Timber had previously been used as the key building material in residential construction. However, the accelerated deforestation resulted in a scarcity of timber and a substantial increase in its price. This led to the greater application of other construction technologies, mainly using masonry and precast concrete systems. This implied increased costs of construction and negative effects on the Costa Rican trade balance.

Against the background of a housing deficit of about 125,000 units in Costa Rica in 1985, the PNB was initiated by a local architect as a part of the "Plan Consolidado de Vivienda" (a housing programme for the lowest income households). The PNB involved an innovative programme with the primary objective of contributing to the alleviation of the shortage of housing facing the lowest income families in a sustainable way through stimulating the application of locally produced bamboo as a building material, as a substitute for timber, masonry and concrete.

The project included: bamboo cultivation to increase the stock of the natural resource; upgrading of the skills and knowledge of the human resources; R&D on bamboo and its construction technologies; the design and construction of bamboo houses; and the industrialisation and commercialisation of the project activities. These components were effectively integrated. There were rather strong relations between a number of Costa Rican consultants and the national government during the periods of social-democratic policy orientation. However, a lack of capabilities to meet the PNB's objectives was recognised, and these needed to be addressed through local efforts as well as ITKFs.

Technology and knowledge flows, regime and performance

Financial resources and land were made available by the Costa Rican government for the establishment of a 200 ha bamboo forest. The rest of the funds for the project were provided by the United Nations Development Programme (UNDP) and the Dutch government.

ITKFs in several fields, from various sources and through various modes contributed to the achievement of the targets of the PNB project. Bamboo samples were imported from other Latin American countries, and bamboo plantations were established to provide the necessary raw materials for the residential construction programme, since the specific *Guadua* bamboo was originally not available in the country.

Knowledge about bamboo cultivation was transferred from Germany. R&D projects were carried out, resulting in the designs of the houses and development of specific bamboo preservation and construction solutions, in close collaboration between Costa Rican, Colombian, German and Dutch experts (see Figure 16.5).

Work by Dutch researchers on rationalisation and improvement of the bamboo processing activities resulted in recommendations for further industrialisation (involving systematisation, standardisation and mechanisation) of the production plant for prefabricating bamboo wall elements with modest means. The Dutch researchers also contributed to studies on the diversification of the production of bamboo-based products to make items including furniture and timber substituting panels as intermediate products for other industrial sectors such as packing, since the bamboo harvest was greater than what was actually needed for the planned housing construction. US expertise contributed to the industrial design and engineering of furniture items. Training programmes, short courses and conferences were organised and information services provided to the more than 1000 professionals, technicians and family heads involved in the cultivation, production, preservation and application of bamboo technologies in housing construction, as well as for those involved in the administration of small bamboo-based community businesses. These activities were carried out in collaboration with Costa Rican, other Latin American and European experts.

Figure 16.5 Basic innovative bamboo house in Costa Rica
Source: de Vries, 2002.

The houses, which had facilities for cooking and washing, matched both the Technology Regime and national setting in Costa Rica. Their appearance and physical performance meet the national building standards as well as the requirements of the Costa Rican population. Moreover, the PNB house scores better regarding construction costs, earthquake resistance and environmental sustainability when compared to similar houses built with concrete. However, they still appeared too expensive for the lowest income population, and the support of the government was necessary. As the Costa Rican government was unable to provide substantial funds, only 700 houses were built for the lowest income group in 1994, although the initial target was at least 2800 houses.

The bamboo-based activities could not be continued on a sustainable basis as was planned because of their dependence on foreign and government financing, which, in turn, depended on the economic and political situation in the country. The economic situation was not always favourable. Moreover, not all governmental regimes were favourably disposed towards bamboo-based residential construction projects, or even the provision of support for bamboo R&D, the cultivation of the bamboo or the bamboo-based industries. Commercialisation of the initial public investments in the activities did not succeed.

Major constraints were the linkages of the PNB with other agents in the construction innovation system that were initially strong, but did not last long. The bamboo-based building materials industry that was established within the PNB framework was public-owned. It appeared to lack the necessary linkages and capabilities for the commercialisation of its products. Private contractors, which were deeply involved in concrete systems production, were reluctant to accept and adopt the bamboo construction technologies. The international relations of the PNB were also strong; but these linkages were limited to the duration of the project. Despite these constraints, the project can be considered a relative success, especially in terms of the experience gained in the country both at the national level and at the level of execution of the project regarding the integrated approach towards finding solutions to the housing problem.

Case 6. South–south transfer of technology for residential construction in Bangladesh

Data collection

The focus of the final case study in this chapter is on the opportunities and obstacles for south–south ITKFs of innovative bamboo technologies from Latin American countries to Bangladesh as a solution to the housing problem of the lowest income households in the Chittagong Hill Tracts, and sustainable overall construction industry development.

Evidence of the advantages of using innovative bamboo construction technologies as the main building material for housing in a number of countries such as Ecuador and Costa Rica, and the idea that it would be possible to make use of the experience gained in other countries and to introduce new, more durable and sustainable bamboo technologies in Bangladesh, influenced the decision to carry out the ex-ante case study (De Vries and Van Egmond, 2002).

Field studies were carried out in four villages in the Chittagong Hill Tracts and data were gathered by means of interviews with the target group and with local construction experts. A review of the literature provided information on the residential construction sector as well as the national setting in Bangladesh.

National setting

The national setting in Bangladesh in which the ITKFs were expected to take place was one of a developing country, reflected in many aspects such as its politics, economy, health, education, technology development, physical infrastructure and housing for a large proportion of the population. Inadequate policy-making and the frequent floods that affect the country seem to hinder significant development. The Chittagong Hill Tracts is a hilly region in the south-east of Bangladesh where most of the country's natural bamboo grows. The Chittagong Hill Tracts population is predominantly tribal, with a culture that resembles that of the neighbouring country, Myanmar, more than those of the people in the plains of Bangladesh. The population of all four villages was quite homogeneous.

Construction innovation system, needs and capabilities

Despite the major nationwide housing problem, the lack of adequate houses in the Chittagong Hill Tracts relates mainly to the lack of quality of the houses. The limited availability of resources means that a large part of the population has to rely on traditional building materials such as bamboo, timber or rammed earth for the construction of their houses. The roof is usually the only exception in the use of local materials: corrugated iron sheets are often used instead of plant-based materials. Over 70 percent of the houses are built with bamboo (see Figure 16.6). However, Bangladesh faces a shortage of bamboo due to poor management, although the bamboo resources in the country could provide an abundant supply of materials for housing, especially for the low-income households.

Moreover, most of the bamboo houses have a temporary character, do not meet the material and other requirements of the households, and the majority of them are either built by the owners themselves, or the construction is controlled by *majis* (local construction experts) who have learnt their trade by family tradition, and operate on a competitive basis. The construction process is basic and requires limited capabilities.

Potential south–south technology and knowledge flows and regime

R&D on bamboo construction technology is necessary to improve the supply of housing, both in terms of quantity and of quality, durability and functionality of the houses. Yet bamboo R&D is not given much attention. Only the Bangladesh Forest Research Institute is engaged in such studies, but it is not part of its core research programmes.

TKFs on innovative bamboo technologies for low-cost housing from Latin America to Bangladesh and their adaptation to suit the local conditions might give an opportunity to improve low-cost housing in less time and at lower cost compared to indigenous bamboo R&D (Table 16.8 illustrates this potential). However, the case study revealed some obstacles in Bangladesh, resulting from the prevailing technological regime and the national setting. Bamboo has low social status as a building material for houses. This constitutes a major constraint for the use of bamboo in general. Moreover, Bangladesh lacks the socio-economic base for initiatives regarding technology development and innovation. Any effort that involves substantial investment in the Chittagong Hill Tracts region in particular may pose a problem as the people in the region are not in a politically strong position in the country. Moreover, the possibility and success of the introduction of innovative technologies requires political stability and political will at all levels of the country. As bamboo is potentially available throughout the whole country, many other regions could be suitable for the

Figure 16.6 Traditional bamboo house in Bangladesh
Source: de Vries, 2002.

application of innovative bamboo technologies, and it would be pertinent to investigate the opportunities in other regions in the country.

Conclusion

The comparison of the findings from the case studies enabled some conclusions to be made regarding TKFs in relation to the technological capabilities in the innovation system of construction in developing countries. The national setting in many developing countries forms a major obstacle to technology development and innovation, which cannot be easily dealt with. The economic situation in many of the countries is far from favourable for

Table 16.8 Crucial factors for ITKFs in construction from Latin America to Bangladesh. Legend: + a promoting factor for assimilation of ITKFs; – a constraining factor for assimilation of ITKFs (source: de Vries, 2002)

Bangladesh			*EC*	*CR*
National context	Economic		–	–
	Political		–	–
	Socio–cultural		–	+
	Physical infrastructure		–	–
Bamboo sector	Innovation system		–	–
	Technology Needs		+	+
	TECHNOLOGY CAPITAL	Technology Stock Products	+	+/–
		Technology Stock Process	–	
		Human Resource Stock	+/–	+/–
		Natural Resource Stock	+	+
Construction sector	Innovation system		+/–	–
	Technology Needs (house)	Functionality	–	+/–
		geometry/dimensions	+/–	+
		Materialisation	–	+
		physical durability	+/–	+
		production complexity	+	+/–
		Costs	+	– –
	TECHNOLOGY CAPITAL	Technology Stock Products	+/–	– +/–
		Technology Stock Process	+	–
		Human Resource Stock	+/–	–
		Natural Resource Stock	+	+

local investment in technology development and innovation. Therefore, technological development has to rely on either foreign financial sources or a political orientation and willingness at various levels in the country to give priority and support to investments in construction industry development. Globalisation, which has also led to changes in national policies whereby there are programmes of liberalisation and privatisation in many developing countries, has generated opportunities and, in some cases, incentives for foreign investments. However, not all developing countries have economic and political stability that makes countries attractive to foreign investors. The physical infrastructure (communication and transportation, energy and water supply network) in many developing countries shows deficiencies that form an obstacle to the diffusion of technologies which flow into the countries.

The innovation system of the construction industry in developing countries shows rather weak or even non-existent linkages among the actors in the network. Thus, ITKFs take place without broad support and acceptance among the actors in the network, with much dependence on single sources of support within a limited timeframe. ITKFs are often isolated actions by a single actor in the sector. Technological capability-building then takes place on a non-durable basis and is easily lost when it is not further diffused and assimilated in the innovation system.

Technological capabilities in the construction industry, in terms of the stock of national resources that are available and can be committed to the construction industry, show deficiencies that cannot all be counterbalanced by ITKFs. The technological regime in the construction industries of developing countries, the expectations, beliefs, norms, values and motives of the local communities, are often different from those of the foreign suppliers of technologies and knowledge. This constrains ITKFs, and can even lead to non-acceptance of the transferred technologies and knowledge. The results of the studies in Ghana and Tanzania show that, through collaboration with foreign contractors, technology development and innovation occurred, although the extent was limited. Collaborations between foreign contractors and local companies occurred mainly in the form of sub-contracting. Sub-contracting has the disadvantage that the local firms are usually involved in only a (small) part of the construction process. The least opportunities for technology development and innovation are in cases where only specialist work is sub-contracted. Sub-contracting, in comparison with joint ventures, offers only a small number of local personnel project management experience.

The transferred knowledge and skills mainly involve construction operations, and, to a lesser extent, the knowledge and skills of project organisation. However, it is not the aim of the foreign contractor to educate the employees in these areas, except for aspects such as health and safety on-site. Most of the training opportunities are provided on-the-job and mainly in the form of learning-by-doing. The local workers of the foreign contractors benefit the most from their involvement in the projects compared to those employed by the local sub-contractors. On most of the projects, local permanent employees of the foreign contractors were involved in management and supervision. Thus, they have a learning opportunity, not only with respect to on-site construction activities, but also regarding organisational aspects.

The best learning opportunities were for those local contractors which had been in business over the longest period, had undertaken more projects, and gained more experience through joint project execution with foreign contractors, and thus have initial technological capabilities which are permanently available in the firm. With this basis of technological capabilities, they are more likely to become involved in joint project execution with foreign contractors and find it easier to understand and acquire the skills and knowledge to use new technologies.

Foreign contractors work with experienced local contractors and tend to collaborate with the same partners on different projects. Also, individual local workers will have a greater chance for a job in a construction project that is carried out with the involvement of a foreign contractor when they have a higher level of initial technological capabilities, particularly in terms of experience. For local personnel who work for foreign contractors, a letter of recommendation at the end of a project can be even more important than a university degree scroll. Foreign contractors have contact with each other and they take each other's employees. However, the chance for such workers to get a job with a local contractor becomes limited since they also ask for higher wages and better working conditions.

The new technologies and knowledge are diffused only on a limited scale throughout the local construction industry. In order to achieve sustainable technology development and innovation, not only should the know-how (skills) be transferred, but also, the knowledge (know-why, when, where and by whom) components should be transferred as well. The acquisition of this knowledge requires dedicated investment in training in countries where the investment capacity is rather low.

The case studies point to a number of strategies that are important to ease the diffusion and implementation of innovative construction technologies in developing countries: the formation and strengthening of networks in the innovation system; and bringing about changes in the technological regime by voicing and shaping expectations about the new technologies and knowledge, and by active technology and knowledge exchange amongst the actors in the innovation system about design and engineering specifications, user characteristics and their requirements, environmental issues, industrial development options, government policies, regulatory framework and governmental role concerning incentives for diffusion and implementation.

However, developing countries differ and so do the innovation systems of the construction industries in these countries. Thus, to manage the processes of innovation and technology flows in order to successfully diffuse and implement innovative technologies, and to fully benefit from their technical and social advantages in construction, the first step to be taken is to gain full insight and understanding of the features of the innovation system: the roles and functional relations of the actors in the actor network; the prevailing technological regime; and the existing and competing technologies.

The case studies provide a broad picture of the major aspects that influence the success of ITKFs leading to technological capability building and improved performance of the construction industry in developing countries. The studies confirm the main threads of the key points in the literature reviewed in the earlier sections of the chapter.

References

African Development Bank (ADB) (2002) *Achieving the Millennium Development Goals in Africa: Progress, Prospects and Policy Implications*, African Development Bank, Abidjan: ADB.

ADB African economicoutlook.org 2010

Betts, M., Robinson, G., Burton, C., Cooper, A., Godden, D. and Herbert, R. (2009) *Global Construction 2020*, London: Global Construction Perspectives and Oxford Economics.

Breschi, S. and Malerba, F. (1997) Sectoral innovation systems: technological regimes, Schumpeterian dynamics, and spatial boundaries. In C. Edquist (ed.) *Systems of Innovation: Technologies, Institutions and Organizations*. London and Washington: Pinter/Cassell Academic. pp. 130–56

Carlsson, B and Stankiewicz, R. (1991) On the nature, function, and composition of technological systems, *Journal of Evolutionary Economics*, 1(2), pp. 93–118.

Central Intelligence Agency (CIA) (2002) *The World Factbook 2002*. https://www.cia.gov October 2002

Dosi, G., Coriat, B. and Pavitt, K. (2000) *Competences, Capabilities and Corporate Performance*. Final Report, Dynacom Project. http://www.sssup.it/LEM/Dynacom/files/DFR.pdf/

Dunning, J.H. (1981) *International Production and the Multinational Enterprise*, London and Boston: Allen and Unwin.

Egmond - de Wilde De Ligny, E.L.C. van (2000) Resource management for sustainable development: The application of a methodology to support resource management for the adequate application of Construction Systems to enhance sustainability in the lower income dwelling construction industry in Costa Rica. In P.A. Erkelens, S. de Jonge & A.A.M. van Vliet (Eds.), *Proceedings of the symposium 'Beyond Sustainability 2000'*, Eindhoven, September, Eindhoven: Eindhoven University of Technology.

Egmond - de Wilde De Ligny, E.L.C. van and Vries, S.K. de (2002) Sustainable construction industry development through international transfer of innovative technologies. *Conference Proceedings CIB W107 International Conference: Creating a sustainable construction industry in developing countries*, 11–13 November, Stellenbosch, South Africa.

Egmond - de Wilde De Ligny, E.L.C. van, Vulink, M. and Benda, S. (2005)

Capacity Building through Joint ventures in the African Construction Industry in *Proceedings 3rd International Conference on Multinational Joint Venture for Construction Works*, Bangkok, Thailand, November 1–2, 2007.

Egmond - de Wilde De Ligny, E.L.C. van (1999) *Technology Mapping for Technology Management*, Delft: Delft University Press

Egmond - de Wilde De Ligny, E.L.C. van and Vries, S.K. de (2002) Sustainable construction industry development through international transfer of innovative technologies. *Proceedings CIB W107 International Conference: Creating a sustainable construction industry in developing countries.* 11–13 November, Stellenbosch, South Africa.

Egmond - de Wilde De Ligny, E.L.C. and Kumaraswamy, M.M. (2003) Determining the success or failure of international technology transfer, *Industry and Higher Education*, 17(1), 51–9.

Egmond - de Wilde De Ligny, E.L.C. van and Vulink, M. (2005) International technology and knowledge flows in the African Construction Industry, Africa Seminar, TU/e, February, Eindhoven.

Egmond - de Wilde De Ligny, E.L.C. (2006) International Technology Transfer, Lecture notes, TU Eindhoven.

Egmond - de Wilde De Ligny, E.L.C. van, Vulink, M. and Benda, S. (2007) Capacity building through joint ventures in the construction industry. *Proceedings 3rd International Conference on Multi-National Joint Ventures for Construction Works*, 1–2 November, Bangkok.

ENR (2008) *Engineering News Record, Top Lists*, McGraw Hill Construction, http://enr.construction.com/toplists/

Erkelens, P.A. and Egmond - de Wilde De Ligny, E.L.C. van (2005) *Report on the Inception Mission in the NPT Project (NPT/GHA/047) Capacity building in the Sunyani and Cape Coast Polytechnics to improve performance of the building and construction industry in Ghana.*, Eindhoven TU/e – Nuffic.

Geels, F. (2004) From sectoral systems of innovation to socio-technical systems Insights about dynamics and change from sociology and institutional theory, *Research Policy* 33 (2004) 897–920.

Ghana Statistical Service (2000), *Ghana Living Standards Survey*, Report of the Fourth Round Accra, GSS.

Hippel, E. von, (1988) *Sources of Innovation*, New York: Oxford University Press.

International Finance Corporation (IFC) (2010) *Building Africa with Improved Infrastructure*, London: IFC. World Bank Group http://www.ifc.org/ifcext/africa.nsf/ created July 14, 2010. Retrieved August 2010.

Lundvall, B.-Å., Interakummerd, P. and Lauridsen J.V. (eds.) (2006) *Asia's Innovation Systems in Transition*, London: Elgar.

Lundvall, B. and Gu, S. (2006) China's innovation system and the move toward harmonious growth and endogenous innovation, *Innovation: Management, Policy and Practice*, 8(1–2).

Mashamba, M.S. (2002) Industry training overview, *Construction News: Journal of the National Council for Construction in Zambia*, Vol. No. 7, 1st quarter, pp. 12–13, 21.

Malerba, F. (2002) Sectoral systems of innovation and production, *Research Policy*, 31(2), 247–64.

Nelson, R. and Winter, S. (1982) *An Evolutionary Theory of Economic Change*, Boston: Harvard University Press.

Radosevic, S. (1999) *International Technology Transfer and Catch-up in Economic Development*. Science and Technology Policy Research Unit, University of Sussex; pp. 14–29, 96–177.

Schot, J.W. and Rip, A. (1996) The past and future of constructive technology assessment, *Technology Forecasting and Social Change*, 54, 251–68.

Stewart, F. (1979) *International Technology Transfer: Issues and Policy Options*, World Bank Staff Working Paper No. 344. World Bank.

Stiglitz, J.E. (1999) Knowledge as a Global Public Good, Oxford Scholarship Online Monographs, pp. 308–26. http://www.oxfordscholarship.com/oso/public/content/economicsfinance/9780195130522/toc.html/

Teece, D.J. (with G. Pisano and A. Shuen) (1997) Dynamic capabilities and strategic management, *Strategic Management Journal*, 18(7), 509–33.

UN (2010) *UN Millennium Development Goals Report 2010*, New York: United Nations. http://www.un.org/millenniumgoals/pdf/

UNCTAD (United Nations Conference on Tariffs and Trade) (2000) Regulation and Liberalization in the Construction Services Sector and its Contribution to the Development of Developing Countries, Note by the UNCTAD Secretariat. Geneva.

UNESCAP (United Nations Economic and Social Commission for Asia and the Pacific) (1992) *Technology Transfer: an ESCAP Training Manual, Booklets*, Ministry of Science, Technology and Environment, ST/ESCAP/862. Bangkok: UNESCAP.

UNESCAP (1994) *Application and Extension of the Technology Atlas*, New York: United Nations.

Vries, S.K. (2002) Bamboo Construction Technology for Housing in Bangladesh, Opportunities and constraints of applying Latin American bamboo construction technologies for housing selected rural villages of the Chittagong Hill Tracts, Bangladesh, MSc thesis TU Eindhoven.

Wells, J. (2001) Construction and capital formation in less developed economies: unravelling the informal sector in an African city, *Construction Management and Economics*, 19, 267–74.

Index

The letter f indicates a figure and t a table.

9781138381339